The Ecological Community

The Ecological Community
Environmental Challenges for Philosophy, Politics, and Morality

Edited by Roger S. Gottlieb

Routledge
New York   London

Published in 1997 by
Routledge
29 West 35th Street
New York, NY 10001

Published in Great Britain by
Routledge
11 New Fetter Lane
London EC4P 4EE

Copyright © 1997 by Routledge

Printed in the United States of America on acid-free paper.

All rights reserved. No part of this book may be reprinted or reproduced or utilized in any form or by any electronic, mechanical, or other means, now known or hereafter invented, including photocopying and recording or in any information storage or retrieval system, without permission in writing from the publishers.

Library of Congress Cataloging-in-Publication Data

The ecological community : environmental challenges for philosophy, politics and
   morality / edited by Roger S. Gottlieb.
      p.   cm.
   Includes bibliographical references.
   ISBN 0-415-91611-9. — ISBN 0-415-91612-7 (pbk.)
   1. Environmental ethics.   2. Human ecology.   3. Environmentalism.
   I. Gottlieb, Roger S.
GE42.E33   1997
179'.1—dc20
                                                               96-9179
                                                                  CIP

Contents

Acknowledgments vii

Introduction: The Center Cannot Hold ix
Roger S. Gottlieb

Part 1. Environmental Challenges for Political Theory and Philosophy

1 Environmentalism and Human Oppression 3
   Peter S. Wenz

2 Time, Narrative, and Environmental Politics 22
   John O'Neill

3 The Rationale for Environmental Restoration 39
   William Throop

4 Empathy, Society, Nature, and the Relational Self: Deep Ecology and Liberal Modernity 56
   Gus diZerega

5 Is Liberalism Environment-Friendly? 82
   Avner de-Shalit

6 Be-wildering Order: On Finding a Home for Domestication and the Domesticated Other 104
   David Macauley

Part 2. Environmental Theory and Moral Questions

7 A Sleepless Ethicist and Some of His Acquaintances, Including the Monoculturalist, the Poetic Naturalist, the Very Famous Biologist, the Happy Scientist, and a Few Heroic Antimodernists 139
   Roger S. Gottlieb

8 Imperialism and Environmentalism 163
   Eric Katz

9 Habermas and the Ethics of Nature 175
   Steven Vogel

10 The Problem of Knowledge in Environmental Thought:
    A Counterchallenge   193
    Robert Kirkman

11 Feeding People versus Saving Nature?   208
    Holmes Rolston, III

III. Struggles Up Close: Current Conflicts in Environmental Theory and Practice

12 Ecofascism: A Threat to American Environmentalism?   229
    Michael E. Zimmerman

13 Materialists, Ontologists, and Environmental Pragmatists   255
    Andrew Light

14 Challenging Pluralism:
    Environmental Justice and the Evolution of Pluralist Practice   270
    David Schlosberg

15 Environmental Justice, Neopreservationism,
    and Sustainable Spirituality   292
    Mark I. Wallace

16 International Justice and Wilderness Preservation   311
    Mark A. Michael

17 Solidarity Across Diversity: A Pluralistic Rapprochement of
    Environmentalism and Animal Liberation   333
    Brian Luke

18 The Sustainability Question   359
    Carl Mitcham

    Contributors   381

Acknowledgments

This book began as a special issue of *Social Theory and Practice* (Volume 21, Number 2, Summer 1995). Co-edited by myself and Peter Dalton, the issue was entitled *The Environmental Challenge to Social and Political Philosophy*. I am indebted to Peter for his valuable work in choosing and suggesting changes for the seven essays that appeared in that issue; and to *Social Theory and Practice* for permission to reprint the pieces by de-Shalit, diZerega, Katz, Light, Luke, Michael, and Zimmerman.

The secretary of the Humanities and Arts Department of WPI, Penny Rock, helped with much of the mailing that goes into a book like this.

Once again the staff at Routledge and my editor Cecelia Cancellaro have done terrific work and been a pleasure to work with.

—Roger S. Gottlieb

Introduction
The Center Cannot Hold

Roger S. Gottlieb

At times, human thought and social practice creep forward or ebb backward, making incremental changes for an infinite number of small, hidden reasons. At other times a shock wave of historical change, with an epicenter in some particular locus of concern, devastates the human landscape and rapidly renders existing beliefs controversial, doubtful, or irrelevant. During recent centuries, the development of science and technology, the rise of mass political movements, the Holocaust, and the struggle for women's rights have all challenged widely held values and long-established social theories. In the contemporary world, it is the environmental crisis that poses the most profound, and the most inescapable, rupture of any and all attempts to continue with "business as usual."

The material aspects of this crisis have been widely documented. Toxic pollution of air, water, and land poses threats to the future habitability of the earth for human beings and has already caused extensive illness and death. In terms of people's experience of their surroundings, the loss of wilderness, alteration of global climate, expansion of genetic engineering, and constant tinkering with local flora and fauna raise the depressing prospect of a world in which everything is man-made. The result is a pervasive sense of loneliness and a lurking fear that we are making a terrible mistake. For those who love the nonhuman as well as the human, our casual devastation of other species is felt as a crippling moral failure.

The intellectual traditions of western philosophy, political theory, and morality are deeply challenged by this crisis.

This challenge takes two forms. On the one hand, old perspectives are now seen as deeply flawed. On the other, every valued resource we have must be utilized in trying to understand the mess we've made. Thus recognizing that we in fact face an environmental crisis creates (among other things) a host of theoretical problems and areas for intellectual inquiry.

Failures of the Past

Like the insight that western thought has been distorted by racism and sexism, the realization that our manner of life leads so easily to the elimination of other species and a kind of slow suicide should make us deeply suspicious of the beliefs our culture has held. Until recent decades, most western thinkers and traditions have been silent about the evolving rape of nature and its consequences for humanity. An occasional Romantic poet, a tradition of conservationism very much outside of mainstream social and ethical theory, early (largely ignored) glimmerings in Heidegger and the Frankfurt School, the first attempts at environmental ethics from Aldo Leopold and Rachel Carson—these exceptions prove a rule of profound intellectual blindness. Theorists for the most part took it for granted that how humanity treated the nonhuman world was not a serious problem. We might wonder how we could know that world, or how to divide up pieces of it most justly among ourselves. But concerns for the autonomy or inherent value of the nonhuman world, its dialectical role in shaping our own intellectual or emotional capacities, its function as a mediator between human beings—these were for the most part not to be found. Only the intrusion of disaster—brought to light largely by people outside the fields of social and political thought—has made a substantial fraction of social theorists begin to think environmentally.

It should also be noted that earlier perspectives are of limited value just because the world is now simply a very different place from what it was. Whatever environmental degradation occurred in the past (and in many cases it was considerable), ecological problems have now reached a point far beyond what could have been conceived of by prior thinkers. When else has a theorist of political power had to confront a government that could poison mothers' milk with pesticides or make sunlight significantly more dangerous by altering the composition of the atmosphere? What other society has had to create norms for economic activity when that activity could affect the daily lives of people thousands of miles away? What morality has had to face a banality of evil in which the most common of everyday actions (driving an automobile, putting fertilizer on the lawn) could contribute to devastating effects on future generations or people at the other ends of the earth?

However it is not just that the vast majority of social and ethical thinkers has been silent on environmental questions or did not foresee their importance. It is also the case that a number of reflective philosophers expressed a profound and damaging estrangement from the nonhuman world. I will confine myself here to brief discussions of themes from two thinkers who are still essential to the basic orientation of a great deal of modern social and political thought: Locke and Marx.[1]

Consider Locke's view of property, which in essence, is remarkably simple: if you alter nature by laboring on it, the result is yours. You may take more than you personally need, for the improved nature—which can now be traded for nonperishable tokens—is a benefit to others. However, you may not waste, because ultimately nature, including your own life, has its origins in the will and beneficence of God.

> Reason . . . tells us, that men, being once born, have right to their preservation, and consequently to meat and drink, and such other things, as nature affords for their subsistence. . . . God, who hath given the world to men in common, hath also given them reason to make use of it to the best advantage of life, and convenience. The earth, and all that is therein, is given to men for the support and comfort of their being. And . . . all the fruits it naturally produces and beasts it feeds belong to mankind. . . . Whatsoever [a man] removes out of the state that nature hath provided, and left it in, he hath mixed his labour with and joined to it something that is his own, and thereby makes it his property. . . . As much as any one may make use of to any advantage of life before it spoils; so much he may by his labour fix a property in. . . . And thus came in the use of money, some lasting thing that men might keep without spoiling, and that by mutual consent, men would take in exchange for the truly useful, but perishable supports of life.[2]

The lasting contribution of Locke's view is to undermine claims that accord property by virtue of social position. However, it also privileges the high-impact agriculture of English colonists and delegitimizes any potential claims of Native Americans, whose agriculture tended to be low impact and frequently shifting. This privileging ultimately rests on a repeatedly stressed image of nature's (for all practical purposes) inexhaustible resources. "Locke's theory depends upon the existence of a New World with an endless supply of space and resources ripe for colonization and plunder."[3] Clearly he is not thinking of England, most of whose wild lands were under cultivation by his own time. Nor does he face the way accumulation of money allows individuals to gain title to so much land that a great many others can barely feed themselves.

A deeper, but not unrelated, problem is the difficulty Locke faces in distinguishing between a reasonable and a wasteful appropriation of nature. In general, a liberalism whose goal is the protection of individual rights is unable to take a substantively critical view of people's preferences. And we see in the brief quotation above how Locke moves from defining our entitlement as "subsistence" to providing us with the rights to "convenience" and "advantage." The notion of "spoilage" is relatively clear in regard to agricultural products; but it is very hard to distinguish between spoiling and useful production in a world (such as our own) that is poisoning itself by the production of junk.[4] When we take land out of the undeveloped commons and move it into the producing private sphere, are we thereby always doing something that (in the deep sense) is "productive"? Locke cannot say.

Finally, Locke does not recognize the possibility that nature might have a destiny outside of human interests. For him, its ultimate value derives from being a gift from God to humanity. Even though animals also labor (say, to collect food or build nests), only humans can acquire property. And this is so because God gave nature to the rational and purposive beings—i.e., to us. Property arises when labor creates value in what is valueless until it is adapted to human interests.[5]

Thus Locke's view, and any contemporary philosophies that accept its premises, legitimizes values and practices that contribute to the environmental crisis. Opposed to his uncritical reliance on people's desires and productivity, contemporary environmental thought is forced to challenge the origins and extent of consumerist dispositions. Production of "advantages" or "conveniences" has revealed itself as a process with no psychological limits and crushing material effects. The notion that human beings are always benefited by the cultivation of wilderness is similarly doubtful. The existence of national parks, wildlife refuges, and the Endangered Species Act, as well as the grief many feel over the countless extinguished species, express the view that numerous areas of the earth are more valuable left alone than adapted to "production."

In essence, the claim that "the earth, and all that is therein, is given to men for the support and comfort of their being" has been deeply challenged. We now have assertions of (and political movements for) animal rights, investigations of the "legal standing" of trees and ethical declarations that human beings are "plain members" of (rather than lords over) ecosystems.[6] People mobilize to save wildlife damaged by oil spills and obstruct clear-cutting of forests in which they have no "personal" stake. And there has been a tremendous upsurge of a spiritual sense that the environmental crisis spells not only the pollution but actually the *desecration* of the world.

Marx's view of nature is more complicated than Locke's; and I cannot do it full justice here. Moreover Marx (and western Marxist descendants such as Horkheimer, Adorno, and Marcuse) actually has much to offer an analysis of the environmental crisis. Marx's concepts of alienation and fetishism describe how human powers and products are being used against humanity and often appear to be uncontrollable. His analysis of commodification captures quite well the way natural entities essential to complex ecosystems are reduced to objects whose only meaning is their value in the marketplace. Marx's account of the dynamic of capitalist accumulation provides what is still the best explanation of the spread of industrial society. And he did write about the destructive effects of capitalism on the natural world.

What *is* problematic is that Marx's ultimate value—the development of human freedom—involves for Marx a separation from and mastery over humanity's natural origins. While emerging from nature, humans differ from animals in their ability to act in accord with a previously formulated plan and to evaluate and alter their ways of life. Their innate sociality and rationality point toward a possible future in which both social life and our exchanges with nature are ordered by conscious mastery. In such a future, social relations become nonexploitative and nature subject to our will. "Marx . . . shared the faith of a spectacularly successful nineteenth-century capitalism in material and technological progress" and "called for a social mastery of nature for the sake of man."[7]

It is this last goal that has rendered Marx and his followers subject to considerable criticism. The point is that for Marx, as for Locke, the nonhuman world derives its ultimate value by being a resource for humanity's creative self-development. The nonhuman finds its true fulfillment only when we humans end up humanizing it.[8] "Through [man's] production nature appears as his work and his reality" thus "he sees himself in a world he has created."[9] In such a world humans are unbound by natural restraints. Both powers and limits derive only from humanity's own condition. "What is wealth? The full development of human mastery over the forces of nature, those of so-called nature as well as of humanity's own nature.The absolute working out of [man's] creative potentialities, with no presupposition other than the previous historic development."[10]

There are two profound limitations in this view. First, as Andrew McLaughlin has observed, any proper understanding that humanity and nature are engaged in a dialectical process of mutual support and transformation cannot at the same time "assume a separation of humanity from nature. [T]his means that the goal of control, mastery, or domination of nature is ultimately incoherent." A contrasting view, asserts McLaughlin, "images nature as an interconnected network—the web of nature—with humanity as simply a part of that net.... It differs from [Marx] in recognizing epistemological and practical limits to human mastery of the rest of nature."[11]

Like Locke, Marx also misses not only the mutual determination of humanity and nature but also the possibility that ultimate values might lie in nonhuman nature precisely insofar as it is *not* humanized by us. Or that human beings might have worth for other reasons than developing our powers and Doing Great Things. Both modes of human existence and modes of interaction with the nonhuman that are based on appreciation, communion, and "letting be" are foreign to Marx's creative, driving, accomplishment-oriented spirit. Similarly, the notion that nature—which does *not* possess freedom in the way that humanity does—might have its own value is simply not something Marx can grasp.

Challenges of the Present

For the kinds of intellectuals who have written and will read and teach this book, the first challenge of the present is to acknowledge the severity of the environmental crisis and to alter our research and classroom conduct accordingly. Unfortunately, however, denial is as prevalent in the academic community as it is in government, media, corporations, and the general population. Many scholars continue to talk about theoretical questions that presuppose that our children will have breathable air, that the sunlight is still safe, or that it is not that important that the United States produces tons of toxic waste per person per year. It is also true that many of us, including myself, have taken quite a long time to come around.

Yet in the last twenty-five years or so there *has* been a realization—not widespread enough but still real—that fundamentally new questions are in order. Environmental philosophy, political theory, and ethics have been born; and this volume is a contribution to their joint enterprise.[12]

What are their common concerns?

Most profoundly, there has been a shattering of traditional claims about the centrality of human beings: that we are ultimately more important than other forms of life; that we can measure our worth or wisdom without assessing our damage to the rest of the world; or that we can expect to have justice within the human community if we do not consider what that (or a comparable) concept might mean when applied to our relations with nonhumans.

Four of the essays below focus on this central issue. Peter Wenz argues that there is a causal connection between our treatment of nature and the oppression of fellow humans. Both institutionally and ideologically, he suggests, the means we use to bring the earth under control require the domination of other people. Eric Katz stretches the usual notion of "imperialism" to apply to relations between people and their nonhuman surroundings. By using this term—with its component elements of power, domination, and intervention—Katz is attempting to get us to see our intrusions into ecosystems and species life as subject to moral criteria. A major theme in my own essay has to do with the ethical standing of our civilization. How can we pretend to moral competence, either collectively or as individuals, while we pursue a form of life that is at the same time suicidal for humanity collectively, deeply aggressive to those communities who suffer most from environmental degradation and the source of the most serious elimination of species in 65 million years? I also investigate uses of "nature," offering an alternative to the supposition that nature is either that which is isolated from the human or simply a social construct. Finally, Holmes Rolston probes the notion that we could really have to choose between "feeding humans or saving nature," and rejects the mind-set that takes for granted the social arrangements that might compel such a choice.

Even as environmental thinkers challenge those kinds of anthropocentric premises that have animated most of Western thought, they cannot confront the environmental crisis from scratch. We could not—and need not—discard Marx's critique of capitalism or Locke's support of democracy because their accounts of nature are lacking. No matter how dire our situation, intellectual traditions are evolutionary.

Environmental philosophy and political theory thus do not comprise some miraculously newborn understanding of humanity and the world, but instead several variants of existing perspectives that have been challenged and changed. Marxism has given rise to ecoMarxism, feminism to ecofeminism, struggles against racism to concerns with environmental justice, and liberalism to examinations of the rights of animals, species and ecosystems. Extensive discussions exist of how these perspectives might have contributed to the environmental crisis and how they might be altered to help solve it.

In this book we find Gus diZerega claiming that liberal moral and political theory can actually be supportive of environmental concerns. diZerega argues that the account of sympathy for others found in Hume, and of the role of principles in self-regulating social systems in Polonyi, allow a sense of personal morality and social organization that is surprisingly compatible with the ecocentric ethics of Deep Ecology. Avner de-Shalit, by contrast, finds political liberalism to be much more problematic. On the one hand, its tradition of free inquiry and expression allow for the promulgation of environmentalist ethics and politics. Yet its insistence on individualism in its understanding of preference formation and social policies makes it hard for liberalism to face the necessary tasks of social reorganization. Steven Vogel questions a different contemporary perspective, that of Jurgen Habermas. After examining criticisms of Habermas for his categorical separation of nature from the human realm of communicative ethics, Vogel suggests that our relation to the nonhuman should best be understood not as the confrontation with a mysterious Other but as an extension or continuation of the social itself. Nature, he believes, is always a socialized presence. His position here provides an interesting contrast both to my own viewpoint and to that of Wenz. Like Vogel, Robert Kirkman is suspicious of certain highly general claims about nature that have arisen in environmental philosophy. He disputes the necessity of positing an organic holism in order to make assertions about what we in fact are preserving; and defends science against charges that its reductionism distorts the reality of nature.

Much of social and political theory is more particular and less grand but nevertheless essential for the ongoing task of trying to understand our current situation. Accordingly, several essays below do not attempt to make large statements concerning entire ideologies or major thinkers but instead confine themselves to trying to understand particular concepts or issues.

Carl Mitcham, for instance, analyzes the concept of "sustainability." Aside from the obvious significance of a term with such wide popular usage, the notion is important for theoretical reasons. If we are to be critical of Locke's inability to distinguish between valid production and spoilage, we need some criteria to help identify social practices justifiably compatible with the nonhuman world. Mitcham traces the history of this idea of sustainable development and describes some problems that can arise when it is used without a sufficient understanding of accompanying social context. John O'Neill and William Throop, though approaching rather different subject matters, share a common concern with understanding why nature is valuable to us and what in fact gives a particular natural setting its identity. O'Neill is working with a general question of the ultimate importance of nature and the role of narrative in assessing that importance; Throop is trying to make philosophical sense of the current practice of restoring ecosystems after social degradation. Both find something essential in the notion of time: that our sense of the essence of a natural setting has to do not only with its characteristics at a given moment but also with the processes that led it to that condition. The violation of nature is thus not just the replacement of one set of entities by another, but also the imposition of a discontinuous, human-oriented "time" on the rhythms of the nonhuman. David Macauley also investigates a boundary area between the human and nonhuman: our relations to domestic animals and the concept of the "domestic" itself. Macauley's goal is neither to eliminate nor indict domestication, but to support greater knowledge of and empathy for those Others who share our dwellings.

Along with challenges to tradition, reworkings of existing theories, and concrete studies, environmental efforts in philosophy, politics, and morality are situated in the realm of concrete social struggles. Environmentalism (thank heavens) is an international political movement and not just a topic for theorists! Fraught with the same differences in social position and strategic inclination as are other mass movements, environmentalism is thus prey to comparable ideological and practical conflicts. Six of the essays included here address questions that arise from those conflicts.

Mark Michael focuses on the difficult question of justice and international efforts at wildlife preservation. He argues that demands for wilderness preservation by the First on the Third World must be accompanied by corresponding redistribution of assets; that is, concern for the nonhuman makes sense (or "justice") only when people are cared about as well. This argument is as pointed as recent attempts to end the killing of African elephants by a European/American led-ivory boycott. The only really successful wildlife preservation strategies, it turns out, must be concerned with local communities as well as with local wildlife.

Michael Zimmerman's disquieting essay (responded to in part by that of Peter Wenz) shows that an environmental rejection of modernity can combine with fascism. Himself a committed environmentalist, Zimmerman nevertheless sketches the historical experience of Nazism, one element of which combined a love of the natural with a genocidal hatred of particular human groups. Like a Marxist after Stalin, who must be suspicious of blanket claims to represent "the workers," Zimmerman knows a nature-loving attack on technological civilization will not always lead to something better.

Essays by Andrew Light, Mark Wallace, David Schlosberg, and Brian Luke share a common concern: how to mediate theoretical and political conflicts within the environmental movement itself. Not surprisingly, some of these conflicts resemble those that occur in other ideological settings. Thus Andrew Light's focus on differences between those he calls environmental "ontologists" and "materialists" might remind us of earlier disputes between those who privileged intellectual as opposed to those who privileged institutional determinants of social life. Light sketches an "environmental pragmatism" in which ecological concern allows for a metatheoretical compatibility between seemingly opposed schools of political ecology. Brian Luke, David Schlosberg, and Mark Wallace confront more focused political disagreements within the environmental community and seek to find ways beyond or through those conflicts. Luke seeks a "solidarity across diversity" among the sometimes (surprisingly) intense disputes between environmentalists and animal liberationists. He suggests that the two movements cannot be reduced to a common thread, but are indeed distinct value perspectives. They do share, however, a common enemy in anthropocentrism. Mark Wallace elaborates a form of "sustainable" spirituality that he thinks may help bridge the gap between those two groups: those who are mainly concerned with environmental problems in urban settings and the direct effect of toxic wastes (the "environmental justice" movement), and those who focus on wilderness or species preservation. David Schlosberg, confronting much the same divide, aims for a reworked political pluralism that will enable different environmental groups—as well as citizens with very different views about ecological issues—to be heard and represented.

It is perhaps a mark of maturity that environmental philosophy, politics, and morality can have serious and extensive internal disputes. Such disputes confirm that these intellectual efforts are rooted in a real social world in which disagreement and conflict reflect a reality of inequality and opposing interests. As Karl Korsch remarked more than half a century ago, theory divorced from political movements becomes ideology. Environmental thought can avoid becoming merely ideological if it continues to be rooted in people's vital concerns—and in those of other forms of life and being as well.

Notes

Miriam Greenspan and Bettina Bergo made helpful comments on earlier drafts of this Introduction.

1. Obviously, other examples could be discussed. Plato's sense of the "natural" part of the soul as needing control by the "rational" and the postmodern tendency to see only "nature" and never nature, are two. Exactly where the problem begins is the subject of some debate. Heidegger sees it as all going wrong with the Greeks. Some cultural ecofeminists locate the problem in 5,000 years of patriarchy. Environmental thinker Paul Shepard sees the even earlier end of hunter-gatherers as the start of our long decline. Contrast Buddhist deep ecologist Joanna Macy, who in a surprisingly Hegelian vein sees our millennia-long alienation from nature as essential for the cultivation of valuable resources such as science and human rights.

2. John Locke, *Second Treatise of Civil Government*, paras: 25–27, 47.

3. Robert Frodeman, "Radical Environmentalism and the Political Roots of Difference," *Environmental Ethics* 14, (4) (Winter 1992), 262. See also Eugene C. Hargrove, *Foundations of Environmental Ethics* (Englewood Cliffs, NJ: Prentice-Hall, 1989), 67–73.

4. Kathleen M. Squadrito, bolstered by Locke's pious writing on education and the virtues of charity and modesty, offers a more sympathetic reading in "Locke's View of Dominion," *Environmental Ethics*, 1 (3) (Fall 1979).

5. Robyn Eckersly, *Environmentalism and Political Theory: Towards an Ecocentric Approach* (Albany: SUNY Press, 1992), p. 23.

6. See, for example, writings by Tom Regan, Christopher Stone, and Aldo Leopold. Lest it be argued that such a view of nature was not historically available to Locke, consider Maimonedes' (twelfth century) statement: "It should not be believed that all the beings exist for the sake of the existence of humanity. On the

contrary, all the other beings too have been intended for their own sakes, and not the sake of something else" (*Guide to the Perplexed*, part 3, chapt. 13).

7. This from Howard Parsons, who believed in Marx's ecological wisdom. Howard Parsons, ed., *Marx and Engels on Ecology* (Westport, CT: Greenwood Press, 1977), p. 69.

8. Steven Vogel explores this attitude (approvingly) in "Marx and Alienation from Nature," *Social Theory and Practice*, 14 (3) (Fall 1988). "For Marx . . . recognizing our cennectedness to the environment means recognizing its *sociality*" (p. 376).

9. John Clark, quoting Marx from the *1844 Manuscripts*, in "Man's Inorganic Body," *Environmental Ethics* 11 (3) (Fall 1989), 252.

10. Karl Marx, *Grundrisse*, trans. Martin Nicolaus (New York: Vintage, 1973), p. 488.

11. "Marxism and the Mastery of Nature: An Ecological Critique," in Roger S. Gottlieb, ed., *Radical Philosophy: Tradition, Counter-Tradition, Politics* (Philadelphia: Temple University Press, 1993), pp. 177–179. Clark (cited above) argues that Marx had the elements of a truly dialectical view, but did not pursue it.

12. Both theology and institutional religion are engaging in the same enterprise in a more comprehensive way than are philosophy and political theory. See Roger S. Gottlieb, ed., *This Sacred Earth: Religion, Nature, Environment* (New York: Routledge, 1996).

Part 1

Environmental Challenges for Political Theory and Philosophy

I

Environmentalism and Human Oppression

Peter S. Wenz

Introduction

Peter Carruthers writes in *The Animals Issue*, "I regard the popular concern with animal rights in our culture as a reflection of moral decadence. Just as Nero fiddled while Rome burned, many in the West agonize over the fate of seal pups and cormorants while human beings elsewhere starve or are enslaved."[1] Others have argued similarly that environmentalists who wanted to curtail logging in old-growth forests of the American Pacific Northwest were insensitive to human needs and that logging jobs are more important than spotted owls.

These are just two examples of a common antienvironmental attitude that opposes concern for human beings and concern for nonhumans. Playing upon the premise, unquestioned in our society, that human well-being takes precedence over other concerns, antienvironmentalists attempt to discredit environmentalism by claiming that it is misanthropic.

The opposition between human well-being and environmentalism is strengthened conceptually by the association of environmentalism with Nazi ideology. Much Nazi thought resembles environmentalist bioregionalism,[2] and Martin Heidegger remained sympathetic to Nazi thought while offering environmentalist critiques of technology.[3] Completely unrelated to the Nazis, J. Baird Callicott, a leading exponent of Aldo Leopold's Land Ethic, has been labeled an ecofascist. In his seminal article "Animal Liberation: A Triangular Affair," Callicott hinted that culling the human population to promote ecosystemic health may be desirable.[4] This prompted Tom Regan to associate environmentalism with fascism.[5] So environmentalism has been accused not only of insensitivity to human needs, but also of tendencies toward racism and fascism. Anyone who values humanity should oppose environmentalism.

However, qualifying the value of human beings out of concern for nonhumans does not lead to human oppression, according to philosopher Roger Gottlieb, because human flourishing and environmental concern are mutually supporting rather than opposed. Human oppression results largely from technologies and institutions developed under the guidance of mainstream anthropocentric views. Gottlieb notes that "the devaluation of nature" that began about 5,000 years ago is related not only to "the development of more advanced agriculture, increasingly complex social divisions of labor and relations of exploitation," but also to "the desire for *control* in and of itself, as a good thing." The result is "interlocked systems of military, religious, economic, and ideological domination" in societies in which "hard labor, power, status, and wealth are unequally and unjustly distributed, and in which "man-made" poverty and exploitation supplant droughts or floods as the greatest threats to material well-being."[6]

This suggests the thesis of the present paper: The best way to help human beings avoid oppression is to value nonhuman nature for itself. I support this position in two ways. First, I look at some indigenous societies where human beings are treated better than we treat each other. I note a connection in these societies between respect for nature and for human beings. I then examine western society's attitude toward, and history of, dominating nature (in the supposed human interest) and explain how a great deal of human oppression results from the mind-set that legitimizes such domination.

Because the relativizing of human moral standing inherent in any nonanthropocentric environmentalism has been associated by its detractors with fascism in general and Nazi ideology in particular, I conclude by considering the Holocaust, the deliberate killing of six million Jews under Nazi control, as a prime example of human oppression that results from a purely instrumental attitude toward nature. I do not explore all the reasons for the Holocaust, or for episodes in human history similar enough to be called "holocaust-like events."[7] I am concerned only to show that such events are in large part a by-product of the ideology that nature exists merely to serve human needs and wants. Antienvironmentalists who associate the Land Ethic or Deep Ecology (two popular nonanthropocentric environmentalist views) with racism, fascism, or any other serious form of human oppression have matters backward. *Paradoxically, in the western context anthropocentric thinking and associated social structures were important preconditions for inhumanity.*

Foraging Societies

Traditional foragers, who are often called hunter-gatherers, live in small bands and travel by foot in a regular sequence throughout the year to obtain food by hunting and gathering. They are discussed here for several reasons. They often express the kind of respect for nature that environmentalists champion. They also live in relatively good harmony with the environment, as environmentalists prescribe. They do not seek successively more powerful technologies that overpower nature in the supposed human interest. Finally, they live reasonably satisfying, fully human lives without much social hierarchy or human oppression. In short, they belie the antienvironmentalist claim that, regardless of the society, environmentalists' prescriptions exact unacceptable tolls on human beings.

Of course, we cannot become foragers, if only because our population densities are some orders of magnitude too high. But if foragers show that some people can have their environmental cake and human well-being too, their example may help to clarify what has gone so drastically wrong with our society that a Holocaust could occur. They suggest that one way to prevent such oppression may be to adopt environmentalist attitudes and prescriptions insofar as these both resemble those of traditional foraging people and can be adapted to our circumstances. However, the work of adaptation is beyond the scope of the present paper.[8]

Although today's few remaining foragers live within the jurisdiction of various nation-states, traditional foragers did not have a state in the Weberian sense of an organization in society that claims monopoly on the right to decide how force is used. The result was not a Hobbesian war of all against all. People used informal pressure along with socialization to handle deviance.

Socialization promoted moral, social, and material equality among group members. No one was permitted a great surplus of valued items while others lacked sufficient means for survival and moderate well-being. Relative equality functioned to promote needed social stability, because people allowed to become desperately poor and deprived would likely display more deviance than a stateless society could handle. They lacked police and jails.

Consider, for example, the Ju/'hoansi (whom anthropologists formerly called !Kung) who hunt and gather in southwest Africa's Kalahari Desert. Equality is enforced partly by social rules that require the sharing of meat, so that when a large kill is made, everyone shares in the feast. "Ownership" of the meat involves the duty of dividing it among all members of the group. This is a position of importance that might have tended to give social advantage to good hunters. So egalitarianism is encouraged by making the person who owns the arrow that killed the animal the owner of the meat. Because arrows are traded freely in this society, the owner may not be a hunter at all, much less the hunter who killed the animal.[9]

Jealousy and personal animosities resulting in quarrels, violence, and even murder existed among the traditional Ju/'hoansi. Adjusted for population size, the murder rate between 1920 and 1955 was about as high as Detroit's in the early 1970s.[10] This may seem too high for decent living until it is realized that these deaths are the only ones deliberately inflicted by human beings on other people. In civilized society, by contrast, homocide is not the only way people deliberately kill one another. War is also common.

Lewis Mumford noted that a consistent characteristic of civilization (i.e., city-oriented societies where people dominate nature) is the war of one city against another. Organized armed conflict is abetted by the power monopoly within each city that allows its rulers to command residents to make war on people inhabiting other cities. "As the activities of the city became more rational and benign within, they became in almost the same degree more irrational and malign in their external relations.... Is it merely by chance," he asks, "that the earliest surviving images of the city, those on the pre-dynastic Egyptian palettes, picture its destruction?"[11]

Anthropologist Richard Lee makes the same point about conditions in the area inhabited by the Ju/'hoansi.

> In the nineteenth century the Botswana chiefdom imposed its order on the band-level San hunters in Eastern Botswana, only to wage intertribal warfare on a much larger scale against neighboring chiefdoms such as the Matabele and the Kalanga-Shona. Then at the end of the nineteenth century, the British industrial state brought the Pax Britannica to the warring chiefdoms of Southern Africa. But a generation later, the British mobilized thousands of Tswana warriors' sons to fight in the Mediterranian theater against the German and Italian states.[12]

In sum, although traditional foraging people lack police control of human behavior, the overall level of deliberate killing of human beings is lower than among "civilized" people because they lack collective killing. Genocide is unheard of and other forms of human oppression, such as slavery, are unknown, because they are impossible in societies where people do not dominate nature. No foraging group has enough physical force at its disposal for large-scale collective intrahuman violence and oppression.

People in foraging societies also enjoy affluence without hard work, according to anthropologist Marshall Sahlins.[13] Lee's studies of the Ju/'hoansi support Sahlins's view: "Ju/'hoansi appear to have the happy combination of an adequate diet and a short workweek."[14]

Distributive justice is generally assured because the entire way of life depends on sharing, according to Lee. "Each Ju is not an island unto himself or herself; each is part of a collective. . . . The living group pools the resources that are brought into camp so that everyone receives an equitable share."[15] Again, this serves to forestall deviance that a stateless group could not handle.

It is no surprise that, habituated as they are to plenty, foragers typically view the world's goods as abundant and sufficient, rather than scarce and insufficient.[16] They do not seek new ways to glean additional resources from nature because they believe that traditional methods suffice to assure not only survival but also reasonable comfort, both for the group and for each individual in it.

Foragers typically view their relationships with nonhuman constituents of the environment as reciprocally social rather than purely exploitational. The Other is valued for itself. Anthropologist Nurit Bird-David summarizes the views of three foraging societies, "the Nyaka of South India, the Batek of Malaysia, and the Mbuti of Zaire."[17] According to Bird-David, "Each group has animistic notions which attribute life and consciousness to natural phenomena, including the forest itself and parts of it such as hilltops, tall trees, and river sources."[18] They believe that people socialize with nonhumans. For example, the Mbuti believe that "the Forest visits the . . . camp, plays music, and sings with the people."[19] It also gives "food and gifts to everyone, regardless of specific kinship ties or prior reciprocal obligation."[20] Sharing among human beings is just a special case of "a cosmic system of sharing which embraces both human-to-human and nature-to-human sharing. The two kinds of sharing are constituents of a cosmic economy of sharing."[21]

Bird-David reports also that forest-dwelling foragers typically attribute to the forest traits that we associate with the divine. The forest is parent as well as provider.

> For example, . . . the Mbuti often refer to the forest as "father" and "mother". . . but also . . . describe it as the source of all spiritual matter and power, including the vital essence of people's lives. . . . In a similar vein, . . . the Nayaka not only refer to natural agencies (especially hilltops and large rock formations) by the terms *dod appa* (big father) and *dod awa* (big mother) and to themselves correspondingly by the terms *maga(n)* and *maga(l)* (male and female children) but also say that dead Nayaka become one with forest spirits.[22]

People with such beliefs tend to behave in environmentally responsible ways. Lacking belief in natural scarcity, they lack motivation to seek and implement new ways to overuse nature. Believing that many natural constituents are friendly beings with whom they engage in reciprocal exchange, they are inclined to treat nonhuman beings with the kind of respect that we typically prescribe for (but often fail to practice with) human beings. Finally, associating nature in general with their parents, and endowing it with qualities we associate with the divine, they are more reverential than exploitative. Thus, their ideas (in some ways) and their practices resemble what many environmentalists advocate. Aldo Leopold, for example, recommends viewing humans and nonhumans as members of the same land community: "A land ethic changes the role of *homo sapiens* from conqueror of the land-community to plain member and citizen of it. It implies respect for his fellow-members, and also respect for the community as such."[23]

At the same time, foragers are very people-friendly. Starting with the belief that nature provides sufficiently for everyone, traditional foraging people share with one another and achieve relative equality. No one is desperately poor, so no state is needed to keep people in line. Although lethal violence occurs, it is not collective and therefore is not as widespread as among us. There is no systematic exploitation of human beings as can occur only in a hierarchical society and nothing like the systematic killing that characterized the Holocaust.

Traditional foraging people show, then, that good intrahuman relationships are compatible under their circumstances with views resembling those of committed, nonanthropocentric environmentalists. Deep ecological commitments that relativize human worth do not lead necessarily to human oppression. I argue next that human oppression is a typical result of the western prescription that people exploit earth maximally in the supposed human interest.

The Eight-Part Pattern

There are eight aspects to the process whereby anthropocentrism in the West leads historically and understandably (but not by logical necessity) to human oppression.

1. People separate themselves conceptually from nonhuman nature and maintain that only human beings are ends in themselves.

2. People believe they are in jeopardy due to scarcity, because nonhuman nature does not automatically provide sufficiently for human wants or for what people have come to consider human needs.

3. People combat scarcity by seeking ever greater power over nature so it can be more thoroughly exploited in the human interest.

4. Technological innovation and specialization of function are used to increase human power over nature and to improve efficiency.

5. Such innovation and specialization increase human jeopardy rather than decrease it as people become, and perceive themselves to be, less self-sufficient and more vulnerable to one another.
6. Due to this perception, people increasingly relinquish power to nation-states in order to reduce jeopardy and coordinate specialized activities, and then compete for access to state power.
7. Many people experience increased insecurity, poverty, and discontent.
8. Strategies used to obtain and maintain centralized government power often result in human oppression.

I call this the eight-part pattern. Its aspects do not occur in strict chronological sequence, but the first three are foundational. Civilizations in general—and modern western civilization in particular, on which I focus—differ in these three respects from the foraging societies discussed earlier. Also, the first three aspects result in the other five.

Because my focus is human oppression in early modern, modern, and contemporary western societies, and in societies importantly affected by modern commercialism, most of my examples to illustrate the eight-part pattern primarily are drawn from these societies and time periods. I do not mean to suggest that the eight-part pattern does not apply, for example, to ancient or medieval China. But such application is beyond the scope of this essay and the range of my expertise.

The examples illustrating the eight-part pattern are not given in strict chronological order, because I want to illustrate the continuing, and often the current, influence of a given part of the pattern before moving on to the next part.

Finally, the association made here between human oppression and increasing control of nature does not preclude associations of such oppression with, for example, patriarchy, racism, and class exploitation. In fact, all three have been integral to the history of human oppression in and from the West, and all appear in current patterns of oppression. My thesis is that attempts to dominate nature in the supposed human interest are at least as fundamental as patriarchy, racism, and classism in catalyzing human oppression. If this is correct, attempts to reduce oppression that focus exclusively on patriarchy, racism, and/or classism, without addressing anthropocentrism, will fail, as a fundamental source (even if it is not *the* fundamental source) of human oppression will not be addressed, and so will continue unabated.[24]

1. People separate themselves conceptually from nonhuman nature and maintain that only human beings are ends in themselves. The most influential understanding of Christianity in modern western society holds that nonhuman nature exists to serve human beings who alone are made in the image of God. This view is reinforced by the Cartesian mind/body dualism, which attributes mind and reason to human beings alone, and by Kantian noumenal/phenomenal dualism, which attributes free will and morality to human beings alone.

2. People believe they are in jeopardy due to scarcity, because nonhuman nature does not automatically provide sufficiently for human wants or for what people have come to consider human needs. For the purposes of predicting human choice, economists have no need to distinguish needs from wants. Scarcity exists, according to mainstream economics, whenever people cannot obtain all that they want. Since wants have no limit, scarcity is endemic to the human condition, as there will always be too few resources, goods, and services to meet human needs and wants. This view can be found in the introduction to almost any economics textbook. Economist Clement A. Tisdell, for example, writes,

> The basic economic problem is considered to be how to manage or administer resources so as to minimize scarcity, that is, to minimize the "gap" between individuals' demand for commodities and the available supply of them.[25]

This view is a major justification for the widespread belief that the economy should grow. Only a growing economy that turns out more and more goods and services, it is thought, can supply the needs people perceive themselves to have. How many politicians have been elected in recent memory on a platform calling for a smaller, rather than a larger, economy?

3. People combat scarcity by seeking ever greater power over nature so it can be more thoroughly exploited in the human interest. Francis Bacon's works are the *locus classicus* of this view. He believed that expulsion from the Garden of Eden deprived people of the power over nature that was theirs by divine bequest. People must regain this power in order to flourish.[26] John Locke also maintained that people should extend their power over nature so as to meet human needs. Nature provides little of what people need, according to Locke, until people mix their labor with it. When people do mix their labor with nature, most of the value of the result is due to the human contribution. In light of scarcity, people have a duty to improve nature for human use. Because seventeenth-century Native Americans failed to do this, Locke denied that they had any property rights in the New World.

The same beliefs concerning human needs and wants are emphasized in our own day by commercial interests that promote development and oppose environmental protection. For example, antienvironmentalists typically equate monocultured trees with old-growth forests to legitimize increased human power over timber resources. The Bush administration adopted minimizing definitions of wetlands to make more areas available for human exploitation.

4. Technological innovation and specialization of function are used to increase human power over nature and to improve efficiency. Francis Bacon was among the most influencial advocates of using technological innovation to gain power over nature. He recommends "digging further and further into the mine of natural knowledge" in order to "establish and extend the power and dominion of the human race itself over the universe."[27]

Few people in the West today question the desirability of scientific research that leads to technological innovations that give people increased power over nature. We feel the need to conquer distance with trains, automobiles, planes, telephones, and televisions, and eagerly embrace new models and versions better suited to meet human needs and wants. Few complain, for example, about fiber optics and communication satellites. The value of technological innovation to give humanity inexpensive reliable sources of nonhuman energy is shared by advocates of both solar and nuclear technologies. Governments support research designed to delay death by curing such diseases as cancer and emphysema. Genetic research is used to help short people become taller and fat people become thinner.

Efficiency and human power over nature are also increased by the division of labor and specialization of function. Although Adam Smith saw the limitations of exclusive emphasis on efficiency, his illustration of efficiency in pin manufacture through division of labor is a classic. Henry Ford's assembly line is another popular example of people getting more of what they want from limited resources through division of labor. Today the theory of comparative advantage in international economics employs this premise. People in different parts of the globe, it is argued, are better at producing different things. Whether this is due to differences in cultural or natural resources, the result is that more of what people want and need is produced when each area concentrates on what it does best and trades with people in other parts of the world for everything else. Economist Tisdell summarizes the situation this way: "The scarcity problem explains the interest of economists in the efficiency of alternative social mechanisms for resource allocaton and their interest in economic growth and in development."[28]

5. *Such innovation and specialization increase human jeopardy, rather than decrease it, as people become and perceive themselves to be less self-sufficient and more vulnerable to one another.*

Specialization

Specialization and the division of labor make people more interdependent, as they depend on others to meet needs outside their own specialties. Carpenters, for example, depend on farmers and butchers, as they in turn depend on carpenters.

Vulnerability is the other face of interdependence. People on whom we depend for life's necessities are people who have power over us. Foragers, for example, depend on others in their group for cooperation in life's basic tasks. But because everyone has similar skills and available natural resources are adequate, no subgroup can control or limit the rest of the group's ability to accomplish their goals. Dependence does not make individuals vulnerable when cooperation and needed resources are so widely available. In addition, foragers know personally and individually the people on whom they depend. This fosters interpersonal bonds and builds confidence that needed cooperation will be forthcoming.

Similarly, although to a lesser extent, peasants in the Middle Ages were self-sufficient relative to people in the modern world because they knew how to do for themselves more of life's basic tasks than we do. Specialization was not as central to their way of life as it is to ours. But without electricians, plumbers, car mechanics, farmers, truckers, and food retailers, to name but a few, we would be in real trouble. Because the goods and services of specialists are available by and large only to those who have money, we are vulnerable to those who control our access to money, to our means of livelihood—that is, a subgroup, whether bosses, clients, or competitors, who effects our access to compensated work.

Not only does specialization make individuals vulnerable to other individuals in society, it also makes entire societies vulnerable to other societies. Our transportation requires not only car manufacturers and mechanics but also natural resources, such as oil and rare metals, that are not readily available in sufficient quantity in our society. So we are vulnerable to people in other societies where those resources exist in abundance, as the imposition of oil embargoes in the 1970s illustrated. We are vulnerable also to whatever forces may disrupt transportation of needed materials from distant lands. Shipping lanes and major canals must be made safe and kept open, hence the strategic importance of areas surrounding the Suez and Panama Canals.

Distant lands are also vital as markets for goods produced domestically. Just as individuals must have employment to earn money to buy life's necessities, societies need exports to get money to buy what they need from foreign lands. Britain's control of India in the ninteenth century, for example, was motivated not only by the need of cotton and other raw materials needed in British manufacture but also by the need of markets for these manufactured goods. Workers who fear for their jobs support foreign policies that keep foreign markets open, but often resist at the same time openning domestic markets to foreign competition.

The example of home construction illustrates how specialization motivated by the desire for greater efficiency in production can make whole societies more vulnerable to one another. These days, most homes are constructed on site, with most materials coming from the society where the home is situated. It is possible, however, that specialization of function and factory manufacture could give an advantage in efficiency to prefabricated housing, which would then be less expensive. This, however, would make countries where housing manufacture was concentrated more vulnerable than they are today to countries where raw materials exist. Timber resources sufficient for domestic production would not necessarily suffice for worldwide production. Shipping would become increasingly important to the housing industry, as would good trade relations that open distant markets. And people in those distant markets, because they would be no longer employed constructing housing on site for others in their society, would have to compete in some other income-producing endeavor to get the money necessary to live decent lives—become world suppliers of computer chips, or something. In short, increased efficiency would entail increased jeopardy for individuals and whole societies.

New Technologies

New technologies are related dialectically to specialization of function and add insecurity to the world in other ways as well. Technological change is more rapid in a world where people have specialized tasks, because some people's specializations are related directly to science and technology. New technologies, in turn, make additional specializations possible. For example, the invention of steamships, railroads, and refrigeration enabled some areas, such as parts of California and Central America, to specialize in the agricultural export of vegetables and fruits.

Developments in the technology and manufacture of war-making materials enabled European powers to conquer and impose colonial rule on much of the world. This rule, as already noted, facilitated global specializations and trade, as colonies supplied raw materials for industry in "mother countries." The need to control colonies catalyzed additional advances in transportation, communication, and implements of war.

Because specialization tends to increase human jeopardy, technological advances that facilitate specialization tend to increase human jeopardy as well, although indirectly. But many new technologies also augment human jeopardy directly. Means of destruction, often developed for use in war, become more powerful, more plentiful, less expensive, and easier to conceal. At the same time, specialization and the use of other technologies make the social and international fabrics increasingly complex and fragile. People who live in high-rise apartments, for example, depend on electricity for the elevator to get to and from their homes. Trade in perishable food items is vulnerable to terrorists, who may disrupt ship, train, air, and truck traffic with modern explosives. Due to this kind of vulnerability, an entire economy can be crippled by a very small group of people who have passive support from a large minority, such as occurred in Northern Ireland between 1969 and 1994.

6. Due to the perception of increased vulnerablility, people relinquish power to nation-states in order to reduce people's jeopardy and coordinate specialized activities and then compete for access to state power. When people are in jeopardy, they tend to be willing to forgo freedom and privacy in favor of safety. George Orwell saw this, so the Party in the novel *1984* kept a war going. It gave people a sense of jeopardy and justified the loss of personal liberties. The McCarthy era in the United States, which began just after Orwell wrote, was a real-life example of the same thing. Today, because we perceive terrorists as threatening air safety, we insist on and applaud searches of our luggage that previously would have been considered gross violations of privacy. And when our health is in jeopardy, most people are still willing to follow "doctor's orders," although they may now seek the opinion of a second doctor.

The nation-state's power has grown simultaneously with the increased jeopardy accompanying specialization and more powerful technologies. The state claims the right to monopolize decisions about how force is used in its jurisdiction. People who would hijack shipments, steal merchandise, blow up communication links, or otherwise jeopardize commercial and industrial activities that are central to our forms of specialization are sometimes caught and prosecuted by agents of the state.

The state also maintains the infrastructure needed for commercial and industrial specialization. It maintains a common currency; regulates the money supply; promotes employment through the provision of government jobs; supplies courts for the peaceful resolution of commercial disputes; makes treaties to keep foreign markets open; trains the work force through public education; and forestalls disruptive behavior through lessons during that education on the glorious history and promising future of the nation.

Because many industrial processes and products are dangerous, the state regulates these to protect the public. In a world of specializations, workers cannot be expected to know how dangerous their working conditions may be, and the public cannot be expected to judge for themselves the value or safety of many goods and services. So, for example, the state licenses doctors, dentists, airline pilots, and many other professionals.

The state profoundly affects employment opportunities and the distribution of wealth through its various laws and regulations, even in relatively free-market societies. Progressive taxation favors a more even wealth distribution, whereas regressive taxation favors greater gaps between rich and poor. Luxury and "sin" taxes discourage the consumption of certain items and reduce related job opportunites. Laws that enable unions to organize increase the price of labor, as do laws prohibiting hiring permanent replacement workers during strikes. In the absence of such legislation, the price of labor is lower, other things being equal. Treaties that allow foreign goods to compete with those produced domestically reduce the price of goods to consumers, but may also reduce job opportunities. Tariffs on foreign goods have the opposite effects.

Thus, impersonal forces of supply and demand do not alone determine one's success in the marketplace. Instead, a relatively small number of government officials can make all the difference. Philosopher and sociologist Ernest Gellner writes:

> Certainly the notion of a just price, inscribed into the nature of things, is a superstition. . . . But, alas, there is no market price either. . . . Once it is fully clear that the market operates only in a political context, . . . it also becomes obvious that the verdict of the market, the price, is not issued by the oracle of "the Market" alone. In fact it can only be the ventriloquist's mouthpiece for the particular political situation which happens to underlie it.[29]

When people's life chances depend so centrally on the money they can obtain, as is characteristic where people are specialized and interdependent in a commercial economy; and when the state profoundly affects job opportunites and wages, there is an enormous motivation to control the state's decisions and direct them toward one's own benefit. So competition for government power is intense. In societies where the views of the majority affect the distribution of government power, there is competition for the hearts and minds of the largest section of the population. This applies not only to representative democracies but also to any country where crowds of protesters and strikers, or large groups sympathetic to a terrorist minority (as in Northern Ireland), can undermine the social infrastructure needed for commercial life.

7. Many people experience increased insecurity, poverty, and discontent. Where people attempt to master nature in the human interest by finding innovative ways to exploit nature efficiently, and technological change is promoted as a means toward this end, the conditions of life change rapidly. New industries are created and old ones destroyed. People are uprooted when the raw materials they have been extracting or the products they have been making are no longer needed. They are encouraged to move to wherever workers are needed. This breaks the bonds of extended families and communities, leaving many people anomic as well as insecure.

International competition can reduce earnings in societies whose wages have been high or whose products can no longer compete. Because earnings are the life blood in commercial societies, insecurity and discontent increase with poverty.

Discontent arises also from increasing gaps between rich and poor in many countries. Where large industrial manufacture dominates production, large capital investments are needed to set up work for oneself and others. Because most people lack needed capital and expertise, they depend on others for employment. A glut of workers competing with one another for jobs tends to cause lower wages in these situations unless the government intervenes. Government intervention is weak where wealthy people have more political power than workers. So gaps between rich and poor, and therefore envy and discontent, increase.

8. Strategies used to obtain and maintain centralized government power often result in human oppression. The government maintains the social infrastructure by educating people in nationalism, by promoting the fiction that impersonal market forces dictate economic results, by blaming poor people for their plight, by promising future abundance for all who are diligent, and/or by blaming subgroups inside or out of the country for problems that many people are experiencing. Mark Twain said that patriotism is the last refuge of a scoundrel; however, he did not anticipate the proliferation of such refuges. Attention is often diverted from desperate circumstances by highlighting the dangers of various segments and "ills" of society. During the Great Depression, for example, National Socialists stressed the culpability and dangers of communists, homosexuals, Gypsies, and Jews.

While the Holocaust was unique, as are all particular historical events and movements, it displays a pattern related to increased jeopardy that results from attempts to subdue nature in the human interest. Market forces, government decisions, technological developments, and concentrations of power in those who offer employment combine to exacerbate human insecurity and discontent. Because the physical and social infrastructures are vulnerable to the use of new technologies of violence, governments must mollify the population. One strategy is to blame difficulties on either foreign enemies or racial, religious, or ethnic minorities.

Xenophobic appeals serve not only to maintain domestic peace in permanently difficult times but also to attract popular support to particular leaders and parties who wish to increase their power. For example, in the 1980s Argentina's military rulers played upon long-standing dislike of the British presence in the Faulkland Islands to galvanize support for their regime and divert attention from a faultering ecconomy and abuses of human rights. Their attack on the islands yeilded only temporary relief from domestic criticism, however, because they lost the ensuing war.

Leaders of Serbia and of Bosnian Serbs appealed similarly to xenophobic tendencies to increase their power domestically. They engaged in a concerted propaganda program to convince Serbs that they should not continue to live in peace with their Croatian and Bosnian neighbors. This gave the Serbian leaders in question the power that comes with leading a popular cause, which included war and "ethnic cleansing." The xenophobic tendencies relied upon by these leaders would not have resulted in violence except for their exploitation in the service of a political power grab.[30] Such a grab for power requires as its context the kind of highly centralized state that evolved to help people dominate nature in the supposed human interest.

Politicians' attempts to maintain and increase their power in a state context also underlie ethnically related killing in Rwanda. Violence started with the assassination of Rwanda's president who had agreed to inter-ethnic sharing of state power. The assassins made xenophobic appeals on the radio, resulting in possibly one-half million deaths.

The pillage, slavery, and starvation of the Dinka people that began in the 1980s in Sudan is another case in point. According to David Keen,

> although the central government (and some international observers) have sought to portray the violence between the Baggara and the Dinka as the latest manifestation of "natural" tribal hostilities, the conflict was by no means "inevitable", "natural" or "traditional". "Tribal" distinctions had been reinforced in divide-and-rule tactics that were part of a long and inglorious tradition embracing the earlier years of British governance. Like the British, post-Independence governments have resorted to the manipulation of conflicts in civil society in order to retain some semblance of control.[31]

Many other forms of human oppression result from western attempts to subdue nature in the supposed human interest. People who assume that nature cannot adequately meet human needs and wants promote specialization and trade. So, for example, much land in Central America is used to grow fruits and beef for export. Peasants are thrown off the land; few can find remunerative employment, so many go hungry. According to food activists Frances Moore Lappe and Joseph Collins, "By 1984 in El Salvador, with countless small farmers losing their livelihood as they were pushed off their land, 72 percent of all Salvadoran infants were found to be underfed."[32] Similarly in Brazil per capita food supplies in urban areas dropped by one-fifth during the 1970s while agricultural exports increased dramatically. Between the 1960s and the 1980s, as the gross domestic product soared, the fraction of undernourished people in Brazil rose from one-third to two-thirds of the population.[33]

Another consequence of attempts to subdue nature is popular indifference to oppression. Hannah Arendt called this the banality of evil and considered it central to the Holocaust.[34] Following Arendt, Douglas Porpora considers such indifference characteristic of what he calls "holocaust-like events,"[35] many of which involve the United States. When peasants in Central America revolted against their destitute condition and the nation-state was too weak to subdue them, the United States aided the beleaguered state. Most Americans seemed indifferent to their government's support in recent decades of state-sponsored terror in Guatemala, El Salvador, and Nicaragua,[36] thereby making these "holocaust-like events" possible.

So why are people indifferent to the suffering of other human beings? Indifference to human oppression can be expected when people believe that nature provides for them insufficiently and so must be mastered to serve humanity better. This belief leads to the division of labor, the predominance of often anonymous exchange relationships in economics, and the association of increased economic activity with improved human well-being. In an industrial society where there is the constant risk of overproduction and unemployment, the economy's growth is fueled by consumerism. People are convinced that their lives are not as good as they could and should be, and that they must buy some good or service to improve their lot; most people feel too deprived to care much about what is happening others. Thus, the banality of evil and the indifference that allows holocausts to happen are tied through the eight-part pattern to consumerism and the despoilation of the earth.

Conclusion

Many Nazis valued local environments, which they wanted to protect from greedy capitalists. They rejected emphasis on the property rights of developers. They believed human flourishing required people to cherish local landscapes and steep themselves in local traditions that reflect interconnections between a people and their particular land.[37]

Although these positions superficially resemble Deep Ecology and the Land Ethic, their ideological context is opposite to what I believe is foundational to deep environmentalism. Deep environmentalism rests importantly on a sense of the possibility of plenty, peace, and diversity, rather than the inevitability of scarcity, conflict, and winner-take-all. It does not view the human condition as inherently one of jeopardy calling for competitive struggle. People can be plain members and citizens of their ecological communities because there can be enough for everyone, regardless of species, so long as no one tries to dominate nature and thus all others. Respect for diversity means respect for others. Reciprocal exchange with others is therefore preferred to competitive struggle against them.

The Nazis were entirely different. They sought and attained power in a state whose nature reflected the political and social evolution integral to western attempts to master nature in the supposed human interest. In other words, the first seven parts of the eight-part pattern were already in place. Although many Nazis had genuine affection for their landscapes and ecosystems, this affection was mixed with belief in continuing jeopardy. To the jeopardy of scarcity that characterizes western economic thought, the Nazis added jeopardy due to international competition and struggle (which is also a characteristic consequence of western attempts to dominate nature), and jeopardy resulting from supposed Jewish conspiracies. Anti-Semitism served the Nazis much as anti-Bosnian sentiments have more recently served Serbian leaders. Thus, because the Nazi ideology emphasized competitive struggle to overcome jeopardy from capitalists, communists, Jews, and so forth, their entire mind-set was opposite to the perspective common among nonanthropocentric environmentalists.

In sum, the Nazis illustrate the thesis that separating ourselves conceptually from nature and seeing scarcity and struggle as endemic to the human condition underlie both much human oppression (from colonial conquests to the Holocaust) and much environmental destruction. This suggests that environmentalist prescriptions do not necessarily disfavor humanity. When people can be more self-sufficient in their local communities they will be at greater peace with one another and with the earth, as they will have no further need to conquer either one.

This need not require the end of civilization. Even though human oppression has generally increased in the last 5,000 years since the advent of city life, civilization also carries benefits, such as writing and symphonies. The challenge is to reap such benefits in self-assured societies in which both people and nature are respected and revered.

Notes

1. Peter Carruthers, *The Animals Issue* (New York: Cambridge University Press, 1992), p. xi.
2. Michael E. Zimmerman, "The Threat of Ecofascism," *Social Theory and Practice* 21 (2) (Summer 1995); pp. 207–238, especially 216–221. "Bioregionalism" refers to the view that human cultures should reflect ties to the particular geographic regions where people live. Human lifestyles and values may differ from place to place to reflect human integration with local ecosystems.
3. For issues regarding Heidegger see Michael E. Zimmerman, "Rethinking the Heidegger-Deep Ecology Relationship," *Environmental Philosophy* 15 (1993): 195–224.
4. J. Baird Callicott, "Animal Liberation: A Triangular Affair," *Environmental Ethics*, Vol. 2 (1980): 324.
5. Tom Regan, "Ethical Vegetarianism and Commercial Animal Farming," in Richard Wasserstrom, ed. *Today's Moral Problems*, 3rd ed. (New York: Macmillan, 1985) pp. 543–544.
6. Roger S. Gottlieb, "Spiritual Deep Ecology and the Left: An Effort at Reconciliation," *Capitalism, Nature, Socialism* 6 (3) (September 1995); 1–21, at 6–7.
7. For this concept see Douglas V. Porpora, *How Holocausts Happen: The United States in Central America* (Philadelphia: Temple University Press, 1990), especially pp. 6–10.
8. For a longer treatment of some of the themes in this paper that includes prescriptions for our society see Peter S. Wenz, *Nature's Keeper* (Philadelphia: Temple University Press, 1996), especially chapter 10.
9. Richard B. Lee, *The Dobe Ju/'hoansi* 2d ed. (New York: Harcourt Brace College Publishers, 1984, 1993), pp. 55–56.
10. Richard Borshay Lee, *The !Kung San* (New York: Cambridge University Press, 1979), p. 398.
11. Lewis Mumford, *The City in History* (New York: Harcourt, Brace and World, 1961), p. 51.
12. Lee, *The Dobe Ju/'hoansi*, p. 399.
13. Marshall Sahlins, "The Original Affluent Society," in *Stone Age Economics* (Hawthorne, NY: Aldine Publishing Co., 1972), pp. 1–39, especially pp. 36–37.
14. Lee, *The Dobe Ju/'hoansi*, p. 60.
15. Ibid.
16. Nurit Bird-David, "Beyond 'The Original Affluent Society'," *Current Anthropology* 33 (1) (February 1992); 31–32.
17. Bird-David, "Beyond," p. 25.
18. Ibid., p. 29.
19. Ibid.
20. Ibid.
21. Ibid., p. 30.
22. Ibid., p. 29.
23. Aldo Leopold, *A Sand County Almanac* (New York: Ballantine Books, 1970), p. 240.
24. In my longer treatment of some of these matters cited in note 8, I argue additionally that the failure to address the problem of anthropocentrism perpetuates

conditions that foster sexism, racism, and classism. So if anthropocentrism is not adequately addressed, human oppression will not only continue but will also continue largely through perpetuation of sexism, racism, and classism.

25. Clement A. Tisdell, *Natural Resources, Growth, and Development: Economics, Ecology, and Resource-Scarcity* (New York: Praeger, 1990), p. 1.
26. My treatment of Bacon is drawn from Carolyn Merchant, *The Death of Nature* (New York: Harper and Row, 1980), pp. 168–72.
27. Merchant, *Death*, pp. 170 and 172.
28. Tisdell, *Natural Resources*, pp. 1–2.
29. Ernest Gellner, *Plough, Sword, and Book* (Chicago: University of Chicago Press, 1990), pp. 186–187.
30. For this view see Tom Sjelton, *Sarajevo Daily* (New York: HarperCollins, 1995).
31. David Keen, "The Benefits of Famine: A Political Economy of Famine in Southern Sudan," *The Ecologist* 5 (6) (November/December 1995); 214–220, at 216.
32. Frances Moore Lappe and Joseph Collins, *World Hunger: Twelve Myths* (New York: Grove Weidenfeld, 1986), p. 86.
33. Lappe and Collins, *World Hunger*, pp. 86 and 87.
34. Hannah Arendt, *Eichmann in Jerusalem: A Report on the Banality of Evil* (New York: Viking Press, 1963).
35. Porpora, *How Holocausts Happen*, p. 9, where he accepts Hannah Arendt's view about the banality of evil.
36. Ibid., pp. 83–117.
37. For information on Nazi thought on these matters see Zimmerman, cited in note 2.

2

Time, Narrative, and Environmental Politics

John O'Neill

Environmental problems raise two major theoretical challenges to orthodox ethical and political thought.[1] The first concerns the place of nonhuman nature in our scheme of values. The second concerns the place of time, history, and narrative. Both challenges raise questions that existed prior to environmental problems, but both have become stark in the new context.

Discussion in environmental ethics has tended to focus almost entirely on the first problem, the ethical considerability of the nonhuman world: the conflict between anthropocentric and biocentric perspectives has been at the center of the debate. It is not my purpose in this paper to deny that there is anything of value in this exchange. However, one consequence is that the second challenge has been largely ignored; where it has been raised, it has been neither adequately formulated nor addressed. Yet in the "old world," at least—and much of the globe is an old world for someone—it is quite as important, if not more so, as the first challenge. Moreover, it highlights more clearly the institutional and political dimensions to the debate around the environment that, in the discussion of the values of nature, often seem to get lost in the thought that if we got our values right our problems would be solved.

I. Justice and Obligations to Future Generations

To say that the temporal dimension of the environmental challenge has been largely ignored might seem a bit of an exaggeration. One aspect of that challenge, our relation to future generations, has been at the center of the discussion of environmental problems. The philosophical debate on this follows now well-worn paths that apply existing ethical and political theories to the new context. On the one side of the debate stand moral theories that entail strong obligations to future generations. The standard theories are those that take morality to begin from an impartial perspective. There are two favorites. The first is utilitarianism, which holds that the best action is that which maximizes total well-being, where well-being is characterized in the classical theory hedonistically in terms of pleasure and the absence of pain, and in modern versions in terms of preference satisfaction. The utilitarian view involves no temporal indexing of pleasures or preferences and entails that total well-being should be maximized across generations, be this by increasing individual well-being or by increasing future populations.[2] The second impartial perspective invoked is some version of Rawls's theory of justice. Rawls ensures impartiality by specifying that the principles of justice be those that would be agreed upon by self-interested individuals in conditions of ignorance of their position in society, of their dispositions to take risks, of their beliefs about the good. Rawls assumes in his own account that those in this original position belong to the same generation, and introduces obligations across generations by piercing the veil of ignorance and granting the agents in the original position the "knowledge" that each cares about someone in the next generation.[3] Later theorists, however, avoid this ad hoc move and assume that some or all generations are represented, where representatives do not know to which generation they belong.[4] Both the utilitarian and Rawlsian perspectives have generated voluminous literatures that address the paradoxes both are taken to involve.[5]

Ranged against these views are another standard set of ethical and political theories that point in the opposite direction toward weaker obligations to future generations. Again, two kinds of theory are particularly influential here. First are those in which the point of moral rules is to serve as a means by which individuals of limited altruism can realize their long-term interests in conditions where they are roughly equal in power and vulnerability—Humean and certain contractarian theories of justice and obligation provide standard examples.[6] The theories are taken to entail that obligations to future generations are either absent or restricted: since there is an inequality of power and vulnerability (we can harm them, they cannot harm us), the circumstances of justice or obligation do not exist. Second, there are those historical and procedural theories of justice whose requirements of justice are satisfied if we have acquired existing goods by legitimate procedures: justice is not concerned with realizing some end state in which goods are distributed according to some ideal pattern.[7] Again, the consequence is that questions of justice do not arise in our relations to future generations: since justice is not concerned with the legitimacy of future outcomes but the legitimacy of the historical acquisitions, there is nothing in justice we owe to the future.

Thus go the standard approaches to the temporal dimension of environmental problems. The challenge is seen in fitting existing theories to deal with the future obligations. None of these approaches is adequate for reasons I outline later. However, I begin my criticism by looking at problems the theories have from the opposite direction. The theories deal with only one-half the temporal dimension of environmental concern, and while that problem is of great significance, it is that which raises the least serious theoretical challenge to standard ethical and political thought. The other half of the temporal dimension of environmental problems concerns our relation to the past. Not only do current ethical theories deal badly with this, but their failure to do so also points to difficulties in their account of obligations to the future.

2. History and Environmental Evaluation

What then is the second part of the temporal challenge to standard ethical and political theory? It lies in the way that the history of a location matters in environmental evaluation. This is true even where we are concerned with "natural history"; the geological history of mountain and valley, the depth of history embodied in the ancient forest, and other such features of nature's history inform our evaluations of natural objects. It is because natural history matters that we cannot "fake" nature. However, the historical dimension is still more evident in the old world in the weight of human history that is impressed upon both natural and human environments. Little in old-world nature conservation is concerned with the conservation of wilderness—there is little, if any, wilderness in the sense of ecosystems unmodified by human intervention. Rather, the central issues in nature conservation are about specific parts of nature that have a particular history of human use that brings with it a particular habitat. The problems that have arisen are often due to abrupt changes in the use of the land: the loss of meadowland and pastures with the use of nitrogen and intensive farming techniques, radical shifts in patterns of grazing, the disappearance of hedgerow and copses to make way for efficient use of new machinery, and so on. Most nature conservation problems are concerned with flora and fauna that flourish in particular sites that are the result of a specific history of human pastoral and agricultural activity, not with a site that existed prior to human intervention. History is also still more clearly evident in the specific problems of conservation of human artifacts that have been left behind by past generations, such as stonewalls, thingmounts, and quarries. Finally, history enters into the value of place.[8] The value of specific locations is often a consequence of the way that the life of a community is embodied within it. Historical ties of community have a material dimension in both human and natural landscapes within which a community dwells. The city is an environment that has suffered its own dislocations: hence the complaint "I no longer know the place." Place is also valued on a more individual level—the local familiar walk, stream, pond, or landscape that invokes a very specific past of one's childhood, parents, friends, or whatever. Hence the strong reactions roused by external changes that transform those places.

In all these cases, the specific history of a place matters in the evaluation. That a particular place embodies a history blocks the substitutivity of one place for another. Place, like nature, cannot be faked. What matters is the story of the place. It is this dew pond, constructed by people long ago to water their livestock, which is where we used to picnic, where I looked for frogs as a child, that I want to preserve. Another identical dew pond built last year could never do as a substitute simply because its history is wrong. We want to preserve an ancient meadowland, not a modern reproduction of an ancient meadowland. Time and history enter also as constraints on our future decisions about and actions toward a place. Marx's point that we make our own history but not in the circumstance we choose can have a normative and not just an explanatory point. One failure we can make in decisions and actions about a place is not to respect its past and specific history, the work of the individuals it contains, the memories it embodies, and so on. And again, at the level of popular resistance to changes to a place, this is often the issue at stake. The resistance is not to change per se, or at least it need not be, but rather with the kinds of changes that are thought appropriate. Indeed, one major problem with the heritage industry is the way it often attempts to freeze historical development. A place then ceases to have a continuing story to tell. When you look at the landscape and ask, "what happened then in the year the heritage managers arrived?" all there is to say is that they preserved it and nothing else happened. The object becomes a mere spectacle taken outside of history.

The specific history of a place matters, then, in an evaluation of it and constrains our decisions as to what kind of future is appropriate to it. This feature of environmental valuation is one that modern ethical and political theories have real difficulties capturing. They fail to take account of the way time and history enter into environmental appraisal. The failing is most obvious in the case of utilitarianism, which, as a consequentialist morality, is notoriously forward-looking. Reference to the past cannot as such count for the utilitarian.[9] Hence, while the problem of future generations is grist to the utilitarian mill, the problem of the past presents deeper problems. While Rawlsian theory is not consequentialist, its impartial perspective stretches across time. Evaluations of specific history and processes form part of that body of knowledge of particulars, of which those in the original position are ignorant: agents enter into deliberation devoid of knowledge of the particular time and place in which they exist.

The way history and time enter into environmental valuation is also dealt with badly by ethical and political theories that do have an historical dimension. Contractual theories of justice are inadequate. Reference to specific past acts that create obligations, such as promising, are not relevant here. We are not constrained in our actions toward places in virtue of any promises we have made. The obligations we have to the past, if that is a proper way to speak here, are entirely nonvoluntary. For this reason, reference to contracts is out of place as are our evaluations of and the constraints on our actions from the legitimacy of the procedures that got us where we are. Goodin, one of the few political theorists to note the importance of history and processes as sources of environmental value, spoils his case by associating it with Nozick's historical entitlement theory of justice.[10] This appeal to Nozick to introduce the point about history misses the mark entirely. The value a place may have—say, as an ancient meadowland—has nothing at all to do with the justice of any procedure that handed it down to us.

Standard forms of ethical and political theory fail then to capture the historical dimension to environmental evaluation. However, neither is this feature captured by deep-green critics of such theories. Deep-green theory shares with orthodox theories an ahistorical approach to values. Indeed, from deep-green perspectives, which tend to come from the "new world," it is often difficult to see just what is of value in the old world of Europe. Given that central normative concept in deep-green evaluation is that of "wilderness," understood as nature untouched by humans, it follows that European landscapes, in virtue of the very fact that they embody so much human history, are without value. This appraisal is not only just plainly wrongheaded, it avoids confronting the problem of how to appeal to "wilderness" in new-world environmental landscapes: for its native inhabitants the new world was an old world.

In both Australia and the United States, the treatment of nature as a primitive wilderness has led to a failure to appreciate the ecological impact of native land management practices, especially those involving burning, a fact that has itself been the source of problems in the treatment of the ecology of the new world. Thus consider the history of the management of one of the great symbols of the American wilderness, Yosemite National Park. In the influential report of the Leopold Committee, *Wildlife Management in the National Parks*, we find the following statement of objectives for parks:

> As a primary goal we would recommend that the biotic associations within each park be maintained, or where necessary recreated, as nearly as possible in the condition that prevailed when the area was first visited by the white man. A national park should represent a vignette of primitive America.[11]

What was that "primitive" condition? The experience of the first white visitors to the area is reported thus:

> When the Forty Niners poured over the Sierra Nevada into California, those who kept diaries spoke almost to a man of the wide-space columns of mature trees that grew on the lower western slope in gigantic magnificence. The ground was a grass parkland, in springtime carpeted with wildflowers. Dears and bears were abundant.[12]

However, this "original" and "primitive" state was not a wilderness but a cultural landscape with its own history. The "grass parkland" was in part the result of the pastoral practices of the Native Americans, who had used fire to promote pastures for game and black oak for acorn. After the Ahwahneeche Indians were driven from their lands by Major Savage's military expedition of 1851, "Indian style" burning techniques were discontinued and fire suppression controls introduced. The consequence was the decline in meadowlands under increasing areas of bush. Totuya, the granddaughter of chief Tanaya and sole survivor of the Ahwahneeche Indians, returned in 1929 and remarked on the landscape she found, "Too dirty; too much bushy."[13] It was not just the landscape that had changed. In the giant sequoia groves the growth of litter on the forest floor—dead branches and competitive vegetation—inhibited the growth of new sequoia and threatened more destructive fires. Following the Leopold report, both cutting and burning were used to "restore" Yosemite back to its "primitive" state.

However, to talk of restoring a park to a "primitive," "natural," or "wilderness" condition simply disguises the nature of the problems. Reference to wilderness suppresses one part of the story that can be told of the landscape: the non-European native occupants of the land are themselves treated as part of the "natural scheme," of the "wilderness," without history, and not as dwellers in a landscape that embodies their own cultural history. Moreover, the language also disguises the way in which the history of the landscape is being artificially frozen at a particular point in time: "The goal of managing the national parks and monuments should be to preserve, or where necessary recreate, the ecologic scene as viewed by the first European visitors."[14] To refer to mythical natural or wilderness states avoids the obvious question: "Why choose that moment in which to freeze the landscape?" There are obvious answers to that question, but they have more to do with the attempt to create an American National culture than with ecology. And here there are normative objections to make to the use of wilderness to gloss over part of the darker side of American history. The appeal to wilderness, in this context at least, is a way of avoiding coming to terms with a troubling historical dimension to environmental evaluation.

Neither standard ethical and political theories nor their deep-green opponents provides an adequate way of dealing with the historical dimension of environmental evaluation. Why? The answer lies in the absence in such perspectives of any reference to the role that narrative plays, or ought to play, in our evaluations and decisions. Our decisions about a place are in part determined by considerations as to how best to continue the narratives that can be truly told of it. It is not a question of merely satisfying some tick list of valued items, nor acting on the basis of some particular past act like a promise, but of ascertaining what actions are appropriate given the particular life history a place embodies. The problem we have is how to continue the story of a place into the future, given its past. The question of our relation to the past is tied up with the account of the relation we have to the future. It concerns problems of how we deal with the passage from past to future and the shape our individual and collective lives take. And while this dimension of evaluations is particularly highlighted by issues in nature conservation, it is not peculiar to it. More generally, part of both individual and social decision-making processes should involve notions of narrative unity and continuity: of an individual life—where should my life go from here given the story of my life thus far; of a community—where should we go from here given the history we have experienced.[15]

3. Back to the Future

This lack of reference to narrative unity or continuity lies at the basis of a common failing in the accounts offered by standard ethical and political theory of our relation to future generations. The questions of what kind of future we want and what we owe to the inhabitants of the world that follow us is presented in abstraction from history. The future both of individuals and communities is treated as if it concerned a distant land that has no connections to that in which we live. This is particularly evident in discussions of the practice of discounting in economic evaluation.

To discount the future is to value future costs and benefits of any proposed option at less than those of the present using some specified discount rate, r. Since benefits and costs in economic valuation are taken to be measures of preference satisfaction and dissatisfaction, this entails taking the future preferences of either ourselves or others as less than those of the present. Current preferences count more than future preferences. Given a discount rate r, a preference measured at £n now is valued in t years time at £$n/(1 + r)^t$. This has clear implications for our treatment of future generations for it magnifies the value we place on current benefits and diminishes the value we place on future costs: we are justified in shifting costs into the future for our current benefit. Hence it is open to the objection that it involves a systematic injustice to future generations.

There are a number of different justifications offered for discounting.[16] Here I focus on one, the appeal to pure-time preferences, and the debate this has generated. The argument goes that individuals have pure-time preferences—they prefer benefits now to benefits tomorrow and costs tomorrow to costs now simply in virtue of the time they occur. And the more distant the future, the more costs and benefits are discounted. People are impatient. The defender of pure-time preferences claims we have to respect these preferences.[17] If well-being consists of preference satisfaction, then we must include these preferences in arriving at the policy that maximizes well-being. The opponent of pure-time preferences holds that a concern with maximizing well-being demands that we ignore pure-time preferences, since to include them would be to fail to maximize the satisfaction of preferences over persons' lifetimes. In the words of Pigou, time preferences only show that "our telescopic faculty is defective."[18] One should take a temporally neutral perspective on the satisfaction of preferences both across an individual's lifetime and across generations. Pure-time preferences are irrational.

What is of note in this debate is the shared assumptions protagonists have about the passage through time of a person's life. Both accounts of time preferences make sense only if one pictures a person's life as consisting of a series of momentary acts of desire satisfaction. What is lacking is a strong sense of identity over time that is given by narrative unity or continuity. Our relation to the future is treated like a relation to other persons. For the defender of time preferences at any moment t_0 O'Neill cares for O'Neill$_0$ who exists now more than O'Neill$_1$ who exists at t_1 who is a physically related relative of O'Neill$_0$ who in turn is cared for more than the more distant relative O'Neill$_2$—and so on as time stretches out into the future. The only relation between the various O'Neills is one of physical continuity. The critic of pure-time preferences assumes the same picture but is an economic maximizer who demands equal consideration for future preferences on the basis of an oddly impersonal perspective. A life is still a series of discrete acts of consumption, but the maximizer asks for any future O'Neill, O'Neill$_n$ to be given the same consideration as the current O'Neill, O'Neill$_0$. The life still lacks internal coherence. The relation of a person to the future becomes one of self to others.

This picture of temporal neutrality has indeed been exploited by utilitarians. Sidgwick raises the following objection to the egoist, understood in the uncommonsensical sense as one who refuses the utilitarian principle of universal benevolence:

> If the Utilitarian has to answer the question, 'Why should I sacrifice my own happiness for the greater happiness of another?', it must surely be admissible to ask the Egoist, 'Why should I sacrifice a present pleasure for one in the future? Why should I concern myself about my own future feelings any more than about the feelings of other persons?'[19]

The thought, developed in detail by Nagel and Parfit, is that a self-interest theory of rationality, according to which each person aims at the world that makes his or her life go as well as possible, is vitiated by the fact that it is agent-relative—i.e., it gives reasons that are reasons only for the agent—but not time-relative—i.e., it aims to maximize well-being across time and not just at the present moment. The claim is that the rejection of impersonality and the acceptance of temporal neutrality cannot be happily combined. My relation to my future selves is not in principle different from my relation to others.[20] Any future "self" is an other like any other. This account of the self to its future underlies all the competing perspectives in the debate about time preferences.

This account needs to be rejected. The picture of identity over time that it assumes is in error. What is absent is a view of human lives as ones that have narrative structure—as stories of physical and mental growth and development, of decline, of the success and failure of projects and relationships. It fails to allow any place for narrative order in our identity, and correspondingly in the evaluation of a moment or period in our lives. The future of one's own life does matter for an appraisal of current life in a way that the current or future lives of strangers need not. Our future depends on how well we can say our present is going. Our present life is part of a larger narrative, and the shape of that whole life matters. A personal relationship that begins in contention and ends in reconciliation is preferred to one that begins in apparent harmony and ends in discord. The way in which the earlier moments of contention and harmony are to be understood and appraised depends on the larger narrative. If a person suffers great difficulties in attempting to write a book, to prove some mathematical theorem, to win some social struggle or whatever, and then ultimately succeeds, the suffering is redeemed in a way that is not the case if the attempt fails. The outcome defines how the present suffering is to be evaluated.

The way matters turn out matters. Our lives are not or should not be a series of disconnected events such that at any moment we can say now whether our lives are going well or badly. The future is what determines the appraisal we can give to the present. Hence the truth in Solon's dictum that we can call no man happy until dead.[21] I dissent from Solon only in holding that the point of death is too soon to make an evaluation: a person's death is not the end of the narratives of which they are part. And this points to the place of narrative in our proper understanding of our relations across generations.

It is possible for us to engage in projects and belong to communities such that how well our lives can be said to go depends on what happens to the projects and relationships that occur beyond our lifetime. It can matter to us the way the future will be, and thus we have a stake in creating a particular future. Consider the activity of doing science. The status of scientific works depends on their relation to both a particular past and a particular future. In relation to the past, a piece of scientific work only makes sense within a particular history of problems and theories to which it makes a contribution. Its success or failure depends on its capacity to solve existing problems where others have failed. However, it also depends on a projected relation to the future in terms of its capacity to solve not just existing problems but also problems not envisaged by its author, and in its fruitfulness in creating new problems to be solved and new avenues of research. Correspondingly, it matters for current scientific activity that there exist future scientists educated in a discipline and able to continue work within it. The same points apply in the arts. The greatness of many works of art lies in their ability to continue to illuminate human problems and predicaments in contexts quite foreign to those in which they were originally constructed. Likewise, many of the aesthetic qualities of a work of art may only become apparent by virtue of its relation to future works. In that minimal sense, Eliot is right: "The past should be altered by the present as much as the present is directed by the past."[22] For this reason, it is significant for us now that there be future generations able to appreciate the arts and contribute to them.

Similar points apply to more "prosaic" activities—to politics, for example. The success or failure of major political projects normally becomes apparent only well after the political actors have ceased to be active: witness the Bolsheviks in contemporary Eastern Europe. These points apply also to everyday working activities, where these involve skilled performances that are embodied in objects and landscapes. Consider the hedgerows of Britain: these are the product of the skilled work of laborers that stretches back for centuries. We fail both past and future if we lose our capacity to appreciate the skill embodied in the hedgerows or create a future social world that contains no appreciation of their value and destroys them as mere impediments to more profitable agriculture.

The accounts of the future offered by standard ethical and political theories are vitiated by exactly the same problem as their accounts of the past: the failure to note the place that narrative structure plays in appraisal and decisions. The future is not a distant land inhabited by strangers unrelated to us. The success or failure of our lives depends on future generations, for it is they that are able to bring to fruition our projects and relationships that form the structure of the narratives of our lives. This point is missed in the standard atemporal discussions of future generations. Golding asks, "What obligations do we have to future generations since they may not belong to our moral community?" and answers that we have none.[23] Others have responded that we do have duties to strangers, whatever their nature, appealing either to utilitarian considerations or impersonal principles of justice. Both sides fail to acknowledge that our primary responsibility is to attempt, as far as possible, to ensure that future generations do belong to a community that has a narrative continuity with ourselves—that they are capable, for example, of appreciating works of science and art, the goods of the nonhuman environment, and the worth of the embodiments of human skills, and are capable of contributing to these goods. This is an obligation not only to future generations but also to those of the past, so that their achievements continue to be both appreciated and extended; and to the present—ourselves. The future matters to us now.

To say this is not to endorse a quiet conservative picture of our relation to the future, although it may be that conservative thought has been more sensitive to this dimension than its liberal opponents. Present, future, and past are not linked by some single set of values that the present passes from past to future, but by arguments both within generations and between them. Indeed, what counts as an achievement or a success in the lives and work of those in the past is often contested—consider, for example, the differing status one might accord to the theoretical works of Locke, Hegel, Mill, or Marx, or to the political actions of Cromwell, Jefferson, Luxemburg, and Lenin. Differences in values and interests among the present generation affect not only the past but also more clearly the future. Such conflicts are associated with differing aspirations for the future and hence, indirectly, with the kinds of people that will inhabit the future. It is not, then, a question of passing on a "shared set of values" to future generations, but rather an ongoing argument about the proper transition from past to future. The conservative image of a quiet and contented continuity of past, present, and future is one voice in that argument and one that is itself open to contestation.

4. A Brief Social History of Time

While the problem of narrative order is one that is particularly visible in the environmental context, as I noted at the start of this paper, it existed prior to it. Indeed, the relation of historical narrative and environment was central to the debates at the opening of commercial society. The disruption of historical narrative by market norms lay at the heart of debates about land and commerce in Britain in the seventeenth and eighteenth centuries, and while that debate has seemed to disappear in much of recent political and economics theory, it has never entirely left the scene and has indeed returned in an acute form in recent social theory.

Early critics of emerging commercial society were concerned in part with the effects of the mobilization of landed property by commerce on the conditions of community and projects across time. The eighteenth-century civic humanist criticism of commercial society was founded on the belief that civic virtues had their basis in stable ownership of landed property. The material foundation of a good society lay in "real property recognizable as stable enough to link successive generations in social relationships belonging to, or founded in, the order of nature."[24] Commercial society, by mobilizing land, undermined that link between generations.

This relationship between the market and the mobilization of land has been echoed by later socialist critics of commercial society. It is not just the mobilization of landed property by the market that undermines intergenerational identity, but also the mobilization of labor. Specific commitments to a particular locality and place, to a stable extended community within a locality, and to a particular craft and profession are inimical to and undermined by the workings of a market society. The theme is particularly evident in Polanyi. Both land and labor are fictitious commodities, objects treated as if they were produced for sale upon the market. As fictitious commodities, they are bought and sold in real markets, and the consequence is the disruption of social ties of place:

> To allow the market mechanism to be sole director of the fate of human beings and their natural environment . . . would result in the demolition of society. . . . Nature would be reduced to its elements, neighbourhoods and landscapes defiled, rivers polluted.[25]

Workers in a market society must be prepared to shift location and occupation if they are to achieve the market price for their labor. The ties of "a human community to the locality where it is," ties of place, are undermined. To have a tie to a place is to have a tie with an environment that reveals a particular past history.[26] This tie is, to use our earlier examples, to recognize the skills embodied in dry stone walls, hedgerows and buildings and to have a sense of continuity with those whose skills are thus made public. Hence, to disrupt ties to place is not merely to remove persons from a particular spatial location but also to divorce them from relations to previous generations and a sense of continuity with the future.

Historical continuity is similarly disrupted by the mobilization of labor across occupations. The advocacy of the mobility of labor was central to the defense of the market economy by its early defenders like Adam Smith whose work was aimed against unions and the practice of lengthy apprenticeship as a barrier to the movement of labor. Again, a major response was that ties across generations are weakened by the disappearance of continuity in craft and work. The relationship of craftsman and apprentice is undermined, and with it the sense in which success in craft work was tied to past and future. The view is echoed in Weil's comment:

> A corporation, or guild, was a link between the dead, the living and those yet unborn, within the framework of a certain specified occupation. There is nothing today which can be said to exist, however remotely, for the carrying out of such a function.[27]

The mobilization of labor by the market, like the mobilization of land, has undermined a sense of community across generations.

These critical accounts of the effects of commercial society on our relations to past and future are, I believe, substantially correct. However, to make them is to raise a problem rather than a solution. For there are good reasons to reject a possible corollary that we return to stable ownership of the land and limited mobility in labor. The particular ties of premodern societies were often oppressive, and the dissolution of old identities a liberation from personal servitude and narrow horizons. Moreover, even if it were desirable to limit mobility, in modern conditions it would not be practicable without excessive coercion. The problem of obligations to future generations is a social and political problem concerning the economic, social, and cultural conditions for the existence and expression of narrative identity that extends across generations. At the heart of that issue is the problem that has been the focus of much social and political theory for the last two centuries—that of developing forms of community that no longer leave the individual stripped of particular ties to others, but that are compatible with the sense of individual autonomy and the richness of needs that the disintegration of older identities also produced. The recent debate between communitarians and liberals is a variation on a well-established theme, and itself restates the problem rather than offers a new solution.[28]

One recent turn in social theory that will clearly not do as a solution to that problem is that of postmodernism, in which a radical aesthetic model of autonomy is asserted at the expense of identity and narrative. The postmodernists owe a large debt to the situationist discussions of the problems of narrative and time in the modern commercial society. The situationists, and Debord in particular, were acute critics of the way historical time had been disrupted in commercial society founded upon commodity production.[29] The criticism, somewhat different from that of the earlier critics of commercial society, played on the way that time is sold and consumed: blocks of time are sold, life is presented as a "sequence of falsely individualised moments,"[30] individuals' lives are fragmented such that the story of development is lost, a life of being frozen at youth is promised: "The spectator's consciousness . . . no longer experiences its life as a passage toward self realizations and toward death."[31] Putting aside the possible exaggeration about the extent that individuals' lived lives are like this, the main problem with the situationist analysis is that this criticism of the disruption of narrative is married to an aesthetic account of individual autonomy as playful self-creation that is empty of content: offered "a playful model of irreversible time of individuals and groups" or a "program of total realization, within the context of time, of communism which suppresses 'all that exists independently of individuals,'"[32] I remain not just unconvinced but unclear what it is supposed to be a possible lived experience. The demands remind me of a comment a friend made twenty years ago that the trouble with the situationists is that they *wanted* the advertisements to be true.

The trouble with postmodernists is that they think the advertisements *are* true. Indeed the distinction between appearance and reality is largely rejected. Much in postmodernism is situationism without social criticism. Where the situationists remained critical of the disruption of individual narrative lives in commercial society, the postmodernists see already existent in consumer culture the possibility of a "playful attitude" to self and narrative. The result is the celebration of the disruption of narrative in modern commercial society. The kind of fragmented individual life of merely physically related selves that is central to economists' accounts of time reappears as the fragmented self who is able to play with her or his identity. However, the aesthetic account of autonomy, of a chooser who can put on and take off identity at will, is mistaken. To be able to play with your identity is to have none. To have an identity is to be the possible subject of a coherent life narrative.

The celebration of the disruption of narrative continuity in individual lives is reproduced by postmodernists at the social level. Postmodernism has been characterized in terms of the loss of the credibility of "grand narratives."[33]. Now, there is a truth contained in this skepticism about grand narratives. What it gets right is that, at any moment, there is always more than one story to be told. Thus, in the environmental sphere a place is always the meeting point of a multiplicity of narratives, some of which are silenced. Moreover, it is right to point out the way that some of these are silenced. Consider again the case of Yosemite valley: the hidden human history that is lost in the name of wilderness needs to be resurrected in order to alter our understanding of what is at stake in the landscape. Genealogy, the favorite critical tool of postmodern thought, works by unmasking troubled histories. It is premised on the importance of history in evaluation.

However, reference to incredulity about grand narratives often is not just a way of uncovering the pluralism of stories that can be truly told but of rejecting the very possibility of a coherent narrative either in individual or collective lives. Postmodernism denies the possibility of the possibility of social narratives. This has its basis in a skepticism about certain particular grand narratives—what Lyotard calls the narratives of emancipation and speculation—a skepticism founded largely in disillusionment with the Marxist narratives that writers like Lyotard once accepted. Now whether one is right to reject these particular narratives, and I believe the narratives have more going for them than is often assumed, their rejection does not entail the rejection of the very possibility of social narratives.[34] What the temporal dimensions of environmental problems show is both how an appeal to narrative continuity at both individual and social levels continues to have some kind of evaluative force and correspondingly a need to have institutional forms that allow us to regain a social sense of narrative continuity over time that commercial society has disrupted.[35]

Notes

1. I focus here on the challenges to liberal theory. There are more traditional problems of global justice and poverty that are as least as important to the practice of liberal institutions.
2. For a useful survey see R. Attfield, *The Ethics of Environmental Concern* (Oxford: Blackwell, 1983).
3. J. Rawls. *A Theory of Justice* (Oxford: Oxford University Press, 1972), section 44.
4. See D. Richards, *A Theory of Reasons for Action* (Oxford: Clarendon Press, 1971); and B. Barry, "Justice Between Generations," in P. Hacker and J. Raz, eds., *Law, Morality, and Society* (Oxford: Clarendon Press, 1972).
5. See in particular the debates that have followed D. Parfit, "Future Generations: Further Problems," *Philosophy and Public Affairs* 11 (1982):113–72.
6. For a useful discussion see Barry, "Justice Between Generations."
7. See F. Hayek, *Law, Legislation and Liberty*, Vol. 2 (London: Routledge and Kegan Paul, 1976); and R. Nozick, *Anarchy, State and Utopia* (Oxford: Blackwell, 1974).
8. On this see the work of "Common Ground": S. Clifford and A. King, *Holding Your Ground* (London: Temple Smith, 1985); S. Clifford and A. King, eds. *Local Distinctiveness: Place, Particularity and Identity* (London: Common Ground, 1993).
9. The utilitarian might treat historical valuations simply as preferences for the past. My objection here would be for the reduction of first order values to preferences—we prefer something because it appears valuable to us; is not valuable because we prefer it. And where we believe what is preferred by a person is evil, we

have no reason to consider it. The appeal to the past matters because it is a proper object of preferences.
10. R. Goodin, *Green Political Theory* (Cambridge: Polity Press, 1992), pp. 26–30.
11. A. S. Leopold et al., *Wildlife Management in the National Parks*, U.S. Department of the Interior, Advisory Board on Wildlife Management, Report to the Secretary, March 4, 1963, p. 4. Cited in A. Runte, *National Parks: The American Experience* 2d ed. (Lincoln: University of Nebraska Press, 1987), pp.198–9.
12. Leopold et al., p. 6. Cited in Runte, *National Parks*, p. 205.
13. Cited in K. Olwig, "Reinventing Common Nature: Yosemite and Mt. Rushmore—A Meandering Tale of a Double Nature" in W. Cronon, ed., *Uncommon Ground: Toward Reinventing Nature* (New York: W.W. Norton, 1995).
14. Leopold et al., p. 21. Cited in Runte, *National Parks*, p. 200.
15. For a discussion, see A. Holland and J. O'Neill, "The Ecological Integrity of Nature over Time: Some Problems," *Global Bioethics* (forthcoming). To invoke the notion of unity or continuity here is not to deny the possibility of quite radical or revolutionary change. But it is to deny that one should approach such change as if one was starting from point zero. Thus, to take a nonenvironmental example, in Northern Ireland it seems to me that participants are quite right to reject the suggestion that they should forget history: if there is to be a break with the past, it has to involve concepts like forgiveness, repentance, reconciliation, and the like, which are all concerned with the appropriate transition from past to future.
16. There are four main arguments: pure-time preferences, social opportunity costs, uncertainty, and increasing wealth. For a discussion see J. O'Neill, *Ecology, Policy and Politics: Human Well-Being and the Natural World* (London: Routledge, 1993), chapter 4.
17. The appeal to pure-time preferences to defend a social discount rate has other problems, those of moving from intrapersonal to interpersonal preferences. It is one thing to have pure-time preferences about one's own life, another to have such preferences about that of others.
18. A. Pigou, *The Economics of Welfare*, 4th ed. (London: Macmillan, 1952), p. 25.
19. H. Sidgwick, *The Method of Ethics* (London: Macmillan, 1907), p. 418.
20. T. Nagel, *The Possibility of Altruism* (Oxford: Oxford University Press, 1970); and D. Parfit, *Reasons and Persons* (Oxford: Oxford University Press, 1984), part 2.
21. See Aristotle, *Nicomachean Ethics* (Indianapolis: Hackett, 1985), Book 1, chaps.10–13.
22. T. S. Eliot, "Tradition and Individual Talent," in *Selected Essays* (London: Faber and Faber, 1951), p. 15.
23. M. Golding, "Obligations to Future Generations," *The Monist* 56 (1972): 85–99.
24. J. Pocock, *The Machiavellian Moment* (Princeton: Princeton University Press, 1975), p. 458.
25. K. Polanyi, *The Great Transformation* (Boston: Beacon Hill, 1957), p. 73.
26. Ibid., p.181.
27. S. Weil, *The Need for Roots* (London: Routledge and Kegan Paul, 1952), p. 96.
28. In that debate the role of narrative is raised most clearly by MacIntyre. See A. MacIntyre, *After Virtue* 2d ed. (London: Duckworth, 1985), chap. 15.
29. G. Debord, *Society of the Spectacle* (Detroit: Black and Red, 1977), chaps. 5 and 6.

30. Ibid., para. 149.
31. Ibid., para. 160.
32. Ibid., para. 163.
33. See J. Lyotard, *The Postmodern Condition: A Report of Knowledge* (Manchester: Manchester University Press, 1984), p. 37.
34. The "temptation to give credence to grand narrative of the decline of the grand narrative" (Lyotard *Postmodern Explained,* p. 29) has never been resisted. Moreover the narrative that is told about the rise and fall of modernity is itself implausible once any empirical detail is added.
35. The arguments in this paper owe much to conversations with Alan Holland. An earlier version was read to a symposium on Environment and Society in Tampere, Finland. My thanks to the comments made on that occasion.

3

The Rationale for Environmental Restoration

William Throop

The gray wolf is reintroduced into the Greater Yellowstone Ecosystem. Grassland at the University of Wisconsin Arboretum's Curtis prairie is burned and seeded with native species. Where the Caledonian forest once stood in the Scottish Highlands, conifers are planted by volunteers, and fences are erected around them as protection against browsing deer. President Clinton endorses the huge Everglades Restoration Project, which will remove obstacles to the natural flow of water from Lake Okeechobee to the Everglades. Each project attempts to reverse some human alteration of the land; each restores part of an ecosystem.

Environmental restoration has emerged as one of the central aims of environmental policy. It has been proposed as a paradigm for human/nature relations, touted as the new rival to the goal of sustainable development, and acknowledged as an obligation.[1] Restoration also raises some of the most challenging questions in environmental philosophy. The question of whether the restoration of natural ecosystems is conceptually possible has spawned a rich literature.[2] How could a restored ecosystem be natural? If one grants that restoration is possible, as I do, then a variety of issues about the aims of the restoration and its evaluation arise: To what state should we restore degraded landscapes—to a pre-Columbian state or simply to some healthier state? Should restoration seek to integrate humans into ecosystems or limit their impact? How disruptive can the means of restoration be? How should injury to current species inhabiting a region be weighed in the evaluation of a restoration project? What features of landscapes make them suitable candidates for restoration; is the presence of exotics sufficient to require restoration, or must they have a significant (and negative) effect on the ecosystem?

Careful answers to such questions seem to require a theory of the value of ecosystems and their components that is sensitive to possible value changes resulting from human activity in the ecosystem. Although some of the broad outlines of such a theory are emerging,[3] much remains obscure. Fruitful theories are usually developed in response to clearly defined problems (Laudan 1977). I elaborate a problem about the rationale for restoration that, to my knowledge, has not received attention, and propose a solution that throws light on an important feature of an adequate theory of value for ecosystems.

I

I will focus on restoration as a practice rather than as a paradigm or research strategy. With minor qualifications, I will follow Morrison's (1987) widely cited definition of restoration as "returning a site to some previous state, with the species richness and diversity and physical, biological and aesthetic characteristics of that site before human settlement and the accompanying disturbances," but I will understand "before human settlement" to mean before *a particular* human settlement and its accompanying disturbances.[4] This permits one to restore an ecosystem to a state that it had prior to some anthropogenic disturbance, but not prior to all such disturbance. An important feature of this definition is that modification of an ecosystem is a restoration only if it is a return to a state that earlier characterized that site. In this sense, restoration of an ecosystem is like restoration of a work of art; only if one alters a damaged painting in a way that closely resembles a state that it was in can one be said to restore the painting.[5]

This narrow construal of restoration is quite common (Bratton 1992), though in practice it may function more as an ideal than a criterion of success. For example, it seems implicit in the 1990 definition of restoration adopted by the Society for Restoration Ecology: "The process of intentionally altering a site to establish a defined, indigenous, historic ecosystem."[6] An indigenous, historic ecosystem is presumably one that existed in the region prior to human alteration. A variety of broader notions of restoration are also current, some of which would include any case of humans returning an ecosystem to health after degradation or repairing damage to an ecosystem.[7] The problem I raise applies only to restoration in the narrow sense.

The above definition leaves open a range of states to which an ecosystem might be restored. Does returning the ecosystem to any predisturbance pattern in the region count as restoration, even a pattern that has not been instantiated since the previous interglacial period? Presumably not. We should countenance a range of appropriate goals for restoration, but the range should be fairly limited. With this in mind, I circumscribe that range by specifying that the goal must be a pattern of ecosystem constituents and processes, hereafter called *a historic pattern*, that (a) characterizes the ecosystem in the region immediately prior to some significant human modification or (b) constitutes a succession stage typically leading to a pattern of type (a).

The goal of restoration is a *general* pattern of species in characteristic relations to each other and to the abiotic elements of the ecosystem. A successful restoration need not duplicate all features of the predisturbance ecosystem. In a restored forest, for example, the trees and boulders need not be situated in the same places as they were before human degradation. The pattern is close enough if roughly the same proportion of tree species returns with the same kind of understory and nutrient cycles that characterized the original. In this respect, ecosystem restoration differs from art restoration, where exact fidelity to the original is the goal. Specifying the goal in this fashion enables us to avoid the metaphysical issue of whether a restored ecosystem can be numerically identical to the ecosystem that humans degraded.[8] A pattern may be restored whether or not the ecosystem is numerically the same. It seems doubtful that our concepts of ecosystems are sufficiently precise to provide informative answers to questions about how much change is compatible with the ecosystem retaining its identity.

Arguably, one might restore an ecosystem to a pattern that would have characterized the area now if the significant human modification had not occurred but that never in fact existed. This would broaden Morrison's definition in order to permit restorationists to adjust their goal to accord with naturally occurring changes in the surrounding land. Although I think such a goal is sometimes reasonable, justifying it raises problems beyond the scope of this paper.

The decision about which historic pattern should serve as the goal of restoration in a particular case will depend on our empirical knowledge, on the feasibility and cost of reaching the goal, on the impact on species in the region, and on the invasiveness of the means for successfully reinstating the pattern, among other things. It should now be clearer that the range of restoration goals includes patterns that result from some significant human modification, as long as they existed prior to some later modification. Activities like the rehabilitation of oak savannas in the northern Midwest would clearly count as restoration, even though the characteristic pattern may have depended on anthropogenic burning (Packard 1993).

II

Many environmentalists make the following assumption:

> A: With respect to a wide variety of cases, restoration of an area is preferable to further humanization of the area, and to letting the current system evolve on its own, and to building a new ecosystem.[9]

This assumption creates a problem, for it is not *obviously* correct, and thus it requires some rationale. If restoration is the preferred response to human degradation of an ecosystem, then we should be able to explain why it is preferred. A standard approach to such an explanation would appeal to the differences in value between the options available. One might anticipate an explanation of why a predamage pattern is particularly valuable and how that value is returned by restoring the ecosystem to that pattern. Unfortunately, the usual way of explaining the former, by invoking the value of the "natural," seems to conflict with the latter. How could the process of restoring the ecosystem, which involves increased "nonnatural" human activity, return the value that a historic pattern had by virtue of its being natural, i.e., relatively unchanged by humans? Here the worry is not about whether one *could* restore the ecosystem, but rather about why one would think that one *should* restore it.

The problem of justifying restoration can be structured as an inconsistent set of claims, each of which seems plausible. Seen in this light, it appears as "a paradox of restoration." To formulate the paradox crisply, I must introduce a bit of terminology. A *diachronic property* of an ecosystem is any property relating the ecosystem to something that exists before or after it. The diachronic properties of most interest to us will relate an ecosystem pattern to its causes, to the processes that brought about the pattern in the region. The *synchronic properties* of an ecosystem are its nonrelational properties, e.g., its being composed of particular species, and those properties relating the ecosystem to other things existing at the time. Although these characterizations are vague in significant respects, they are clear enough for my purposes. Now for the paradox:

(1) Either A is warranted by virtue of the value of the diachronic properties of ecosystem or by virtue of the value of their synchronic properties (or both).

(2) The value of diachronic properties of ecosystems is not sufficient to justify A.

(3) The value of synchronic properties of ecosystems is not sufficient to justify A.

(4) A combination of the value of synchronic and diachronic properties is not sufficient to justify A.

(5) But, A is warranted.

Clearly, one of these claims must be rejected. Each, however, has much to be said in its favor. I will briefly sketch the motivation for believing each thesis. Most of the arguments attempt to show that standard bases for valuing ecosystems do not promise to warrant A. Inevitably these arguments will be incomplete, for they cannot survey all possible rationales for A. Nonetheless, they do identify the challenges that any rationale must meet and show why such challenges are especially hard to meet.

If one is going to justify policy in terms of weighing the likelihood of value being gained and/or lost in each alternative, then one should accept (1). Since the class of synchronic and diachronic properties is exhaustive, these properties will determine the value of an ecosystem. Whether one has an objective or subjective account of values, consistency seems to demand that value supervene on empirical properties. In other words, the empirical properties—relational and nonrelational—of an entity determine its value. There could be no value change without a change in properties, at the very least relations between the object and our psychological states must change.[10] One might try to justify policy on grounds other than weighing value, but it seems reasonable to press for a deeper explanation in terms of the values that are served. If one refuses to give any, one courts the suspicion that one's policies are unfounded. Below, I will consider a specific proposal that denies (1) and show why it seems inadequate.

The prospects for rejecting (2) dim considerably when the difficulties of justifying restoration on the basis of the effects and/or the causes of an historic pattern become evident, for these are the diachronic properties most likely to bear value. One common set of reasons for valuing an ecosystem emphasizes the beneficial effects that it has for us or for other things we value. Although some anthropocentric reasons for restoration depend on synchronic relations between us and certain ecosystems, e.g., on our current use of an ecosystem, many others depend on diachronic relations, that is, on the effects an ecosystem has on our well-being at another time. I will treat both here since the same arguments apply to each. Though justifications for restoration based on effects will work in particular cases, they do not have much promise as a general rationale for A, because it is doubtful that, in general, historic patterns will have beneficial effects to a greater extent than do modifications to historic patterns, which may be easier to create.

Sometimes the recreational values of hunting and fishing, or the scientific value of understanding natural processes, or the increased productivity of consumables, do provide strong reasons to restore ecosystems. However, in many cases the good of restoration outweighs anthropocentric considerations, which favor the creation of other ecosystems. For example, fishing may be enhanced by stocking lakes with nonnative species, but many, including some of those fishing, would prefer a restored lake. Productivity would often be enhanced by humanizing of a landscape, but where such productivity is not essential to our survival, we sometimes prefer a return to a natural pattern. Sometimes people have preferred turning strip mines into recreational areas; however, for many, restoration continues to be the default option even in cases where human welfare would be increased by creating other ecosystems. Appeal to more indirect effects, e.g., the effect of a historic pattern on climate, is unlikely to help here, for often there is no reason to believe that effect cannot be achieved by patterns involving exotic species. It is noteworthy that restorationists do not typically justify their policies primarily on anthropocentric grounds, though these often play a subsidiary role.

As noted above, another common set of reasons for valuing ecosystems focuses on the processes that led to their present patterns. Ecosystems are often valued because they have "naturally evolved," that is, because humans have played little or no recent role in the production of their patterns. It is a well-worn theme in the philosophical literature on restoration that even a perfectly restored ecosystem would not have the value of the original ecosystem because they have different origins. This is often defended by appeals to analogy. Restored ecosystems have been compared to art forgeries (Elliot 1982) and to artifacts (Katz 1992). If restorations are "fake" natural ecosystems or artificial constructions, then they lack much of the value of the historic ecosystems. Such analogies are vulnerable to charges that the analogies fail in relevant ways. Thus Katz argues that restorations are not like forgeries because there is usually no intent to deceive in the former case, and Sylvan adds that forgeries do not typically involve modifying the original work. Sylvan (1991, 21) also criticizes Katz's analogy on the grounds that in restoration most of the work is done by nature, with humans adding a helping hand, unlike the construction of artifacts, where humans do the essential work.

These analogies are not required to make the basic point that motivates Elliot and Katz. Let us call any diachronic property that relates an entity to the processes that caused it to have its character a *historical property*. Thus being painted by Picasso is a historical property of *Guernica*, and having evolved from australopithecenes is a historical property of humans. The claim that ecosystems have value by virtue of their historical properties is strongly supported by consideration of a range of counterfactuals pertaining to ecosystems that have the same structure and composition as our natural ecosystems but that were largely created by humans. Suppose we were to discover that the Greater Yellowstone Ecosystem had been created and was being maintained by some very intelligent humans in a style reminiscent of Jurassic Park, that in effect Yellowstone was a large zoo. We would no longer see it as the fantastic product of eons of natural forces. We might still appreciate it for its beauty and grandeur, and in addition we might see it as one of the great works of humanity, but intuitively it would lack a significant value that it currently has, a value that can only be attributed to its historical properties since these are the properties that distinguish it from Yellowstone as we know it in this world. The relevant historical properties seem to involve having a structure and composition that evolved "naturally," i.e., largely independent of human activity.

But can an appeal to this kind of diachronic property justify restoration? It seems highly unlikely. After restoration the ecosystem lacks the property of being naturally evolved, which seemed to give the earlier unmodified ecosystem some of the value it had above and beyond the values based on its synchronic properties. Of course, after restoration, natural process will take over and the humanization will wash out of the system (cf. Rolston 1994), but that will also occur if a new ecosystem is created or if the damaged ecosystem is left alone. Indeed, if the value associated with being naturally evolved is sufficiently important, it would seem preferable, other things being equal, to allow a humanized ecosystem to evolve on its own rather than to tamper with it further, thereby further diminishing the value associated with it being a result of natural processes. Neither of the diachronic bases for valuing ecosystems seems sufficient to warrant A; thus (2) seems hard to reject.

But (3) is no easier. Ecosystems are often valued in terms of synchronic properties, such as health, biodiversity, beauty, and perhaps dynamic stability. These, however, are not sufficient to justify restoration in the appropriate range of cases; for often these properties will be instantiated to the same or greater degree by nonhistoric patterns that could be created in the region. In such cases they cannot account for the preference for restoration. For example, in some strip-mining sites, significant ecosystem health, beauty, and biodiversity could be conserved by creating wetlands so as to produce more biodiversity in a region rather by returning the site to a previous state. Although sometimes considerations may warrant wetland creation, there is a presumption in favor of returning the site to a historic woodland state. Somehow the woodland would be more natural, and hence more valuable, than would a wetland that did not "belong" there. Similar cases could be imagined for other synchronic properties, like health and beauty.[11] Thus, at least sometimes, the value of returning to a historic state outweighs the synchronic value of alternative patterns. In such cases, clearly an appeal to synchronic values cannot explain the preferability of restoration.

Little is gained by combining the values based on synchronic and diachronic properties, for the value of historical properties appears to add nothing to the case for restoration, and any combination of other values will justify alternatives to restoration in too wide a range of cases. The above arguments make (4), the claim that diachronic and synchronic values are jointly sufficient to justify restoration, as difficult to reject as (2) and (3). This puts significant pressure on (5); perhaps the restorationist's assumption is not warranted. But the cost of rejecting (5) is abandoning restoration as an important policy goal. One might still believe we have an obligation to enhance degraded ecosystems, but one would have no reason to prefer an indigenous pattern over some other pattern unless the former happened to satisfy some other goals. This runs counter to the deep intuition that restoration *is* good practice; other things being equal, we should put back what we have destroyed. Only if there is no alternative should we be driven to reject (5).

Again, I emphasize that the above considerations are far from knockdown arguments for (1) through (5); in this respect the problem is not quite a paradox. Other synchronic or diachronic properties might plausibly be used to justify restoration in the appropriate range of cases, and other ways of warranting restoration might not appeal to value theory. I suggest, however, that an aura of paradox emerges from the above arguments because the value of a historic state seems to depend on its historical properties. Other value-bearing properties will often warrant preferring the creation of nonhistoric patterns. Unfortunately, the historical properties that seem to bear value are precisely those that we cannot recreate, values associated with being naturally evolved. Before outlining my own solution to this problem, I will consider several responses that are likely to attract immediate attention.

III

It might be argued that biodiversity and/or ecosystem health would warrant restoration in the appropriate cases if they are complimented with one or more plausible empirical assumptions. Thus (3) can be rejected after all. One might assume that the speed and ease of enhancing ecosystem health are usually maximized through restoration, or that the cost of renewing health is usually minimized. One might accept a "nature knows best" principle; nature put this here before, so trying to put it back is probably going to create the best, most viable, and most healthy ecosystem in the region. This might be coupled with an argument from ignorance. We do not have enough knowledge to create other ecosystems. The best we can do is to try to put back what we modified. Although these empirical claims are sometimes accurate, they will not apply in the full range of restoration cases. The quickest and easiest route to health or biodiversity may be the establishment of alternative ecosystems, as in the creation of wetlands in old strip mines or the preservation of prairies containing some exotic species. In at least some of these cases, restoration still seems preferable to the alternatives.

Furthermore, as our knowledge and technologies are enhanced, the force of the empirical assumptions is likely to diminish. With increased knowledge, we will be able to ascertain that some constructed ecosystems will be as healthy and have as much biodiversity as the original ecosystem; and with enhanced technology, we will have better means of creating new healthy ecosystems. However, we do not think that the virtue of restoration is a function of ignorance or lack of ability. If anything, increased knowledge should enhance our obligations to restore as well as enhance our ability to do so.

In some cases, considerations of convenience, speed, and cost do justify reconstructing degraded ecosystems on some new pattern, rather than restoring them. In other cases, they contribute to the case for restoration, but this is not so in all cases where we take restoration to be preferable. We must seek elsewhere for a sufficient rationale for A.

It might be argued that a principle of restitution justifies restoration without any assumption of the original pattern having more value than other patterns that might have been instantiated there. Perhaps restorationists simply accept some defeasible deontological principle to the effect that we owe it to nature to provide restitution for harms humans have inflicted. If so, then they can reject (1). It is not clear, however, why restitution should be in the form of returning an ecosystem to a historic state, given that other states that are easier to produce will often eliminate the harm. Such a principle certainly justifies doing something, but not necessarily restoration. The appeal to general deontological principles, applicable to a wide range of cases, seems unlikely to justify restoration over alternative ways of redressing harm. If one narrows the principle to specify restoration as the obligation, then explanation of the preference of restoration will seem ad hoc. One would like a justification of restoration that enhances understanding of why it is warranted, but a straightforward appeal to a principle that says that we owe it to nature does little to enhance understanding; it simply cuts off the question. This kind of move is sometimes reasonable, but it is unattractive in this application, for it is solution by fiat.

In light of the difficulty of warranting A, we might be tempted to abandon it, to reject (5), and turn toward a value theory that makes little of the distinction between the created ecosystem and the naturally evolved ecosystem (Turner 1994). This solution might be motivated in part by an explanation of A that showed it to be a result of indefensible sentimental values, say nostalgia. We were attached to the old ecosystem, and perhaps we will feel better if it were restored, but these feelings are not grounds for a theory of value. Or it might be argued that A's plausibility is a result of a poor analogy between the restoration of a work of art and of ecosystems. One might be justified in restoring a painting because it was supposed to look a particular way, which was determined by the intentions of the artist. Ecosystems are not supposed to be any particular way, so there is no reason to return them to any particular historic pattern.

Perhaps indefensible values and analogies or empirical misconceptions do underlie A, but awareness of these explanations does not diminish A's intuitive appeal. Normative judgments that remain in increasingly idealized situations, i.e., under conditions of increased reflection and empirical knowledge, have an increasingly strong claim to truth (cf. Blackburn 1984, chapter 6). Only if we find no other acceptable solution to the puzzle should we allow such explanations to undermine A. Other potential rationales for restoration are bound to occur to a thoughtful reader. The pattern of my arguments should suggest how these would be addressed. Although other values may contribute to a rationale for restoration in some cases, in other cases they are likely to justify other responses to humanization over restoration, though intuitively restoration is preferable. I cannot here rule out the possibility that some complex combination of values justifies restoration in the appropriate cases, but I believe that a simpler account can solve the paradox adequately.

IV

I suggest that the paradox results from a failure to see how a valuable historical process transfers value to the contingent product of that process. To substantiate this hypothesis, I must further develop the common view that at least one historical property of ecosystems bears significant value; and I must argue that at least some of the value associated with having this property is also held by a historic pattern which does not now have the relevant historical property. If this argument is plausible, then restoration may be justified, because it preserves a value that is a function of the historical property without reproducing on ecosystem with the property itself. Thus, I will show how it is reasonable to reject (2).

Let us say that an ecosystem is wild to the extent that its processes and components are caused by factors that have not been recently and significantly altered or controlled by humans. Although "wild" is being used here as a technical term, the above definition reflects one standard usage of the term. The definition contains several dimensions of vagueness; this should not be surprising, for important values are often linked to rather amorphous properties like integrity, courage, friendship, etc.

Wildness is a matter of degree. Fine discriminations of wildness may seem arbitrary, but a great many comparative judgments will be clear—

typically, a wilderness area is wilder than a farmer's woodlot, which is wilder than an urban park. An ecosystem's degree of wildness is a function of both the extent of humanization and its temporal proximity; over time, humanization will wash out of an ecosystem. Eventually, even a significantly altered ecosystem will return to a high degree of wildness; Dartmoor, U.K., seems to be a good example of this.

Wildness-decreasing humanizations result from human alteration or control or both. An ecosystem is altered by humans to the extent that it is changed from the way it was or would have been independent of human activities. Some human activities attempt to maintain an ecosystem in some historic state rather than to alter it; for example, pine barren ecosystems may be preserved by anthropogenic burns. Such burns involve human control but not, in my sense, human alteration of the ecosystem, since natural fires would have produced similar results. Highly controlled ecosystems would be the equivalent of biological zoos. Such an ecosystem may be characterized by natural patterns and processes, but the result is not very wild.

Some human alterations do little to affect wildness because they leave most of the ecosystem components and processes in place. Human trails make wilderness areas only slightly less wild, whereas strip mining and the introduction of aggressive exotics like kudzu in the southeastern United States significantly reduce wildness. The speed at which humanization washes out of an ecosystem over time depends on the extent of alteration and on the resilience of the ecosystem, among other things. Since aspects of an ecosystem may be significantly humanized while others are not, one may distinguish various respects in which an ecosystem is wild. A lake in northern Maine recently stocked with nonnative game fish can be wild in a great many respects, even though it is not wild with respect to some fauna.

Callicott (1991) and Baldwin, De Luce, and Pletsch (1994) have argued that since Darwin we have recognized that humans are a part of nature, and thus it makes little sense to isolate the historical property of not being altered by humans and to value that property. Clearly, in the biological sense of "natural," human alterations are natural, but there are other senses in which they are not. Darwinian theory seems to support a conceptual pragmatism according to which concepts are selected to serve different purposes of the organism (Rorty 1979). I suggest that a human/nature distinction serves an important purpose in formulating norms for our behavior in a world overrun by humans.[12] This distinction does not conflict with biology; it is simply not a scientific distinction.

A full defense of the value of wildness would involve showing that valuing wildness is in wide reflective equilibrium with empirical beliefs, other general evaluative principles, and specific intuitions about value. This cannot be undertaken here. At the very least, the intuitions tapped by

Elliot (1982) and Katz (1992)—that even a fully restored landscape lacks some of the value of the original—strongly suggest that wildness is valuable, as does the intuition that biodiversity crafted by the genetic engineer lacks much of the value of wild biodiversity.

V

Suppose wildness is a significant value-bearing property, how does that contribute to the rationale for restoration? I propose that when something has value in virtue of its having a particular history, the instantiation of its nonhistorical properties in the appropriate circumstances acquires some value from that history. I will clarify and support this proposal by applying it to the restoration of works of art. Arguably, works of art have some value because of their historical properties. A canvas painted by Monet has more value than a qualitatively identical canvas painted by an amateur (Radford 1978). Suppose a Monet were damaged in one corner, and one wished to repair the damage, which would require repainting significant parts of the corner. Even if a board of aesthetic experts were to convince us that the painting would be more powerful and more aesthetically pleasing were the corner to be repainted in a slightly different way from how Monet left it, most of us would think it should be restored to the predamage pattern. The result of the restoration would have more value than the alternative in question, though perhaps not as much value as the painting had prior to damage. But why? The restored area does not have the historical property of having been painted by Monet. In all likelihood, it does have the historical property of being as Monet intended it, which may be sufficient to return value to the restoration. But imagine a more fanciful case where one has good evidence that the painter did not intend the particular kind of brushwork he left in the area. Perhaps he intended to simply fill in the area and return to it later. Still, we would think the painting should be restored to the original pattern.

One plausible explanation of this is that the contingent product of the valuable historical process acquires value from the process. Monet might have used somewhat different brushwork in an area; in that case, *it* would have been valuable since it was the product of a process (being painted by Monet) that we value. Once a pattern has value in this way, then in the appropriate contexts its reproduction has some of the value of the original due to its history, even though the reproduction lacks the historical property of the original. It is as if some of the value of the historical property rubs off on the contingent result of that history.

The example of art restoration involves complexities beyond the scope of this paper. It may make a difference how much of the painting was damaged and restored; continuity with the original work may carry significant value. The value of a restoration may depend in part on synchronic relational properties of the painting, such as how the restored painting sheds light on other paintings by the artist. The identity of a work of art may be linked to its history in a way that partially explains the value of restoration. I cannot show that no other account of the value of art restoration is satisfactory, but my hypothesis about the transfer of value from process to product has the virtue of simplicity. It seems to account for our intuitions, and it provides a deeper justification for restoration than the deontological principle that we have a basic obligation to restore damaged paintings. Moreover, when applied to ecosystem restoration the hypothesis promises to solve the paradox in a direct and natural way.

Restoration of ecosystems cannot replace the property of wildness, at least not in the short term. However, restoration can approximate the pattern of components and processes that was originally brought about in a more wild fashion. If this pattern acquires value from its having been brought about largely by nature, then such value is returned through restoration even though the historical property is not. Nature might have created ecosystems with very different patterns in this region, and if so then these would have acquired some value as a result. When deciding what to do with a degraded ecosystem, we take into account the relative values of the synchronic properties of the alternatives, as well as their costs, but we also factor in the value that attaches to an alternative because of its similarity to a historic state. As more ecosystems become degraded or reconstructed on nonnatural patterns, the value of historic patterns tends to increase. Not just any natural pattern will do; we prefer patterns that characterized the region in question.[13] The hypothesis of value transfer from process to product explains this.

As noted above, sometimes the goal of restoration is a historic state that is only moderately wild. In a world where so much has been humanized for so long, we must often settle for returning patterns that are merely less humanized. The above account can justify such activity if the degraded ecosystem being restored is more humanized than the goal pattern. To justify Packard's oak savanna restoration project in this way, one must maintain that restoring the savanna returns the ecosystem to a more valuable pattern that is more wild.[14] Naturally, other values, such as the value of enhancing the diversity of kinds of ecosystems in a region, also play a role in the justification of savanna restoration.

One may wonder whether enough value is transferred from process to product to make much of a difference in the selection of responses to degradation. After all, wildness is just one of many values that an ecosystem might have, and the value a restored ecosystem would have thanks to the return to a historic pattern is less, prior to the washout of humanization, than the value of a wild ecosystem with that pattern. Elliot (1994a) suggests that wildness is *intensely* value adding. If so, then one should expect that the contingent products of wild nature also carry significant value. Even if that may be doubted, this solution seems preferable to the alternatives. The challenge was to justify restoration in the range of cases in which restorationists take it to be warranted. My proposal posits a locus of value that can help to justify restoration, but it does not tell us how much value a historic pattern in a region has. In this sense, it is a formal solution to the puzzle that must be developed as it applies in individual cases.

Unlike the solutions using only the values of synchronic properties or of the effects of ecosystem patterns, my solution is not blocked by examples of ecosystems that intuitively should be restored but that such values fail to justify. Any solution appealing only to nonhistorical properties will be threatened by counterexamples, because the connection between the goal of restoration, a historic pattern, and the values justifying restoration will be coincidental. Only a solution positing a necessary connection between the goal and the justification avoids such counterexamples. Such a solution can sidestep the arguments against justifying restoration in terms of historical properties only by identifying a value that is a function of causal origins without requiring a specific kind of causal origin. I suggest that only a solution of this sort will permit one to retain (1) and (5).

Without specifying weights for competing values, including the value of having a historic pattern, I cannot claim to fully meet the challenge I have set. I have, however, removed a major obstacle to meeting that challenge, and sketched an important element of a value theory that seems required if restoration is to be warranted. I do not expect agreement of all knowledgeable parties on the weight of the value of a historic pattern, nor on wildness value itself, but I believe that this argument can alter such weights. My larger aim has been to contribute to a reflective understanding of a value system that is accepted by a number of environmentalists and in doing so to facilitate the defense of this system and the conversion of others to it. I believe that many people do value wildness in certain degrees and contexts, even if they would not initially acknowledge it. By making salient contexts in which their preferences are plausibly explained by wildness value, environmentalists can put pressure on others to achieve a greater reflective equilibrium between their avowals of value and those preferences, and often this will result in an increased appreciation of wildness value. If I am correct, restoration can be defended against its critics by highlighting the value of wildness and making salient the ways in which valuable processes confer value on some of their contingent products.[15]

Notes

1. For examples of a variety of perspectives on restoration, see Jordan, Gilpen, and Aber 1987; Baldwin, De Luce, and Pletsch 1994; and articles in *Restoration Management and Notes*.
2. Important philosophical treatments of the issue include Elliot's seminal article (1982) and Gunn 1991, Sylvan 1991, Katz 1992, Bratton 1992, Cowell 1993, Attfield 1994 and Elliot 1994a.
3. See especially, Elliot 1994a.
4. Cited in Elliot 1994b, 37.
5. See Gunn 1991 for a nice discussion of limitations of the analogy between art restoration and ecosystem restoration.
6. Cited in Cowell 1993, 30.
7. For example, Jackson, Lopoukhine, and Hillyard 1995 define restoration as "the process of repairing damage caused by humans to the diversity and dynamics of indigenous ecosystems" (71).
8. See Gunn 1991 for a discussion of this issue.
9. This assumption is not often articulated, but it is implicit in the practices of restoration ecologists and in a variety of U.S. environmental laws and policies. For an excellent review of the latter, see Berger 1991.
10. For the purposes of this paper, I need take no stand on the metaphysics of value.
11. Nonphilosophers—and many philosophers—would prefer that real cases be used to make such points. I agree. Unfortunately, real cases are often sufficiently complex and messy that one cannot be sure what features of the cases are driving the intuitions about the preferability of one option over another. Such features can only be reliably isolated in cases described highly abstractly. Naturally, there are other dangers in abstract cases; for example, one's intuitions about what one should do may not be reflected in practice. In order to highlight the structure of the puzzle, I risk the latter dangers. I believe that intuitions in real cases will lead to similar conclusions. A case in point is a dump outside of Madrid that has created controversy because it has been very good for biodiversity in the region (*New York Times*, 8/10/95). Returning the dump to its original pattern will reduce that biodiversity, but it seems preferable nonetheless. Of course, as with any real case, there are other relevant variables, such as the pollution from the dump.
12. Rolston 1994, chap. 1, explores several distinctions between the natural and the nonnatural and replies in more detail to the charges that humans are parts of nature.
13. Elliot 1994a suggests that returning an ecosystem to *a* "natural design" increases the value of the restored area. Perhaps this is so, but it seems far preferable to return it to *the* natural design that characterized it before degradation. Creating an ecosystem with a natural design not indigenous to the region seems little better than creating a new ecosystem.
14. Some of the controversy surrounding this case revolves around a concern that the woodlands currently occupying areas that were once oak savannas are as wild, perhaps more wild, than the original states to which Packard wishes to return the areas. See letters to the editor in 1994 issues of *Restoration Management and Notes*.

15. My thinking on restoration has been deeply influenced, over a period of several years, by numerous interchanges with Ned Hettinger. Conversations on Adirondack trails with Steve Schwartz significantly sharpened my thought on key points and enabled me to avoid some errors.

References

Attfield, R. (1994). "Rehabilitating Nature and Making Nature Habitable." In R. Attfield and A. Belsey, eds. *Philosophy and the Natural Environment*. Cambridge: Cambridge University Press.

Berger, John (1991). "The Federal Mandate to Restore: Laws and Policies on Environmental Restoration." *Environmental Professional* 13: 195–206.

Baldwin, Dwight, Judith De Luce, and Carl Pletsch. (1994). "Introduction: Ecological Preservation versus Restoration and Invention." In Baldwin, De Luce, and Pletsch, eds. *Beyond Preservation: Restoring and Inventing Landscapes*. Minneapolis: University of Minnesota Press.

Blackburn, Simon (1984). *Spreading the Word*. Oxford: Oxford University Press.

Bratton, Susan (1992). "Some Alternative Models of Ecosystem Restoration." in Robert Costanza, Brian Norton, and Benjamin Haskell, eds. *Ecosystem Health*. Washington, DC: Island Press.

Callicott, J. Baird (1991). "The Wilderness Idea Revisited: The Sustainable Development Alternative." *The Environmental Professional* 13: 235–247.

Cowell, Mark (1993). "Ecological Restoration and Environmental Ethics." *Environmental Ethics* 15: 19–32.

Elliot, Robert (1982). "Faking Nature." *Inquiry* 25: 81–93.

——— (1994a). "Extinction, Restoration and Naturalness." *Environmental Ethics* 16: 135–44.

——— (1994b). "Ecology and the Ethics of Environmental Restoration." In Robin Attfield and Andrew Belsey, eds. *Philosophy and the Natural Environment*. Cambridge: Cambridge University Press.

Gunn, Alastair (1991). "The Restoration of Species and Natural Environments." *Environmental Ethics* 13: 291–310.

Jackson, Laura, Nikita Lopoukhine, and Deborah Hillyard (1995). "Ecological Restoration: a Definition and Comments." *Restoration Ecology* 3: 71–75.

Jordan, William (1994). "'Sunflower Forest': Ecological Restoration as the Basis for a New Environmental Paradigm." In Baldwin, De Luce, and Pletsch. *Beyond Preservation*.

Jordan, William; Michael Gilpen, and John Aber (1987). *Restoration Ecology*. Cambridge: Cambridge University Press.

Katz, Eric (1991). "The Ethical Significance of Human Intervention in Nature." *Restoration and Management Notes* 9: 90–96.

——— (1992). "The Big Lie: The Human Restoration of Nature." *Research in Philosophy and Technology*: 81–93.

Laudan, Larry (1977). *Progress and Its Problems*. Berkeley: University of California Press.

Morrison, Darrel (1987). "Landscape Restoration in Response to Previous Disturbances." In Monica Turner, ed. *Landscape Heterogeneity and Disturbance*. New York: Springer.

Packard, S. (1993). "Restoring Oak Ecosystems." *Restoration Management and Notes* 11: 5–16.

Radford, Colin (1978). "Fakes." *Mind* 87:66–77.
Rolston, Holmes (1991). "The Wilderness Idea Reaffirmed." *The Environmental Professional* 13: 370–77.
Rolston, Holmes (1994). *Conserving Natural Value*. New York: Columbia University Press.
Rorty, Richard (1979). *Philosophy and the Mirror of Nature*. Princeton: Princeton University Press.
Sylvan, Richard (1991). "Mucking with Nature." Unpublished manuscript.
Turner, Frederick (1994). "The Invented Landscape." In Baldwin, De Luce, and Pletsch. *Beyond Preservation*.

4

Empathy, Society, Nature, and the Relational Self
Deep Ecology and Liberal Modernity

Gus diZerega

To a substantial degree, environmental advocates are motivated by an ethical concern for nature's nonhuman dimensions. They regard a human-centered ethic as morally questionable. Many have concluded that modern society, with its Promethean mind-set, is ecologically unsustainable and morally impoverished.[1] In return, their critics often claim these views are misanthropic and subvert the principles of western civilization.[2]

This paper holds both sides to be more wrong than right. Ecocentric ethics need not be as alien to liberal principles as they often seem. In fact, that liberal tradition beginning in the eighteenth-century Scottish Enlightenment is unusually hospitable to ecocentric insights. The work of David Hume and Adam Smith, and their contemporary heirs F. A. Hayek and Michael Polanyi, offers an important link between modern and ecocentric thought. Understanding this link offers an opportunity more deeply to ground the foundations of western modernity, and in doing so harmonize them with the world of nature.

Copyright 1995 by *Social Theory and Practice,* Vol. 21, No. 2 (Summer 1995).

What is "Deep Ecology"?

"Deep Ecology" is a term coined by Norwegian philosopher Arne Naess. Its general perspective also applies to other environmental thinkers who broadly share Naess's view of humanity's proper relationship with the nonhuman world, although arriving there by different ways. Deep ecological perspectives have also been described variously as "biocentric," "ecocentric," or "transpersonal." All contrast themselves to "anthropocentric" points of view.[3]

"Anthropocentric," or human-centered, ethics argues that value in the natural world originates in the attitudes human beings take toward nature. Trees, waterfalls, and butterflies alike possess no moral standing beyond their capacity to serve human ends, or perhaps as repositories of values important to humans. In themselves, they have no meaning or ethically significant value.

Most anthropocentric views regard the nonhuman world in instrumental terms. Eugene Hargrove argues that this need not follow, and that deeming a belief anthropocentric simply "refers to a human-oriented perspective, seen from the standpoint of a human being."[4] Such beliefs need not regard everything nonhuman in instrumental terms. Granting Hargrove's point, anthropocentric ethics still focus on the ethics of people considered in some important sense as separate from the natural world, with value projected onto nature by human consciousness.

By contrast, a Deep Ecological or ecocentric view holds that plants, animals, and sometimes even natural forms of land and water merit respect independently of their utility for or regard by human beings. Hargrove sympathetically but skeptically identifies the chief challenge confronting a Deep Ecological argument: "To succeed [it needs] to go beyond valuing based on the human perspective, which seems impossible."[5]

Hargrove's own solution for how to deal ethically with the natural world is to reject the widespread equating of ethics with utility, adopting what he terms a "weak anthropocentrism" that can recognize noninstrumental values in nature. I agree with Hargrove that even for the human world, ethics do not reduce to utility. But even so, his approach does not touch upon that dimension of experience that, more than anything else, motivates deep ecological sensibilities: direct sensuous contact with the natural world as a source of meaning.

From what has been said so far, biocentric ethics appear incompatible with two major intellectual pillars of liberal modernity: economics and ethics.

Economics

Contemporary economic theory takes human preferences as given, meaning that their content is separate from the economic process. Preferences are also taken as abstract, that is, they are devoid of any particular content.

Economic theory makes no distinctions in the relative value of our preferences. Rather, it analyzes how exchanges among people can enable each, on balance, to attain more of their preferences than would otherwise be the case. Because economists take desires and wants as given, and assume the market process tends to meet them efficiently, economic theory necessarily treats everything except the final consumer as a pure means. Economic value stems solely from its ability to efficiently meet consumer desires. Other kinds of value are treated as subjective preferences, indistinguishable from preferring jam over jelly.

Traditional economic theory is inescapably anthropocentric. Further, it is anthropocentric in the narrowest sense, conceiving value solely in terms of utility. It is hard to imagine how an ecocentric view could be harmonized with economic theory.

Ethics

Dominant western ethical perspectives rest upon various philosophical arguments seeking to establish the case for human rights, equality, or well-being. This is so whether their foundation is Kantian or utilitarian.

Theories of human rights translate very poorly into the natural world. For example, the so-called animal rights argument has been effectively criticized from many ecocentric perspectives as confused, naive, and absurd.[6] Theories of human rights seem inescapably connected to human beings. In my view, this is because the rationale for human rights is connected to having the capacity for relationships characterized by political equality. Animals are unable to enter into political relationships with human beings, and so the logic of human rights fails to apply. The right to life does not exist within a vacuum, but leads one to recognizing a right to freedom of speech, assembly, and the like. Animals have no use for such rights, for they are incapable of exercising them.

Utilitarian ethics are no better suited to incorporate ecocentric perspectives. For the natural world to maintain itself, the overwhelming number of living beings must serve as food for others. Predation is a problem for utilitarians. Yet without predation it is unlikely that intelligent or even complex life would have evolved. Nothing would have arisen capable of thinking about animal liberation.[7]

If neither natural rights nor utilitarian ethics are reasonably applicable to the natural world, the challenge of ecocentric thought seems radical indeed. If it prevails, the ethical foundation for liberal modernity would appear to be undermined. On the other hand, if liberal views can make the stronger case, ecocentrism appears intellectually bankrupt.

I think either conclusion is in error. Market processes and liberal democratic society can both exist within an ecocentric framework.

Natural Beauty

I begin my argument by probing an experience probably common to us all: encountering natural beauty. Most philosophers argue that beauty does not exist independently of the perceiver. I have no quarrel with this judgment, as far as it goes. But most modern philosophers (and almost all economists and political scientists) go farther. They argue that perceiving beauty is a subjective reaction located completely "inside" the perceiving individual.

This judgment is mistaken. To appreciate why, we must accomplish two tasks. First, show that subjectivist theory cannot reasonably account for our experience of certain values. Second, we must show that not only is nature one of these values, in addition its value cannot be shown to be derived from what we usually consider to be the human perspective.

The claim that natural beauty is a purely subjective preference does not even remotely fit our experience. When we encounter natural beauty, we do not congratulate ourselves for having, by the power and quality of our minds, created a new instance of value in the world. We do not take pride in having a mind that effortlessly creates beauty. More importantly, we would regard someone who did think this way as not really understanding beauty as we experience it. What we experience is the discovery of beauty, which is perceived as in some sense separate from who we are.

A subjectivist would hold that, regardless of our experience, "beauty" simply refers to feelings of pleasure elicited by encountering certain kinds of stimuli. As such, it can be situated somewhere within a person's "utility function." To be sure, differences exist between the pleasure of eating chocolate ice cream and the pleasure from seeing the Grand Canyon, but ultimately they are quantitative. Both are subjective judgments expressing our preferences. Subjectivists point out that not everyone likes either the Grand Canyon or chocolate ice cream. Neither argument is adequate.

Consider the difference between clearly instrumental values depending upon personal desires and natural beauty. Here I want to be as unambiguous as possible. Because many things serve more than purely instrumental roles, I will use money and entrepreneurial action as clear examples of instrumental value and instrumental action. When I part with something of purely instrumental value to me in order to gain something else I value more, I am unambiguously pleased. Such is the case with an entrepreneur selling something for a profit. Had the entrepreneur not anticipated making a profit, he or she would not have obtained the goods to be sold in the first place.

Now it is certainly the case that many of us would sell something we regard as intrinsically valuable if offered enough money, which itself has only instrumental value. But in doing so we will not feel unambiguously pleased. Usually we will feel compromised, somehow dishonest with ourselves, and rationalize what we have done. Significantly, we will often respect a person who holds out against enormous temptation, even when we would not ourselves do so—whereas we would regard a person who held what seemed to us a pure instrumental value, such as money, as crazy if she or he did not exchange it for a resource of greater instrumental value. Why?

A clue comes from Hargrove's suggestion that

> when . . . feelings of pleasure [from natural beauty] are . . . compared with other instrumental values that can be obtained, for example, by clear-cutting or strip-mining, the value of the aesthetic experience then appears trivial, ridiculous, and indeed indefensible.[8]

Something that has only instrumental value is without worth except as it meets a human want. When we contrast a value that serves a physical need for housing, food, or clothing, with one serving "psychological pleasure," the latter always appears less substantive. In practice, pure subjectivism turns into utilitarian judgments based upon a physical conception of need.

According to pure subjectivists, being subjective, values cannot be compared between people. The value of a dollar to Ross Perot cannot be compared to the value a dollar holds for a starving man. In practice no one really believes this is true. And for good reason. If it turned out that Perot valued every dollar as much as a starving man, we would rightly conclude that there was something weird about Perot but not about the starving man. Nor would parents agree that their child always knows better what he or she wants than do they, or that they arbitrarily impose their values when they believe they know what is best for their child. This is true even if the parents are economists.

We are not so isolated from understanding our fellow human beings as claimed by pure subjectivists, who hold that intersubjective comparisons of value are arbitrary. We are similar enough to one another to make rough but not arbitrary comparisons. We generally rank the value of one person's physical need above those of another whose physical needs are luxuriously met, particularly when the need is great and the person is regarded as minimally responsible for his or her condition.

Instrumental values serving more vital needs are given preference over those serving needs less vital. A dollar spent to buy milk for a starving infant is better spent than buying jam for Perot's breakfast muffin. Since natural beauty has little capacity to serve vital instrumental needs without being transformed beyond all recognition, it always loses in this kind of comparison. But the comparison is illegitimate.

Beauty in nature provides meaning, something pure instrumental values do not. The experience is its own reward. It does not need to point to something beyond itself.

Instrumental values meet needs that use them up. When my money is spent, that's it. If I am now penniless, my memory of having had money I once spent on instrumental values will, if anything, tend to make my present circumstances more unbearable. The memory of experiencing natural beauty can be fulfilling in itself. Remembering a beautiful sunset is quite different from remembering I once had $20,000.

The perishable character of instrumental values is why economists always urge efficiency in using them, so that more will remain to meet other needs. And when instrumental values are not scarce, as is usually the case with oxygen, they are treated with complete disregard. However, there is an ambiguity in the idea of economic scarcity that confuses our understanding of the distinction between instrumental values and values such as natural beauty.

Economists frequently say that because love is scarce, we should minimize relying on social arrangements that depend on it. In making this recommendation they combine two senses of scarcity. There is the scarcity that comes from there not being enough loving people, and that most people love a fairly narrow range of others, rarely exceeding family and friends. Insofar as this kind of scarcity is meant, economists have a good point.

But there is another kind of scarcity that holds that if a resource can be used for one purpose, it will not be able to be used for another. The value of a resource is realized only by using it up. This type of scarcity is true for instrumental values, and central to economic theory. It is not true for values like love or beauty. I do not have a limited amount of love that I need to "invest wisely" in order to get "the greatest returns," although I may have limited time available to cultivate these relationships. If loving others did not take time to develop, there is no logical limit to the number of people I could love. If I perceive beauty in one place, that does not use up my capacity to perceive it somewhere else.

Attempting to price love and natural beauty automatically assaults their integrity, giving them an inappropriate frame of reference where they cannot be adequately understood. It is like asking the price of your son or daughter. Taking this question seriously changes our experience from something rewarding in itself to being a means to a separate end. This is why when people do succumb to temptation and sell something that is fundamentally of intrinsic value to them, they feel bad.

Beauty cannot be appropriately ranked along a common continuum with purely instrumental values such as money. However, this argument is still rooted in "the human perspective." We have not left Hargrove's "weak anthropocentrism." But what do we mean by "human" here?

The Gestalt Self

Consider the valuing individual. If we look carefully we will find, as David Hume and many Asian philosophical traditions have observed, that no isolated atomic self looks out upon the world from an independent vantage point, grandly bestowing value upon it. Valuing happens, but the valuer is hard to find. Hume wrote:

> When I enter most intimately into what I call *myself*, I always stumble on some particular perception or other, of heat or cold, light or shade, love or hatred, pain or pleasure. I never can catch *myself* at any time without a perception, and never can observe any thing but the perception.[9]

Hume has been criticized because his observation seems unable to account for the continuity and coherence of awareness that we also experience. Kant and others have emphasized the active role our minds play whenever we perceive something, so that we encounter orderly patterns, not simply a turbulent chaos of perceptions. Even so, the self considered as an isolated entity somehow separate from the world and observing it remains as difficult to find as Hume claimed. Kant showed that creativity is taking place, but he did not show who it is that is being creative.

The concept of a gestalt provides a tool for answering this question. What we perceive, gestalt experiments show, is an object constituted from out of its relationships with its surroundings. Michael Polanyi's demonstration that our explicit knowledge and perception rests upon an irreducible tacit component builds upon this insight. "Perception" he writes, "is a comprehension of clues in terms of a whole."[10] Our conscious awareness is rooted in awareness of which we are not conscious. This subsidiary awareness frames and makes possible our conscious perception. When we perceive an object over time, the angle of perception, available light, and many other factors will continually change. Our ability to integrate a shifting series of sense impressions into a discrete object comes from seeing these perceptions in relationship with many others that create the context where we see an apple.

A thought experiment suggests we experience a continuous sense of self without having any particular attribute that is fundamentally separate from our experience of continuity and of shifting sense data. Imagine yourself minus any particular single trait or quality of experience you presently have. What remains is easily recognizable as yourself. No single element of experience is an essential component of our identity. We have a self in that we experience continuity, but Hume is correct in observing that we cannot find it. It is not a particular entity. If we had never interacted with others, and so had no external experience at all, we would have no sense of self.

Our perception of self is a gestalt perception. Its unity is a result of its relationships with its environment. As such, our sense of self can be either expansive or crabbed, depending upon its boundaries. Boundaries are determined by encountering relationships that are fundamentally Other.

Because we can separate our sense of self from particular aspects of our experience, some people gradually wall off ever greater dimensions of their experience from their sense of who they are. As their sense of self shrinks, the realm of Other-than-self grows, becoming more alien and more threatening. Some people even separate their selves from threatening psychological drives and urges. Ultimately, even their minds become alien to who they think they are. Little sense of self is left but fear and confusion. As we shall see, it can also expand. Despite needing to label it with a noun, the self is a verb, not a thing. "It" is akin to the "it" we mean when we say "It is raining."

Our self is real enough, but it is a focal point of relationships. Our sense of our self appears distinct from its environment because it is a gestalt perception growing out of the configuration of all the environmental relationships in which it partakes. Consciousness, as Deep Ecologist Joanna Macy says, "does not exist prior to or independently of its environment, but is called into being and conditioned by that which in turn becomes its object. It is always consciousness of something."[11] Because no configuration of relationships is like any other, our individuality is truly unique, and fundamental to who we are.

Human consciousness is irreducibly creative because it does not passively experience its outlook, but takes a stance with regard to it. It creates a perspective on the world, a perspective that molds but does not fully determine experience. The self creates a world—but is itself created in return. The poet Yuri Yevtushenko captures this insight:

In any man who dies there dies with him
his first snow and kiss and fight.
It goes with him.

They are left books and bridges
and painted canvas and machinery.
Whose fate is to survive.

But what has gone is also not nothing:
by the rule of the game something has gone.
Not people die but worlds die in them.[12]

This world is the gestalt of all the relationships that, when considered together, make up a human life. And each relationship is itself determined in part by the complex of other relationships that exist at the same time. Our individuality is real but no more separable from the world than a whirlpool is from water.

Insights of this kind play an important role in Arne Naess's thinking. Naess suggests we think of the world as a relational field where "there are no completely separable objects, therefore no separable ego or medium or organism.... Within such a field, any concrete content can only be related one-to-one to an indivisible structure, a constellation of factors."[13] Macy notes that "To be a person . . . is to participate, at every level of our being, in a reality wider than that enclosed by our skin or identified with our name."[14]

It is instructive to compare Naess's and Macy's observations with those of F. A. Hayek, a liberal in the Humean tradition. Hayek observed that Mind can exist only as part of another independently existing distinct structure or order, though that order persists and can develop only because millions of minds constantly absorb and modify parts of it."[15] The overarching order Hayek described is embedded within social practice. We can take Hayek's insight a step further. Social practice is itself embedded within and influenced by the natural environment within which it exists. The deep ecological framework simply continues developing insights extending from David Hume through Polanyi and Hayek.

It is one thing to recognize that we are the outcome of our relationships; it is quite another to ask whether we should care about these relationships and, if so, how? A skeptic can always ask why he should care about the flea that bites him, or the person down the block. To explore this question I turn again to David Hume and to Adam Smith.

Sympathy

In Hume's day, as in ours, there were many who said that at bottom all human action was self-interested. In Hume's words, this view held that "whatever affection one may feel or imagine he feels for others, no passion is or can be disinterested; . . . the most generous friendship, however sincere, is a modification of self-love; . . . even unknown to ourselves, we seek only our own gratification while we appear the most deeply engaged in schemes for the liberty and happiness of mankind."[16] Today advocates of purely economic models of social and policy analysis, rational choice, and similar caricatures of human action make similar assumptions.[17] As with the earlier argument that all value is subjective, this egoistic analysis fails to resemble our actual experience. For example, Adam Smith observes that when we experience "fellow-feeling with all the emotions of our own breast . . . both the pleasure and the pain are always felt so instantaneously, and often upon such frivolous occasions, that it seems evident that neither of them can be derived from any such self-interested consideration."[18] An even deeper analysis provided by Hume and Smith further supports common experience against the egoistic model.

Hume argued that rigorously examined, the logic of egoism defeats itself. Acting in our self-interest beyond the spur of the moment requires the ability to anticipate our future situation. To do this, we imagine our anticipated future. On the basis of that projection, we choose whether to pursue actions intended to lead there. Hume pointed out that this hypothetical future self of ours does not yet exist.

Our power to project our imagination into an anticipated future arises from what Hume and Smith termed our *sympathetic* capabilities. By sympathy they refer to a capacity we would today call empathy or, in Naess's words, "identification."[19] Sympathy is not simply a passion or feeling. Hume emphasizes that "we must be assisted by the relations of resemblance and contiguity in order to feel the sympathy in its full perfection."[20] Smith adds that "men, though naturally sympathetic, feel . . . little for another, with whom they have no particular connection, in comparison of what they feel for themselves."[21]

Our intellect, which helps grasp or deny these "particular connections" to make these comparisons, is also required. Our predisposition to sympathy can therefore be cultivated and strengthened. Smith made the important observation that

> we can never survey our own sentiments and motives, we can never form any judgment concerning them, unless we remove ourselves . . . from our own natural station. . . . . But we do this in no other way than by endeavouring to view them from the eyes of other people, or as other people are likely to view them. . . . We endeavor to examine our own conduct as we imagine any fair and impartial spectator would examine it.[22]

The same capacities that enable us to put ourselves into our own future shoes also enable us to put ourselves into the shoes of another. Thus, rational self-interest that depends upon being able to anticipate the probable future consequences arising for us from something we do now requires that we have the capacity to sympathize with others. The capability to act effectively in one's rational self-interest can only exist in beings capable of altruistic concern.[23]

As I consider further implications of this argument, it seems that children's legendary cruelty might be linked with their equally legendary incapacity to act in their own long-term interest. Their capacity for sympathy needs to develop in order to adequately consider their interest and the interests of others. As a capacity linked to intellectual understanding, we would expect it to grow over time.

To be sure, some children also relate to their world as little animists. But their animism is immediate and spontaneous. While displaying genuine sympathy, it is a sympathy without the depth and continuity provided by experience. Indeed, very young children when seeing another get hurt will sometimes touch the equivalent portion of their own body.[24]

Sociopaths, people who are incapable of empathizing with other people, are also unable to act in their own long-term interests. They frequently end up in jail, are incapable of maintaining friendships (even when such friendships would be useful over the long run), and generally fail at living a successful life. This is also what we would expect if Hume's argument is valid.

The perspective I am developing explains why people would regard a system of moral rules as good. But specific moral rules are secondary to sympathy and can be revised. Moral rules are in part judged by their impact—a "utilitarian" position. But their foundation is respect for others with whom we sympathize. We would be bothered by a rule promising good results but not respecting others. In this sense this argument is not utilitarian.

Sympathy with Animals

Sympathy, as Hume observed, "extends itself beyond our own species."[25] Some species are more like ourselves than are others. We can more easily sympathize with chimpanzees than with fish and fish more easily than with insects. But it is not the case that sympathy, even with insects, is impossible. The possibilities for sympathetic identification never fall to zero. Even the simplest forms of life can flourish or not, react to stimuli that are harmful or beneficial, and enjoy good or suffer ill health.

Aldo Leopold explored these deeper levels of sympathy, writing that while we can mourn the demise of the passenger pigeon, which none of us has ever seen, passenger pigeons would not have mourned our own passing. He observed that "for one species to mourn the death of another is a new thing under the sun."[26] For Leopold it was just such an interspecies connection, when he reached a dying wolf "in time to watch a fierce green fire dying in her eyes" that forever changed his attitude toward the natural world. "I was young then, and full of trigger itch; I thought that because fewer wolves meant more deer, that no wolves would mean hunters' paradise. But after watching the green fire die, I sensed that neither the wolf nor the mountain agreed with such a view."[27] Neither the wolf nor the mountain!

The standard objection to this argument is that Leopold was anthropomorphizing. The claim is that there is a barrier to sharing experience between the human and the animal world. But this barrier, as Mary Midgley reminds us, already exists between people. It is the barrier and logic of a solipsist. "The barrier does not fall between us and the dog. It falls between you and me."[28]

We have two reasons for believing we can accurately sympathize with animals to a point. Evolution teaches us that all life is related. The more closely the physical nervous system of an animal approaches our own, the stronger the burden of proof must be on those who say its experience is wholly unlike our own. It makes more sense to say we have important similarities with other forms of life than to wall our experience off from everything else in the world. Descartes could do this because he believed in Genesis and allowed "doubt" to overwhelm his common sense. But we have no excuse.

Second, those who work sympathetically with animals often get along very well with them, whereas those who do not are much less successful. How significant is this? Perhaps as significant as the same observation with regard to human beings.[29] Nowhere are there reasonable grounds for saying of sentient beings that past a given point sympathy is impossible.[30]

Human Uniqueness

The human capacity for sympathy is qualitatively greater than other animals' because we are radically more self-conscious than they. Our awareness of our self creates the capacity to be aware of other selves who are of no use to us and whom we have never seen. This conclusion allows us to unite the liberal western regard for the integrity of the individual and the deep ecological concern for the integrity of the nonhuman world. Egalitarian Deep Ecologists who cannot find a qualitative distinction between human beings and animals and their anthropocentric critics are both wrong. The strongest evidence for our uniqueness in the world requires at least a biocentric ethic in order for it to manifest.

The stronger our sense of our own individuality, the greater our inherent capacity to realize our connections with all life, for our individuality arises from those connections. The greater our sense of self as a being extending over time, the greater becomes our capacity to sympathize with other beings. This is because the farther into the future our self-interest extends, the more developed our capacity for sympathy must become to encompass it. While some relationships will always be outside our conscious perception, the more of them of which I can be aware, the larger and more differentiated becomes the world in which I live. My realm of both awareness and choice enlarges. We can conclude that an ecocentric ethic is more in keeping with uniquely human nature than any perspective trying to base ethics on narrow self-interest or even the interests of the human race alone. The most loving human beings are the most human human beings.

Ethics and Sympathy

We now return to Hargrove's challenge to deep ecologists that their position depends upon "going beyond valuing based on the human perspective, which seems impossible."[31] The human perspective includes an open-ended capacity for sympathetic identification with the nonhuman as an essential part of what it is to be human. The more we develop this capacity, the less our perspective can be identified with the modern notion of self and the "human perspective" it implies. Those failing to develop this capacity are simply diminished human beings.

Naess argues that "the ecological self is that with which the person identifies. This key sentence (rather than definition) about the self shifts the burden of clarification from the term 'self' to that of 'identification,' or rather 'process of identification.'. . ."[32] As people sympathetically expand their awareness of the relationships comprising not only themselves but other selves as well, they see themselves as members of ever larger and more diverse communities. The result of such identification is increasing respect for nonhuman life. Other beings matter. In the absence of compelling reasons, we should not interfere with them.

This argument is not ethical in the sense of telling us what we should and should not do. Conceptions of ethics as inhibiting or obligating rules often place it in opposition to our selves. Ethics as we usually think of it limits us. However, to the extent people agree with my argument, they will want to act in a sympathetic manner toward all their relations. Ethics as a system of rules alleviates the gaps and failures in our understanding. By contrast, my argument gives us a foundation for ethics that we will not experience as a limitation. Sympathetic relationships with others are expansions, not limitations of our selves.

Of course, people can refrain from expanding their awareness, simply turning their backs upon their most uniquely human capacities. Their reasons for doing so will usually stem from fear or ignorance, which are hardly strong foundations for creating an ethical outlook, let alone living a life! The most complete expression of our individuality is a loving relationship with all beings.

From Biocentrism to Ecocentrism

We have dealt with Leopold's wolf, but what of Leopold's mountain? He believed it also disapproved of his attitude toward wolves. I will not here make a case for animism, although in my view that case can be made. It will take us too far afield. Even without accepting animism, what Leopold did was inappropriate from the perspective of a mountain.

Let us return to my discussion of beauty. When most impressed by natural beauty, we encounter it without comparing it to anything else. We do not analyze it. We do not evaluate it by an external standard.

If someone asks us what is most beautiful about a mountainside, to answer we first withdraw from our immediate experience. As soon as we attend to a part of what we regarded as beautiful, perhaps an aspen grove, we lose the whole. What about the grove is most beautiful? Again we distance ourselves, again adopting an evaluative stance, losing touch with the grove as well. Each experience becomes a member of a class. In important respects more is lost than is gained by this exercise.[33] This process is destructive when applied to human beings we love. Try doing it to your son, daughter, or lover. It is equally destructive when applied to mountains.

Michael Polanyi's discussion of using a probe helps deepen our argument. If we use a probe focusing only on the sensations we receive in the palms of our hands, the instrument will be all but useless. If, on the other hand, we focus on the probe's tip, we can learn a great deal from using it. We then experience the probe as an extension of our bodies. We dwell within it.

Polanyi distinguished between what he termed subsidiary and focal awareness. When my focal awareness is on the probe's tip, I am on a subconscious level aware of its feel in my hand, but am not directly conscious of it. As soon as I shift my focus to my hand, I lose touch with the meaning I receive from the probe's tip. They are mutually exclusive forms of awareness. Polanyi concludes that:

> Our subsidiary awareness of tools and probes can be regarded now as the act of making them form a part of our own body . . . we shift outwards the points at which we make contact with the things that we observe as objects outside ourselves. While we rely on a tool or a probe . . . [w]e pour ourselves into them and assimilate them as parts of our own existence. We accept them existentially by dwelling in them.[34]

The same process holds when we experience natural beauty. We dwell within the experience. We are not gaining information about the shape, texture, or depth of something, but rather another kind of meaning—that deeper significance of nature, which we call beauty.

This deeper significance changes us in a way that using a probe does not. Skillful use of a tool enlarges our power to accomplish instrumental ends. What is experienced as outside of me is a target for my manipulation. When experiencing natural beauty, however, my frame of mind is not manipulative. It is open and accepting. My experience is of encountering something intrinsically bigger, deeper, and more valuable than my daily concerns. I dwell within, but do not seek to manipulate the world outside.

Those claiming what I simply describe as subjective experience downplay or ignore three facts. First, this same approach yields reliable information not otherwise available when applied to our instrumental concerns, as with the skillful use of a tool. Indeed, Polanyi argues it is the basis of all our knowledge. Second, if there is some thing or quality in nature separate from our usual sense of self and which is beautiful, there is no other way by which people could experience it. Third, those experiencing beauty in this way all say it is different from a preference of taste. In many cases the experience changes how they live their lives. Just as with sympathy for others, the boundaries of our selves expand. In both cases we dwell within a wider frame of reference.

Returning now to Leopold, the evaluating mind, deciding that wolves are "bad" and out of place in the wilderness where he encountered them destroys the wilderness experience. The boundary between self and nature is strengthened, as crucial elements within it are rejected. The mind sets itself up as judge, prosecutor, and jury for what it encounters. Whatever fails to measure up becomes an obstacle, trash, or vermin, and disposable. These judgments are all intimately connected to narrowly human instrumental purposes. They are quite different from seeking to live in harmony with the land, which does not draw such a sharp boundary and so leaves one open to the experience of its beauty.

When experiencing beauty, we want to preserve, not destroy. When dwelling within the land, its desecration causes pain. Given good reason we can destroy, but always with a sense of regret.[35] Leopold was deeply insightful when he said that the mountain would not agree. To experience the mountain in its entirety he had to give up his calculating attitude toward it. It was more than simply an abode for deer. To maintain his experience, he had to relate in a respectful way, rather than regarding a basic element of its biological community as fit for extermination.

We have moved from a biocentric to an ecocentric position. The natural world is beautiful. Its beauty can only be experienced to the extent that we open ourselves up to it in a noncalculative way. As with sympathy with all life, we transcend what is usually considered the "human point of view" in order to experience the intrinsic value of the land. Since this value can be experienced only to the extent that we do not approach it instrumentally, in an important sense we discover beauty that was there all along. To discover beauty it is we, and not the land, that must change.

The gulf between deep ecology and anthropocentrism has now been bridged by showing the dichotomy to be a false one. It turns out that the source of human uniqueness is our capacity to widen our senses of self to be indefinitely inclusive, to dwell within not only animals but the natural world as a whole.

Our recognition that other entities are not simply tools for our use, in Warwick Fox's words, "amounts to a revolution in our treatment of the nonhuman world that is comparable to the difference for humans of a legal system that operates on a presumption of guilt until innocence is proved beyond a reasonable doubt and one that operates on a presumption of innocence until guilt is proved beyond a reasonable doubt."[36]

Community

We all live in communities. Communities can be distinguished by the kinds of relationships that most typically characterize them. We participate in an enormous number of communities, many of which are themselves composed of smaller communities.[37]

From this vast assemblage, four characterize modern life, are open to all people, and are defined by relationships applying equally to all adults. These communities are the family; the neighborhood or township; the anonymous society of cities, states, and nations; and the world of nature. Different concrete relationships apply within these four communities. In all these communities the character of the relationships defining membership is more central than any conception of abstract individuality. These concrete relationships can impinge on the self-interest of individuals conceived in purely abstract terms.

These defining relationships are rooted in respect in that all members play legitimate though not necessarily equal roles and have legitimate interests that should be recognized. While values of respect are universal, being rooted in sympathy, the ways in which they are most appropriately applied are context-specific. Respect underlies love and affection within the family, friendship and standing within the neighborhood, and liberal rights within the larger society. Liberal rights are the form respect takes among a large and relatively impersonal community of social and political equals. While most relationships have an instrumental dimension, to maintain the community indefinitely members should not be treated only as a means to our ends. Their interests never count for zero.

Because these communities comprise vital elements of a good human life, public policy should seek to provide an environment conducive to harmonizing and sustaining them. Heretofore liberal principles have recognized only purely human communities. This anthropocentric boundary is not justifiable.

Natural Community

What most distinguishes the natural community is that its membership is made up fundamentally of individuals in their capacity as members of species. The common interest of an ecological community is the well-being of the species comprising it. The most frequent objection to this argument is that species are not individuals and so cannot have interests. Biologist Julian Huxley's response to this objection seems to me quite reasonable. Huxley writes:

> Whenever a recurring cycle exists (and that is in every form of life) there must be a kind of individuality consisting of diverse but mutually helpful parts succeeding each other in time, as opposed to the kind of individuality whose parts are all coexistent. The first constitutes . . . species individuality, or individuality in time, while the other corresponds to our ordinary notions of individuality.[38]

We have seen the individual to be more verb than thing, a gestalt of relationships. Individuals cannot adequately be encompassed by what we commonly think of as existing within our physical boundaries. In the natural world the concept of individual as we apply it in the human realm becomes even more problematic. Bees, ants, and termites certainly are not individuals in the human sense. A slime mold is even farther removed, normally existing as thousands of independently living amoebae. In time these one-celled individuals coalesce to create a multicelled being that crawls along the forest floor, halts, and sends up a spore packet that disperses spores, enabling it to reproduce. In the natural world, species individuality seems as firmly rooted as the individuality of single organisms.

Taken as a whole, a species embodies a particular way of living in the world. No individual member can completely manifest this potential. Animals must learn how to live in the world to a much greater extent than we once thought, and their learning creates opportunities for creativity and differentiation. For example, sometimes songbirds of the same species will have different "dialects." Indeed, even bacteria appear capable of learning.[39] In the course of a single lifetime a human being can explore only the tiniest portion of human potential.

Being a member of a community cannot be translated into the interests of an abstract individual because abstract individuals lack embeddedness in the concrete relations that make them individuals in the first place. This observation applies most strongly to members of the natural community. Except for the top of the food chain, most individuals sooner or later end up as meals for others. The well-being of the natural community requires it. Predation ultimately benefits all species comprising the natural community. Deer as a species would suffer if all predators that eat deer were eliminated—however much certain individual deer might benefit. Deer possess a "public interest" that is not the sum of the interests of individual abstract deer. Over time, more individual deer will benefit from appropriate predation insuring the well-being of their species than would be the case in its absence.

Just as we rightly value diversity and individuality within human communities, we should value diversity and individuality within the ecological community. As our understanding of our relationships with others becomes richer and more varied, we ourselves become richer, with more dimensions to who we are. We more fully embody our innate human potential for sympathy and its resulting deep individuality.

Ecocentric Liberalism

I distinguish that variant of liberal thought arising from the work of David Hume and Adam Smith from other varieties of liberal theory. The first I term "evolutionary liberalism," while the second I label "individualist" liberalism.

"Individualist liberalism" treats the individual as the irreducible and atomistic unit of analysis, analytically or ethically or both. It is here in the liberal tradition that emphasis is upon the individual abstracted from its relationships. This tradition includes all those who base their work on rational self interest, natural rights, or utilitarianism. It crosses the "left-right" political divide, including the work of Milton Friedman and Robert Dahl, of both the early Robert Nozick and John Rawls. Its precursors include Hobbes, Locke, Destutt de Tracy, Bentham, and the Mills. Despite their genuine and important differences, all these thinkers can be said to take a Promethean view of humankind's relationship with the world, for it exists to be mastered and controlled. The analysis I have developed in this paper is fundamentally incompatible with the outlooks that characterize this variant of liberal thought.

Evolutionary liberal perspectives emphasize that individuals are always embedded in society. Social processes are constitutive of individuals, and in that sense prior to the individuals themselves. In Peter Berger and Thomas Luckmann's words,

> Society is a human product. Society is an objective reality. Man is a social product.... An analysis of the social world that leaves out any one of these three moments will be distorted ... only with the transmission of the social world to a new generation (that is, internaliza-

tion is effectuated in socialization) does the fundamental social dialectic appear in its totality.[40]

Evolutionary liberals do not believe human beings can reconstruct society anew every generation, picking and choosing the qualities they wish. There is no vantage point outside of society to accomplish such a reconstruction. In a sense, sympathy is "outside" a particular society, but it is too abstract to be anything but a general guide for assisting the evolution of already existing institutions. Sympathy takes a particular form depending upon the community in which it manifests.

People always look at political choices from within a social standpoint they have already to some degree (but never totally) internalized.[41]

Society comprises the independently existing structure that Hayek described as helping constitute our minds. It incorporates generations of experience and knowledge too vast to be deliberately grasped by a human mind. Much of this knowledge is continually reproduced and modified as the unintended consequences of human action, as people pursue their privately ordered purposes within a framework of customs and other institutions.

It was the evolutionary liberals who discovered that institutions evolve by virtue of how effectively they coordinate human cooperation. Cooperation in a complex society requires enabling institutions and practices for it to be enhanced, because no person can have any but the tiniest view of the whole. Nevertheless, if individually perceived knowledge and opportunities are to be useful to the community as a whole, some means for mutual adjustment and coordination must exist. Societies and practices that are most successful in assisting such coordination will tend to prevail over time, because they liberate the creative power of the human mind.

Society is, in Hayek's terms, a spontaneous order that itself incorporates other spontaneous orders, all of which are in a continual process of mutual adjustment. The economy, science, language, and custom and customary law are all spontaneous orders or what is more often called self-organizing systems. I would include democracy in this list, for the greatest intellectual failure of evolutionary liberals has been their general failure to see that liberal democracies are also self-organizing processes.[42] The market promotes economic coordination, science promotes coordination in the search for scientific knowledge, and liberal democracy promotes the public good. Of course all can be much improved over their present performance.

Unlike the bogus "social ecology" promoted by Murray Bookchin, the evolutionary liberal framework delineates a true social ecology, lacking only adequate grounding in the natural world.[43] Even so, the harmony between evolutionary liberalism and natural processes is very strong. In fact, Darwin's theory of evolution grows out of his familiarity with the work of Hume and Smith. Darwin adapted the evolutionary model they identified in society to the natural world.[44]

Any self-organizing social institution can effectively integrate far more information than could be grasped by any person or committee. Further, it allows a great variety of individual projects to be pursued, each in a way that, on balance, generates information and resources useful for other people to pursue their plans. Far from subjecting individuals to unintelligent processes, the more a society incorporates self-organizing processes the more creative and intelligent it will become in its interactions with its environment.

Evolutionary liberalism is often considered politically "conservative," usually because its advocates have opposed programs advocating massive state intervention, long a hallmark of the socialist left. Their reasons for opposition, however, have largely emphasized the limitations on knowledge available to those holding political power, and the ways in which institutions bias and select out the knowledge that will inform action. The result, they argue, ultimately undermines the program's goals. Emphasizing limitations on our knowledge only seems "conservative" to those who prefer not to think we have any significant limitations at all.[45] The evolutionary perspective is not so much "conservative" as it is a different framework from within which to think about social and public policy.

Evolutionary liberals consistently have pointed out that the customary and ethical institutions of a society perform valuable roles in maintaining a self-organizing social order. They have been often misunderstood as simply defending the ethical and customary status quo. This is a misreading. As Hayek observed:

> The social scientist who endeavours to understand how society functions, and to discover where it can be improved, must claim the right critically to examine, and even to judge, every single value of our society. . . . [But] we can never at one and the same time question all its values.
>
> The . . . evolution of a system of values passed on by cultural transmission must implicitly rest on criticism of individual values in the light of their consistency, or compatibility, with all other values of society, which for this purpose must be taken as given and undoubted. . . . Because prevailing systems of morals or values do not always give unambiguous answers to the questions that arise, but often prove to be internally contradictory, we are forced to develop and refine such moral systems continuously.[46]

The evolutionary liberals can teach much to ecocentric thinkers. For obvious reasons, Deep Ecologists are generally very critical of liberal market society. But their criticism often confuses important but ultimately surface problems with what they take to be radical shortcomings. As a result, they often take positions that ignore the evolutionary liberal criticisms of political direction, and insights on how to approach problems, even when these arguments would contribute mightily to their own understanding of why political control has such a darkly checkered record when applied to environmental issues.[47]

Fundamentally ecocentric perspectives are in harmony with the major tenets of evolutionary liberalism. Deep Ecologist Joanna Macy observed that:

> a totalitarian society might seem to be more unified and coherent than a free one, and seem to display in its apparent discipline an ideological "group head." Yet, to the extent it discourages differentiation between its subsystems and hampers the voluntary flow of information, its mentality . . . is on a more primitive level than that of a free one. At the other extreme, a society where individuals and groups are not ready to assume transpersonal loyalties and responsibilities is not integrated enough effectively to self-organize. Fractured and incoherent, it remains on a low level of adaptability and awareness.[48]

Only her use of "transpersonal" distinguishes Macy's terminology from the evolutionary liberal perspective.

How should liberal society reform, so as to relate decently with the nonhuman world? An insight of Hayek's helps orient us. Expediency, when used to evaluate a prospective immediate gain with a prospective future loss, will be biased in favor of the promised gain. In a self-organizing system, specific political interventions within the system that override its constitutive principles will often promise to lead to desired outcomes. But these interventions disrupt the processes of mutual adjustment maintaining the system as a whole, retarding its capacity to adjust to changes and reducing its ability to serve the maximum number of participants. This is why, as Hayek argued, principles should trump expediency in self-organizing social systems.[49]

Hayek's point holds equally well for our interactions in the natural world, which is also self-organizing. But what principles should they be? Clearly, those principles that enable the environmental community to sustain itself.

A major sustaining principle is that the wastes every organism produces are useful to another organism. When human impact upon the natural world was limited by low numbers and small-scale technology, it was possible to ignore this principle. Today it no longer is. There should be a powerful presumption against creating products that neither biodegrade nor recycle.

A second principle is that using a renewable resource should be compatible with its indefinite renewal. Destroying a fishery, polluting ground water, and exhausting soil fertility are examples of activities that would be inadmissible under such a principle. If an old growth forest is to be used, it should be used in such ways as to indefinitely perpetuate old growth ecosystems.

Third, using resources must maintain a diverse flora and fauna. It is illegitimate to exterminate species, even ones occasionally lethal to us such as grizzlies, great white sharks, or mountain lions. Species are the primary component of an ecological community.

These principles do not undermine the well-being of the other basic communities, although they definitely limit certain activities some people currently take for granted. They are, however, central to maintaining the environmental community's indefinite sustainability. Their observance is as ethically justified as the rules forbidding violence and theft within purely human communities.

A fourth principle underlies these three. Nothing living can be ethically treated as a pure means—and very likely nothing in nature as such should be so treated. To some extent this goal can be reflected in law. For example, farming methods should ensure decent treatment for meat animals until they are slaughtered. But its implications extend beyond what the law can be reasonably expected to enforce. It is not of the same nature as the previous three, which are deducible from the conditions required for an ecology to survive. This principle is rooted in our humanness. The first three can be deduced from a relatively narrow self-interest. The fourth cannot.

Implications for Economics

An ecocentric framework's major impact upon the market concerns property rights. Property rights define power relations. We do not own "property" so much as bundles of property rights. Each right determines a type of relationship into which I may appropriately enter with another person or with the thing owned. The richer and more numerous the property rights that may be bought and sold on the market, the more complex the resulting market order becomes.

Nevertheless, not every relationship should be a property right. Slavery has been appropriately banned, even though at one time there was a market price for slaves and, in Virginia, even a slave-breeding industry. When the right to own slaves was abolished the market order did not collapse, though the market for slaves did. This limitation on what could be legitimately bought and sold on the market was a net gain for human well-being, even if it did destroy a way of life thousands of years old. The principle of slavery is antithetical to principles maintaining free human societies.

The same point holds for practices destructive to our wider natural community. Certain contemporary property rights are illegitimate. As human wisdom increases, they should be disallowed. The right to ruin soil, create nonrecyclable waste, or seriously reduce genetic diversity within a species are all examples of inappropriate property rights. If they were removed we would most definitely not see the demise of the market. We would no more distort the price system or cripple entrepreneurial opportunities than did removing slaves from market transactions. We would return neither to the Middle Ages nor to the Pleistocene.

These arguments demonstrate that evolutionary liberalism and deep ecological principles share many points in common. Their foci differ, one being concerned with human society, the other with our appropriate relations with nature. Nevertheless, they are complementary. Each is a necessary element in creating the conditions for a good and prosperous life for all our relations to the seventh generation.

Notes

1. A classic presentation of this point of view is Bill Devall and George Sessions, *Deep Ecology* (Salt Lake City: Peregrine Smith Books, 1985). See also Christopher Manes, *Green Rage* (Boston: Little, Brown, 1990).
2. Anna Bramwell, *Ecology in the 20th Century: A History* (New Haven: Yale University Press, 1989), p. 248.
3. While there are differences between "ecocentric," "biocentric," "transpersonal," and "deep ecological" perspectives, for most purposes of this article they are not central. Among the basic books developing a deep ecology, biocentric, or ecocentric point of view, see Arne Naess, *Ecology, Community and Lifestyle* (Cambridge: Cambridge University Press, 1990); J. Baird Callicott, *In Defense of the Land Ethic* (Albany: SUNY Press, 1989); Devall and Sessions, *Deep Ecology*; Warwick Fox, *Toward a Transpersonal Ecology* (Boston: Shambhala, 1990); Dolores LaChapelle, *Sacred Land, Sacred Sex, Rapture of the Deep: Concerning Deep Ecology and Celebrating Life* (Silverton, CO: FinnHill Arts, 1988); Aldo Leopold, *A Sand County Almanac* (New York: Ballantine,1966); Joanna Macy, *Mutual Causality in Buddhism and General Systems Theory* (Albany: SUNY Press, 1991); Holmes Rolston, *Environmental Ethics* (Philadelphia: Temple University Press, 1988); and Gary Snyder, *The Practice of the Wild* (San Francisco: Northpoint Press, 1990).
4. Eugene Hargrove, "Weak Anthropocentric Intrinsic Value," in Max Oelschlaeger, ed., *After Earth Day* (Denton: University of North Texas Press, 1992), p. 142.
5. Ibid., p. 152.
6. For example, the collection of essays in Eugene Hargrove, ed., *The Animal Rights/Environmental Ethics Debate: The Environmental Perspective* (Albany: SUNY

Press, 1992). The classic defense of animal rights is Tom Regan, *The Case for Animal Rights* (Berkeley: University of California Press, 1983).
7. Peter Singer, *Animal Liberation* (New York: *New York Review Press*, 1975), p. 238. For a telling examination of the problems with this approach from an ecocentric perspective, see J. Baird Callicott, "Animal Liberation: A Triangular Affair," in Callicott, *In Defense of the Land Ethic*, pp. 15–38.
8. Hargrove, "Weak Anthropocentric," p. 159.
9. David Hume, *A Treatise of Human Nature*, bk. 1, pt. 4, sec. 6; and J. Baird Callicott and Roger T. Ames, *Nature in Asian Traditions of Thought* (Albany: SUNY Press, 1989).
10. Michael Polanyi, *Personal Knowledge: Towards a Post-Critical Philosophy* (Chicago: University of Chicago Press, 1962), p. 97.
11. Macy, *Mutual Causality*, p. 129.
12. Yevgeny Yevtushenko, *Selected Poems* (Baltimore: Penguin, 1962), p. 85.
13. Arne Naess, *Ecology, Community and Lifestyle* (Cambridge: Cambridge University Press, 1989), pp. 56–57.
14. Macy, *Mutual Causality*, p. 184.
15. F. A. Hayek, *The Political Order of a Free People* (Chicago: University of Chicago Press, 1979), p. 157.
16. David Hume, "An Enquiry Concerning the Principles of Morals," appendix 2, in Henry D. Aiken, ed., *Hume's Moral and Political Philosophy* (New York: Hafner, 1948), p. 270.
17. See the excellent critique of these models in Steven E. Rhoads, *The Economist's View of the World: Government, Markets, and Public Policy* (Cambridge: Cambridge University Press, 1985).
18. Adam Smith, *The Theory of Moral Sentiments* (New Rochelle, NY: Arlington House, 1969), p. 10.
19. Arne Naess, "Identification as a Source of Deep Ecological Attitudes," in Michael Tobias, ed., *Deep Ecology* (San Diego: Avant Books, 1985), pp. 256–70.
20. David Hume, "A Treatise of Human Nature," bk. 2, sec. 2, in Aiken, ed., *Hume's Moral and Political Philosophy*, p. 7.
21. Smith, *Theory*, p. 125. This quote of Smith's suggests how Alasdair MacIntyre misreads Hume when he suggests Hume denied that people had a natural regard for the public interest. See *A Short History of Ethics* (New York: Macmillan, 1966), p. 174.
22. Ibid., pp. 161–62.
23. Hume: see Henry Aiken's introductory essay, p. xxiii, in Aiken, *Hume's Moral*. It seems to me that Hume's argument implies a radical critique of the subjectivist argument that in principle denies interpersonal comparisons of utility. We need to make such comparisons in order to act rationally in our own interest. Of course the chances of error increase as the self we examine becomes less like ourselves, but individuality does not necessitate incomprehensibility.
24. See Daniel Goleman, "The Roots of Empathy Are Traced to Infancy," *New York Times*, March 28, 1989, pp. 13, 21.
25. Hume, "A Treatise of Human Nature," bk. 3, pt. 2, sec. 1, in Aiken, *Hume's Moral*, p. 52.
26. Aldo Leopold, *Sand County Almanac* (New York: Ballantine Books, 1970), p. 117.
27. Ibid., pp. 138–39.

28. Mary Midgley, *Animals and Why They Matter* (Athens: University of Georgia Press, 1983), p. 130.
29. Colin Tudge, *Last Animals at the Zoo* (Washington, DC: Island Press, 1992), pp. 193–240; see also John A. Fisher, "Taking Sympathy Seriously: A Defense of Our Moral Psychology Toward Animals," in Hargrove, *Animal Rights/Environmental Ethics*. See also Evelyn Fox Keller, *Gender and Science* (New Haven: Yale University Press, 1985), pp. 95–138.
30. Goleman, "Roots of Empathy."
31. Hargrove, *Animal Rights,* p. 152.
32. Quoted in Warwick Fox, *Transpersonal Ecology*, p. 230.
33. Abraham Maslow, *The Farther Reaches of Human Nature* (New York: Viking, 1971), p. 69.
34. Polanyi, *Personal Knowledge,* p. 59.
35. For a view of how one can both make use of and respect the natural world, see Richard Nelson, *The Island Within* (New York: Vintage, 1991).
36. Warwick Fox, "What Does the Recognition of Intrinsic Value Entail?" *The Trumpeter: Journal of Ecosophy* 10 (1993): 101.
37. Indeed, it is not going too far even at the physical level to say that individuality is a community endeavor. See in particular the work of Lynn Margulis, for example: "Microcosmos," in Connie Barlow, ed., *From Gaia to Selfish Genes: Selected Writings in the Life Sciences* (Cambridge: MIT Press, 1992), pp. 57–66; and with Richard Gurrero, "Two Plus Three Equals One Individuals Emerge from Bacterial Communities," in William Irwin Thompson, ed., *Gaia 2: Emergence: The New Science of Becoming* (New York: Lindesfarne Press, 1991), pp. 50–67.
38. Julian Huxley, "Blurred Bounds of Individuality," in Barlow, ed., *From Gaia to Selfish Genes*, p. 73.
39. Evelyn Fox Keller, "Between Language and Science: The Question of Directed Mutation in Molecular Genetics," in *Secrets of Life, Secrets of Death: Essays on Language, Gender, and Science* (New York: Routledge, 1992), pp. 161–178.
40. Thomas Luckmann and Peter Berger, *The Social Construction of Reality* (New York: Anchor, 1967), p. 61. See also Alfred Schutz, *The Structures of the Life-World* (Evanston: Northwestern University Press, 1973), chapter 4.
41. Alfred Schutz, "On Multiple Realities," in A. Broderson, ed., *Collected Papers*, Vol. I., A. (The Hague: Martinus Nijhoff, 1964), pp. 207–59.
42. Gus diZerega, "Democracy as a Spontaneous Order," *Critical Review* 3 (1989): 206–40.
43. Gus diZerega, "Social Ecology, Deep Ecology, and Liberalism," *Critical Review* 6 (1992): 2–3, 305–70.
44. F. A. Hayek, "The Results of Human Action but Not of Human Design," *Studies in Philosophy, Politics and Economics* (New York: Simon and Schuster, 1967), p. 101. See also Stephen J. Gould, *Eight Little Piggies: Reflections in Natural History* (New York: W.W. Norton, 1993), pp.138–52.
45. An excellent discussion of this issue is in David Ehrenfeld, *The Arrogance of Humanism* (Oxford: Oxford University Press, 1981). Unfortunately, Ehrenfeld seems largely unaware of the evolutionary tradition in modern thought.
46. F. A. Hayek, "The Errors of Constructivism," *New Studies in Philosophy, Politics, Economics and the History of Ideas* (Chicago: University of Chicago Press, 1978), pp. 19–20.

47. Two excellent essays that probe these matters more deeply than the usual deep ecological discussions are Neil Evernden, "Ecology in Conservation and Conversation," and Michael E. Zimmerman, "The Future of Ecology," both in Max Oelschlaeger, ed., *After Earth Day: Continuing the Conservation Effort* (Denton: University of North Texas Press, 1992), pp. 73–82, 170–83. Robyn Eckersley's in most ways very good *Environmentalism and Political Theory: Toward an Ecocentric Approach* (Albany: SUNY Press, 1992) dismisses liberal environmentalism in a few pages, but she focuses only on individualist liberalism. In my reading, her own position is in most respects compatible with the position I have been developing.
48. Macy, *Mutual Causality*, p. 200–1.
49. Hayek, *Rules and Order*, pp. 55–71.

5

Is Liberalism Environment-Friendly?

Avner de-Shalit

The state of the environment is a novel issue on the political agenda, yet it is one of the most important. People the world over have begun to realize that the damage to the environment caused by human activities is drastic and at times even irreversible. The pollution of the soil, contamination of water by chemicals, air pollution, the damage to the ozone layer, acid rain, the complications in radioactive waste management, the extinction of certain animal species, deforestation, misguided urban development—all these and other phenomena constitute what is now widely called "the environmental problem."

Some people think that scientists or engineers can solve or at least suggest solutions to all these problems.[1] But although theoretically speaking most environmental problems have scientific solutions, in practice these solutions are thrust aside by economic criteria and considerations. Now, the latter are simply the reflection of social and political ideologies, and so, in fact, environmental policies involve decisions on the allocation of financial resources and time, on the distribution of money and political power, and on public priorities. Consequently, the solution to these problems is political,[2] with the result that philosophers and political theorists have tried to find the moral grounds for environment-friendly policies. This paper deals with this attempt. My argument is twofold: first, that liberalism, for a number of reasons that will be investigated here, has provided a good framework for the evolution of the "Green" ideas and environmental philosophy. But second, that with regard to environmental policies, liberalism nevertheless faces difficulties: while it allows and encourages discussion of environmental issues, it cannot permit its outcome, namely the implementation, maintenance, and justification of environmental policies, and therefore it precludes constructive public action to secure environmental protection.

This ambivalent attitude may confuse the reader, since there is no sharp and unequivocal assertion in favor of or against liberalism in the context of the environment. On the contrary, the intrinsic complexity of liberalism and the gap between the theoretical discussion and praxis reveal the complexity of liberalism's relationship with environmentalism. But this paper may speak to the suggestion that a more "social" liberalism could, perhaps, deliver the goods.[3] In that respect my project calls for further research; however, I am suggesting that theory about the politics of the environment should be placed not merely in its "Green" or "movement" context, but in a wider framework, that of the liberal tradition in political theory.[4]

Liberalism and Environmental Philosophy

It may be argued that at first appearance liberalism and environmental philosophy do not tie in with each other because liberalism's most fundamental feature—the contract—leaves out those who cannot join it, that is, trees, rocks, rivers, animals, and so on. The most prominent of contemporary liberal theorists, John Rawls, for instance, leaves this issue to metaphysics.[5] Others have tried to follow Rawls's path and yet modify it so that it generates an animal-rights ethics[6] or show that his theory implicitly requires extensive environmental policies.[7] But in this section my aim is not so much to show how liberalism and environmental philosophy can be seen as identical twins or as ideal ideological bedfellows, but rather to argue that liberal societies have become a fertile ground for the promotion of ecological attitudes and environmental philosophy. There are four main reasons for this: the first two reasons lie in the sphere of philosophy, or theory, and involve both the content of the liberal idea and the tradition of liberal thought.

Let us start with the former. One of the main components of liberalism is antichauvinism: the moral agent does not automatically exalt its own virtues and discredit those of others. (The moral agent could be a single person or a collective body to which the person belongs, whether it is a voluntary body—for example, a party, firm, or a body into which the person was born—for example, a nation, race, gender, class.) Chauvinists do not consider "others" on equal terms, but liberals have rejected such attitudes and instead propound the idea that all people are equal since they are all human beings cast in the same mold. Hence liberals have contended that all humans deserve equal rights and that we should follow a policy of "respect for others" and "respecting others as equals."[8] In short, liberalism as a social philosophy has rejected all expressions of chauvinism from national to male chauvinism.

At the same time, the situation in which humans deplete resources and damage the environment has been described on several occasions, and quite rightly, as "man (or human) chauvinism"[9] with respect to the illtreatment of nonhuman animals (sometimes even plants or ecosystems) and their exclusion from the ethical community.

It is only natural, then, that many liberals, both theorists and politicians, have adopted Green ideas and ecological attitudes: the essence of liberalism, as indicated above, is the philosophy of "respect for others." An environmental attitude implies extending the notion of "others" to include nonhuman animals, "all sentient creatures," or even "all living objects" or "ecosystems."

> A land ethic changes the role of homo sapiens from conqueror of the land-community to plain member and citizen of it. It implies respect for his fellow members, and also respect for the community as such.[10]

Most liberals would find it difficult to adhere to "holism" or to Aldo Leopold's "Land Ethic," which rests upon the premise that the individual is a member of a community of interdependent parts; but it is arguable that the elements of antichauvinism and "respect for others" that have characterized liberal thought have influenced and contributed to the emergence of Green ideas, at least in their defense of individual entities in the environment. Admittedly, many Greens are keener on collections of living objects, for example, species, rather than on individual entities. This will be discussed below, where I suggest that a more social type of liberalism suits environmental issues. It also remains to develop a theory of humans' relationship with the nonsentient objects; Christopher Stone has perhaps shown the way.[11] Nevertheless, Roderick Nash has drawn an interesting analogy in this connection between the liberal political campaign against slavery in America and the environmentalists' attempt to persuade the public that the circle of the ethical community should be enlarged to include animals, and even plants or ecosystems. Just as the Abolitionists were considered radical when they claimed that blacks were being ill-treated and denied any moral status, so the environmentalists are regarded as radical today. Consequently, Nash may be right in claiming that environmental ethics is a "logical extrapolation of the powerful liberal tradition."[12] Liberals in the nineteenth century passed from male chauvinism and racism to moral universalism; environmentalists today pass from a human chauvinism to a broader moral universalism, arguing that "the conscious suffering of a sentient creature is indeed intrinsically bad from that creature's standpoint." Thus, we look at other species

> [as] we look at ourselves, seeing them as beings which have a good they are striving to realize just as we have a good we are striving to realize . . . [Hence] their lives can be made better or worse by the way humans treat them, and it is possible for humans to take their standpoint and judge what happens to them in terms of *their well-being* [emphasis in original].[13]

Now while liberals in the nineteenth century wished to protect the vulnerable, that is, the potential victims of modernity and progress, environmentalists toward the end of the twentieth century still protect the victims of modernity, but unlike progressive forces in the nineteenth century, they extend the circle of protection to include the natural environment.

So much for the content of liberalism. But there is a second aspect of liberal theory that is built into the tradition of liberal philosophy. Liberals have always subjected their positions, values, ideas, and theories to critical scrutiny, and have been the proponents of openness and tolerance, not only in political life, but in the academic and philosophical debate as well.[14] Even philosophers and theorists who attack the liberal tradition for its inability to tackle, understand, or solve ecological problems,[15] must admit that they can do so because they live and work in a liberal and tolerant society.

Moreover, in order to accept environmental philosophy one must be relatively open to new ideas and tolerant of criticism—not only of one's own theory, but also of one's methodology. This is because environmental philosophy is (mostly) biocentric or ecocentric rather than anthropocentric: that is, it considers nonhuman objects (individual animals, plants, or ecosystems) as moral "clients" as well as human beings, and unlike traditional morality, it discusses the moral relationship between humans and nonhuman entities. So while one should be careful in claiming similarity between liberalism and environmental philosophy (just because the former argues for tolerance), it seems fair to maintain that it is at least partly due to liberalism's rejection of methodological monism and fostering of a philosophical and intellectual milieu in which new ideas can flourish that environmental philosophy has emerged.[16]

The third reason why liberalism became the breeding ground for a flourishing of ecological attitudes lies in the sphere of internal politics. I refer here to a tradition of defending the individual against the church, the state, large-scale industries, and firms. This started in the eighteenth century with the defense of the individual against the state or church, followed by a warning that democracy might still yield despotism. A suspicion of the despotic rule of the majority (thus, liberals defended the minorities—especially the intellectuals—from the masses) grew into the ideology of citizens' rights in the twentieth century. Recently liberalism became the defender of the underdog, a crusader against monopolies and for proper government and universal law.[17]

The same stand is taken today by environmentalists: they must challenge the activities of huge industries and firms, mainly because they are unsustainable. Such firms, for purely profit-making motives, often pollute the air or the water, decimate forests, and in general neglect the right of individuals to a clean environment. The state must impose regulations, but very often the state, for economic and other reasons, also ignores the dangers to individuals. The role of environmentalists, then, is to document the situation, publish warnings, and sometimes take the liberty to act.[18]

Indeed, this stand is taken by many environmentalists. Even Jonathon Porritt, ex-director of Friends of the Earth in Britain and one of the leaders of the British Green Party, who attacks liberal politicians for their lack of support for the Green cause,[19] employs the liberal terminology of rights when presenting his Green philosophy.

> The fact that people's *rights* are being denied is in itself a serious enough problem. . . . And the fact that there are so few . . . who are prepared either to inform people of the denial of their *rights* or to help them to fight for those *rights*, turns a problem of indifference into crisis of inaction [emphasis added].[20]

Finally, a fourth reason why ecological attitudes have taken root in liberal societies lies in the sphere of international relations. Here one particular element inherent in liberal thought should be highlighted: internationalism. Admittedly this notion sometimes stands for "free trade," which is not environment-friendly, according to many environmentalists. But at the same time, internationalism embodies a strong belief in and reliance on multilateral agreements and international organizations, together with the conviction that political problems may be solved by sometimes tiresome negotiations and that national interests do not necessarily run counter to international cooperation. While many people consider the international arena as a place where "might is right," and where those whose interests are harmed can only complain post factum, liberals have regarded international relations as a sphere in which it is possible to foresee problems and apply the treatment before the damage is caused.

All these elements are crucial, because environmental problems can and must of course be tackled only through international cooperation. This is precisely the element that is both lacking and needed in contemporary politics in regard to the solution of environmental problems, which are rarely entirely local in character.[21] Indeed, the idea that prevention is better than cure has been the underlying rationale for the 1972 Stockholm Conference on the Human Environment, the Toronto Declaration, the London Dumping Convention, *Our Common Future*, the 1992 UNCED in Rio de Janeiro, and other notable international agreements.

At this point it may be argued that at least a few of the features of liberalism discussed so far characterize other currents of thought and ideologies as well, for example, socialism. This may be true, although the second feature, that is, the rejection of methodological monism and advancing academic pluralism, is, I believe, characteristic mainly of liberalism. But even if not, my argument is neither that liberalism is a necessary condition for the emergence of environmental philosophy nor that it is a sufficient condition, but rather that liberalism—as a philosophy and a political attitude—influences, stimulates, and encourages the environmental deliberation and the rise of Green thought.

The Political Task

While in recent years much thought has been given to ecology, it seems that it has nevertheless been too little. The damages to the environment are now of such horrifying proportions that many people think that these problems call for radical and urgent solutions, an approach not necessarily compatible with traditional liberal democracy's reluctance to undertake dramatic changes. And yet now, in the 1990s, the world is witnessing a worldwide democratization and the flourishing of liberal-democratic ideas, and no one should be required to abandon this trend.

The extremely complex political mission for the 1990s, then, is twofold: on the one hand, no longer to relieve ecological "suffering," but rather to introduce reforms radical enough to save the environment and reverse what is still reversible; and on the other hand, to ensure this should not come about at the price of imposing regulations that limit or reduce liberties. Thus, there are two dimensions to this mission: to sustain the growing enthusiasm for democracy and liberty and at the same time to save the environment.

As already indicated above, these two dimensions are closely correlated. Many environmentalists, political theorists, and philosophers, in the light of environmental problems, among them scarcity of resources and pollution, have expressed pessimism and anxiety regarding the future of liberal democracy. Lack of resources on the one hand, and selfish, or self-interested, behavior on the other, have caused them to believe that the only solution might be the imposition of regulations and policies against people's wills. William Ophuls, for example, wrote that "the return of scarcity portends the revival of age-old political evils, for our descendants if not ourselves."[22] As Paehlke points out, all these writers fear that the only possible outcome will be "severe economic restraints, self-discipline beyond that which is likely to develop voluntarily."[23]

But Ophuls's prophecy need not necessarily come to pass. For instance, while his suggestion of replacing the market by political action is, I believe, welcome (see below), it is still questionable whether liberty should give way to authority and whether egalitarian democracy should be abandoned for the sake of "political competence and status."[24] Ophuls's appeal for a strong leadership[25]—reminiscent of George Bernard Shaw's *Man and Superman*—is not the only alternative.

It seems that democratic societies should be looking for a middle path, which might include, at least prima facie, noncoercive but planned and consistent policies. These policies must be in line with sustaining and promoting democracy and people's liberties, and yet they may also imply a shift from discussing environmental issues politically to a politics of the environment, and hence state intervention. This leads us to the second main question of the paper.

Environmentalism and Liberalism: Individuals' Economic Behavior

Why does liberalism outside the debate in the universities fail to justify environmental policies? What happens when it comes to real politics? The debate in the universities has been mainly focused on the question of whether there are intrinsic values in nature or on the issue of animals' rights, because it has never been faced with the actual need to convince the other side, for example, developers, vivisectionists, and so forth. But when politicians come to justify their policies to the general public, they must do this in more traditional, general, and popular terms,[26] and it is at this point that liberalism runs into difficulties.

Now, in order to demonstrate this, a distinction should be made between two conceptions of politics that represent two interpretations of liberalism. One, which generally speaking is more common in contemporary mainstream American liberalism, is based on the values of neutrality, minimal state intervention, an opposition to regulations, and a concept of politics as an aggregate of autonomous decisions—all of which are antithetical to environmental policies. The other interpretation, sometimes called "social liberalism," is not hostile to advancing certain ideas of the good (for example, conservation) and is more open to state intervention.[27]

But before we discuss the best conception of politics in the context of ecology, a more radical liberal argument that the state should not intervene at all because individuals' economic behavior is the best foundation for solving environmental policies needs to be refuted. I deliberately refrain from claiming to refute the "market approach" to the environment. This is a much debated issue, and had I set out to rebuke it, I should have written a separate article. But I would like to examine the literature on the premises of the market approach in our context, with special attention to the "correct" (not in the sense of politically correct) role of politics.

First of all, many arguments have been put forward against the economic behavior approach with respect to the environment. According to several ecosocialists, we can no longer trust the magic and the invisible hand of the market to do the work for us; we must plan and initiate. Individuals' economic behavior has proved to be both inefficient and inequitable in coping with ecology.

For instance, Jeremy Seabrook, an ex-member of the British Labour Party and currently a Green activist, contends that the market is the best and most efficient mechanism to ruin the entire universe. In spite of this, people all over the world have been attracted to the idea of the market and its promise of a better material future, and so forth, because "the economy became the arena in which the guilt for what had happened [in WWII] was to be assuaged." The market, he argues, became "the object of a superstitious reverence: if only this could be made to work, to grow, to provide, we would surely gain exemption from any recurrence of the barbarities of the recent past." The market, he contends, was successful in doing just this, but it cannot provide what people really need and long for: a pleasant and harmonious life. In the East, he claims, the imported idea of the market ruined the traditional rural way of life and its social manifestations through the process of urbanization, whereas in the West the market ruined our conception of nature.[28]

But the critique of the economic behavior approach in relation to the environment is also based on economic arguments, using "market" terminology to demonstrate that individuals' behavior cannot tackle the environmental issue. According to those arguments, this will always result in more pollution, because the costs of this pollution are borne by nature as well as by other people who share the environment rather than by the polluter itself.[29] The polluter does not have to be motivated by ill will: this problem is ingrained in the idea of the market and in its very "imperative," as Eckersley calls it—"grow or die."[30] Such private attitudes cannot respect environmental notions of limits to growth and carrying capacity. An environment-friendly product is likely to be more expensive than its rival product, and hence no "rational" (that is, profit-seeking) entrepreneur will consider such products.

At this point, market advocates put forward the idea of penalties for the polluter. Let us intervene slightly in the market, they say, and charge the polluter for what he or she has done.[31] So anybody who pollutes, depletes, or utilizes a certain natural resource beyond a certain degree will have to pay for it one way or another. But how can we assess the damages? Can we do this when the damage is not local, or within national borders, but rather international, for example, acid rain in Canada due to air pollution in the U.S.? How do we assess the damage when only a small and very specific section of the society is hurt (for example, the workers in a factory, or those who used to play golf on a ground that is now to be developed)? Moreover, it is very likely that the consumers themselves have to pay for the pollution that the manufacturer causes, especially in the cases of monopolies or special products that very few firms or even only one firm produces. So if all polluting manufacturers are charged for the pollution they cause, they will transfer the cost to the consumers. All consumers will then pay the real and full price of energy (including cleaning costs). This, in turn, will increase inequality, because the proportion of income spent on energy declines as income increases, although the use of energy increases. Thus, paying the full price of energy seems unfair for the worst off.

But there is a further problem with regard to assessing environmental damages. While perhaps it is possible to determine the cost of a certain illness (for example, it is equal to the cost of hospital treatment plus a certain amount for compensations), how can we determine the "cost" of a life? Indeed the very term is strange. We are thinking of the value of life, but can it be translated into the "cost" of life? To estimate a person's probable future earnings or any other criterion of the value of life is obnoxious, because in that case someone who possesses this amount of money would be able to purchase another person's life. But, if, as some advocates of the individuals' economic behavior approach have maintained, this question is too artificial or even irrelevant (for instance, because most environmental problems do not cause death), there is still the question of assessing the value of the lives of nonhumans. Thus, how much is the life of a sea otter worth? David Moberg comments that

> surveying people about how much the animals are worth to them or measuring lost income if sea otters disappeared may keep a few economists employed, but it does not answer the question.

This is indeed a cynical, although a serious reply.

> If only one respondent said it was of infinite value, that would throw off the survey. If you limit the response to how much a person would be willing to spend, the result would obviously be affected by how much money people have, a standard flaw of market preference analysis.[32]

Yet Moberg's response is only part of the answer. The truth of the matter is that any calculation or assessment of the cost/value of animals' lives is, at the end of the day, anthropocentric, and any value in the world of nature is instrumental,[33] just as in the Lockean theory only by mixing human labor with natural objects could these objects have some (instrumental) value. Preservation, according to this approach, is a problem only because if humans want to enjoy, say, sea otters, they must first "possess" them. What we should have done instead is to ask the animal itself for the value of its life (for itself). Just as, in the case of human lives, we do not ask a murderer what the value of his victim's life is, but rather ask the victim herself or himself; so should we do in the case of animals' lives. Now I do not wish to slip into the heavily discussed question of the degrees of life, the difference between a virus and a horse, and so on. But if, as indicated above, liberalism has enabled biocentrism to flourish, this is not the right moment to retreat to anthropocentrism. Indeed, this is true not only in cases resulting in the death of animals but in any case of environmental damage. Evaluating the cost in terms of how much people would pay is totally anthropocentric, and if liberalism follows this path it will not be able to genuinely emerge as "environment-friendly," because environmentalists' criticism of the market approach to valuing is what makes a certain good valuable is not the state of mind of the consumer who wants that good, but rather something inherent in that good.[34]

Now, some "economic behavior" advocates may remain skeptical with regard to biocentrism. Nonetheless, it is undeniable that these calculations fail to evaluate the cost for some other human beings: those of future generations. The not-yet-born are either ignored or "discounted." The main reason for the latter is that these economists tend to discount the future: a value x in the future is less than x now, because it would be equal to the amount that x would yield in this future if it were invested now and benefited from the interest rate. This, of course, is incompatible with any notion of intergenerational equity.[35]

In addition to the above difficulties with the evaluation of the cost of environmental damages, there is a serious difficulty with the notion that serves the market advocates in their approach to this issue. They assume that all we have to do when we decide on environmental policies is ask people how much they will pay to conserve a forest, save the life of three whales, preserve a certain building, and so forth. Economists call this mechanism the WTP (willingness to pay) test.

But following psychological research that Amos Twersky and Danny Kahaneman conducted in the U.S. and Maya Bar-Hillel in Israel, I conducted a simple experiment that reveals the fallacy of the concept of WTP. I told a group of forty students that they should imagine that an ecological disaster has occurred and that there is an urgent need to clean our country's coast. The first twenty students were given papers in which they were asked whether they would contribute 1 percent of their salaries this month to clean the coast. All of them replied positively. Then they were asked whether they thought it would be possible to know the average amount that their fellow students would be willing to pay. Ninety percent replied that it would be possible; the average WTP predicted was 1.85 percent of one's salary.

The second group of twenty students were told the same story, and yet they were asked whether they would be willing to pay 4 percent of their salaries. They all asserted that they would contribute this amount, and thought that the average WTP would be 6.16 percent of one's salary.

The results are clear: the concept of WTP reflects nothing about "individuals' autonomous wills." Rather, we see that the format of the question and the starting point affect the WTP. Is there no consistent notion of WTP?

So I conducted a similar experiment. This time I told forty students that the department had decided to allow them to use our common room for their coffee breaks. Each student would contribute as much as he or she wished in order to run this "coffee shop." They were asked whether they would pay $6 a month. They all answered "yes." When asked about the WTP of their fellow students, their answers varied from $2 to $16, with an average of $7.42.

The second group of twenty students were given the same story, but were asked to contribute $30 a month; 20 percent agreed, whereas 80 percent disagreed, but the average evaluations of their fellow students' WTP was $15! My conclusion is that WTP is perhaps interesting, theoretically speaking, but it is not a reliable mechanism for revealing preferences.

It goes without saying that this technical question does not exhaust the discussion of the relationship of liberalism with environmental concern, but it does point to the difficulties with regard to the political and practical implications of liberalism in this context. And beyond all these difficulties with the "economic behavior" approach, there is—according to the critique of the market theory—another, perhaps greater, difficulty. This is the belief that individuals' economic behavior yields the best (sometimes defined in terms of being the most rational) results. Do such results include the difficulties that the people of Athens have in breathing every summer, because of traffic pollution? The problem with the market is that even if it corrects itself, it may be too late, because the damage—sometimes to humans—is irreversible.

Some market advocates answer that the question is not one of metaphysical beliefs but rather of an ethical belief in the idea of freedom. Since they are aware of the environmental damages in our world, they have suggested that firms should be allowed to react in their own ways and according to their own methods to the changing needs of the market, including the need to be more environment-friendly. This, as Eckersley notes, "is not simply a defense of economic efficiency; it is also linked to a defense of political freedom."[36] Thus, for example, if there must be less pollution, then a system of selling pollution rights should be introduced. Pollution rights will be distributed to firms: if x is the degree of pollution that is tolerable, and n is the number of firms, each firm will get x/n pollution rights. Those who can continue manufacturing and limit the resulting pollution will do so, and cover the cost of doing so by selling pollution rights to other firms.[37]

But while some firms adjust to the new circumstances by limiting pollution, others buy pollution rights, and the problem that remains is the distribution of exposure to hazardous waste and pollution. Third-world countries have long been disproportionally exposed to dangerous waste. For instance, in the 1970s it was quite common for African countries to "export" land to western firms for burying toxic and radioactive waste. And in the western world the poor find themselves unable to buy houses located in safer areas or to look for safer jobs. Even if taxes are introduced, those who find it most difficult to cope will be, again, the least advantaged.[38] Indeed, why should we wait until the dirt is produced in order to clean it up? Why not prevent pollution before workers breathe in smog, or before people are hurt and the environment is damaged? Why should we let the private sector make the decisions on where to pollute? Are such decisions private at all? Should they be in the hands of those who run the industries and pollute rivers, meadows, and seas?

However, to some Greens and other advocates of the environmental case, the above approach seems too superficial and simplistic. In the final analysis, the world is not ready to give up any form of private initiative or any expression of political freedom in the sphere of economics. The ultimate opposite to an economy that lacks any state action at all would be an economy lacking any profit-making motive, but many people doubt not only whether full-state or public ownership is congruent with liberty, but also whether such a system is better for the environment. As Goodin notes, "the environmental consequences of public ownership in Eastern Europe do not serve as a happy precedent."[39] Thus these people seek to modify, refine, or correct the market, to make it more environment-friendly. They do not reject all that liberalism stands for, but still think that if liberalism implies an "individual's economic behavior" type of economics, then it is inadequate in the sense that it neglects aspects of ecology, wilderness preservation, and so on. The right question to be asked, these Greens argue, is not how to get rid of the market, but what sort of environmental goods it can deliver and what it cannot.

The market economy, to start with, does not compensate or reward individuals and firms that do act to conserve the environment. A person who lives in an historic house and preserves it is not usually paid to do so. However, if she sells the house to a developer she may make a lot of money. Similarly, the capitalist who invests huge sums of money in order to recycle his or her industry's waste is not rewarded by his or her neighbor capitalist who pollutes the river that runs next to this industry.

Admittedly the market finds solutions to environmental problems: the quality of the water is bad, so we switch to drinking mineral water; there are no green spots left in town, so a farmer opens her land to the public and we take our car, drive out into the country, and pay the farmer to enjoy nature; there is too much noise, so there are glass factories that offer us cheaper double windows. But the serious question remains: are we satisfied with these solutions? Is this what we wanted? Can we all afford them? And can the market, even if it does solve one or two problems, become the right solution to the ecological crisis in general?[40]

Moreover, the market may suggest theoretical ways of coping with environmental damages; but can it enforce them? Pollution is a good example. All the more litter and a clean environment is a good that (almost) everybody wants. But some environmental goods (for example, leaving a remote forest untouched) are desired by only a part of the population. Individuals' economic behavior will not save this forest, because its economic value to timber merchants is much greater than the amount that some environmentalists are able to pay to save it for aesthetic reasons.[41]

It seems that this list of goods that individuals' behavior cannot supply in the environmental context is quite long and includes some significant examples. So even if we accept the market as a system relevant to the "environmental era," there is still a strong need for "politics." In other words, we shall eventually realize that the economy is not a genuinely market one. But then, the promarket economists fail not so much in their economic theory as in their political theory. Therefore, instead of dealing with the issue of a market or nonmarket economy, the right question to consider is that of the more suitable type of politics.

So I now want to return to this question, and argue that liberals must come to terms with the political fact that the need to promote environmental policies reveals: namely, that the state of the environment is closely related to our view of "the political" and the political process, including the debate over the good life; and that the issue of the environment involves the goals of our political life rather than merely the means of achieving certain goals.

Environmentalism and Liberalism: Conceptions of Politics

When discussing the inadequacy of liberalism's image of politics in the context of ecological policies, I must refer first to the impossibility of neutrality in the context of conservation or any other environmental policy.

The idea of neutrality, as advanced by many prominent liberal theorists, is that the state should stay out of the debate on the nature of the good. Official policies should not promote or reflect any conception of what constitutes the good life; on the contrary, the state should be indifferent to any discrepancies between those ideas. For example, while discussing distributive justice, Gauthier writes:

An essentially just society . . . does not need to shape individuals in order to afford them justice. . . . In saying that an essentially just society is neutral with respect to the aims of its members, we deny that justice is linked to any substantive conception of what is good, either for the individual or for society.[42]

But can one remain neutral in matters of conservation? A neutral argument in favor of conservation must make a fair political procedure the only criterion for choosing a policy. Suppose a liberal, it is argued, holds a belief about the importance of preservation. This person may hope that a way of life, which in one way or another is related to the environment (for example, to a beautiful forest) will be available to posterity. But this liberal fears that, owing to the destruction of the environment, this way of life will disappear. The destruction of natural objects is, in fact, destructive of the very possibility of certain competing ideas of the good life. The liberal fears that in the future one (that is, future generations) may not be in a position to make a neutral choice among the ideas of the good life, now available.[43]

But this way of reasoning is misleading. If you and I are choosing between x ideas and I assert that we should choose between $(x+1)$ ideas, you will probably ask me why I insist upon idea $(x+1)$. I cannot answer "because it exists," because then you may wonder why I do not advocate idea $(x+2)$. I therefore have no other choice than to explain why $(x+1)$ is especially desirable, significant, important, and so forth. We wish posterity to enjoy a certain object precisely because it has been desired and found satisfying in the past. Moreover, we do not preserve—indeed we sometimes try to destroy—what we think is wrong or bad, for example, nuclear weapons. Ironically, according to the above liberal argument we should conserve nuclear weapons—perhaps the idea of war itself as a way of life—so that future generations will be aware of it and have the opportunity to choose between a larger variety of ideas of the good. This, of course, is unwise, if not outrageous. In short, an argument in favor of the conservation of a certain object must be based on the claim that this object is in itself good, especially when conserving it contradicts the interests of certain people at the present time (for example, higher salaries, more jobs, and so on).

It is possible at this point to defend the "neutral liberal" stand by distinguishing between those things that are good in themselves and those things that are instrumentally valuable, arguing that nuclear weapons fall into the latter category. Leaving aside the empirical question of whether there are people who regard nuclear weapons, or wars, as a good, this is exactly the point: "neutral" liberals usually find individuals' well-being the most proper—if not the only—moral basic consideration for social policies. They then tend to define individuals' well-being in terms of satisfying personal individual wants, interpreted as those individuals' subjective wants. But they could have defined them according to objective wants (preferences a person would hold if [s]he were fully informed, not confused and rational).[44] Otherwise it is enough to assume that there may be someone who thinks that wars are good to imply the preservation of this idea. We must therefore assert that wars are not good, but then we should do this according to objective wants. In other words, the challenge of the environment is such that objective wants should also be considered and well-being should be an account of final goods.

In the above liberal argument, however, the concept of neutrality is, in fact, derived from the more fundamental values of choice and autonomy, which are, it has been suggested, basic to American liberalism.[45] The latter is therefore considered very democratic inasmuch as, according to its doctrines, political decisions should reflect nothing other than the aggregate of people's preferences and respect their choices. For example, environmental policies are congruent with liberalism if and only if a majority (or a winning coalition) is in favor. Thus, private preferences and economic measures, backed by the legitimacy of "rational behavior," have supplanted the debate on political ideals and the image of the good life.[46] This philosophy holds that society is an instrument for the benefit of individuals; all the more, therefore, should nature be subjugated by humans, who through its progressive transformation fulfill their individualistic desires.[47] And according to this idea of politics, everything is reduced to private interests, which are held in balance of a market or exchange.[48]

But if there is any methodological innovation in Green philosophy it is that the wall between what has been considered nature or the environment on the one side and culture on the other side falls. The concept of the environment becomes part of our culture, or political. Therefore it is repeatedly argued and widely accepted that what is now needed in environmental politics (which are more than the political discussion of environmental matters) is something more sophisticated: not merely policies that are responsive, whatever the individual preferences are or whatever the outcome may be, but policies that, while responsive, take into account the good of the community as a whole as well and offer solutions to problems that are rarely considered, and, still less, resolved by individualistic, self-involved, short-run interests.[49] Moreover, there is a good chance that, in the environmental context, individual and private preferences will contradict the general good. Thus Jonathon Porritt writes:

> There may well have been a time, at the start of the Industrial Revolution, when Adam Smith's assertion that the sum of individual decisions in pursuit of self-interest added up to a pretty fair approxi-

mation of public welfare, with the "invisible hand" of the market ensuring that individualism and the general interest of society were one and the same thing. But in today's crowded, interdependent world, these same individualistic tendencies are beginning to destroy our *general* interest and thereby harm us all [emphasis added].[50]

Indeed, economic and self-interested individualistic preferences could easily lead to the continuing depletion of scarce resources, be it oil, clean air, or scenic landscape. As Ted Schrecker writes, the resistance of business to environmental regulations has been "bitter" and firms have fought "long and expensive court battles to avoid conviction," sometimes issuing threats to shut down firms.[51]

Garret Hardin's well-known "Tragedy of the Commons"[52] demonstrates this claim: according to him, the state of the environment resembles a pasture open to all. Each herdsman tries to keep as many cattle on the common as possible, but the carrying capacity of the land is insufficient. Each herdsman seeks to improve his own position. If he asks himself the utility of adding one more animal to his herd, he answers that the advantage is +1 (the herdsman receives all the proceeds from the sale of the additional animal), whereas the disadvantage is only a fraction of −1 (the effects of overgrazing being shared by all other herdsmen). The tragedy, of course, is that all herdsmen reach the same conclusion. Further, the question of how to reduce, say, pollution on the roads, illustrates the fact that environmental issues involve concepts of "public good," "collective action," and "free rider," and that the state is needed to provide the necessary solution: that is, environmental policies.

Thus, one must concede that "environmental imperatives are basically matters of principle that cannot be bargained away in an economic fashion," and that "not all of us think of ourselves primarily as consumers; many of us regard ourselves as citizens as well."[53] And as citizens we are concerned with the public interest, or the good of the community rather than with our own personal interests. As citizens, we have obligations that are not always compatible with our private preferences as individuals, and these obligations must receive priority.

> [The] cost-benefit analysis . . . prevents us from achieving a certain kind of self-determination. . . . It prevents us from deciding who we are, not just what we are. . . . There is a right and a wrong way to manage those national parks. . . . The wrong thing to do is to make a big drive-in for Winnebagos. This has nothing to do with what turns a social profit or maximizes wealth.[54]

Many Greens claim that to live in rural areas, for instance, is "living more in harmony with nature" than living in an urban jungle of cars and industries.[55] In other terms, these Greens may be arguing that there is less alienation in village life, or none at all. So they actually claim that one sort of life is better than another, or, in other words, constitutes the good life. Is it? May they be kept poles apart, the Futurists and George Sorel thought that this type of harmony was merely a form of degeneration and degradation. Although I subscribe to the Greens' view here, this is not relevant to my purpose. The point to be established is that a debate on whether this sort of life is good is perfectly legitimate. One must realize that there *is* a debate here on the idea of the good. In the final analysis, this should be obvious. Any Green or counter-Green argument must make some assumption about the idea of the good, since the argument rests on a theory of value, that is, on the idea of an intrinsic, noninstrumental value.[56] And such a theory of value is simply a theory of the good.

Thus the state of the environment calls for politics of the common and a debate on the good. At this point the objection might be raised that, all the same, the Greens' environmental concern represents nothing but an individual's private preferences. The argument would be that the Greens' claim that the world will be destroyed if their suggestions are not implemented is similar to the warnings issued by an almost unknown candidate for the 1992 American presidential campaign, John Huglin, that if he were not elected and his proposals not taken seriously, there would be a worldwide holocaust or, more seriously, any religious fanatic's prophecy, and his desire that we should all attend churches, synagogues, mosques, and so on. Moreover—the argument goes—a contractor's wish that a certain valley should be inhabited and developed (which is, for that matter, another private preference) is in no way inferior to the environmentalist demand that all work should stop and the beautiful valley should be preserved.

But this criticism is deceptive. With regard to the last point, the question is not a matter of which standpoint is inferior or superior; both developers and environmentalists express ideas of the good life, images of how this world should be and how humans should live. They even sometimes use the same arguments: for instance, developers put forward the psychological argument that if there are more roads and more jobs, then people will be less tense. And environmentalists argue that since tension is caused by noise, traffic, and the fast rhythm of our lives, if we wish to reduce tension, we must limit growth, build less roads, and so on. But the point is that these two programs are not preferences that can be bargained over until a compromise is reached. The two sides represent opposite conceptions of a good world, of what is and what is not desirable.

Now with regard to the fanatics, the challenge is easily answered: environmentalism is based on rational evaluations and scientific, empirical knowledge, whereas religious fanatics do not appreciate such evaluations. Their system is irrational in essence; the environmentalists, on the other hand, base their call for change on scientific grounds and empirical—though controversial—data. So environmentalists do not simply express private preferences, but put forward ideas of the good, based on scientific knowledge and phrased in moral terms.

The next question to ask, then, is the following: isn't there a price to pay for the liberals' insistence on regarding politics as a matter of individuals' autonomous decisions? There must be if it limits a consideration of the good of the community, welfare policies, and so on. For the ecologist, politics must provide the framework in which common and general interests are discussed and protected, if not promoted.

We are speaking, then, of state intervention.[57] Now many liberals would argue that one can remain neutral with regard to the idea of the good, but nonetheless advocate interventionism. You don't have to debate the nature of the good in order to justify state intervention. The Rawlsian theory of justice is neutral, it is argued, but is nevertheless in favor of state intervention. I subscribe to the view that in fact Rawls cannot put forward the idea of interventionism without being committed to some idea of the good.[58] This debate, however, is beyond the scope of this paper. For our purpose it is sufficient to claim that even if Rawls (or any other liberal), basing himself on the idea of neutrality, does justify a certain kind of interventionism, it cannot be the one that suits the case of the environment. As we have seen, where the latter is concerned, interventionism is based on and justified by the debate on the good. Any less weighty reason for interventionism would not justify the dramatic policies that are necessary to deal with the ecological disasters with which we are faced nowadays and the political difficulties that accompany them. Indeed, if such dramatic policies were implemented without being based on a genuine debate concerning the good, they would turn out to be nothing more than what Ophuls fears they would be: that is, regulations imposed on a large section of the population without this section understanding why they are needed.[59]

So although liberalism has been a fertile ground for environmental philosophy, it has a basic difficulty with regard to a public environmental policy and its justification: most liberals adhere to neutrality and regard liberal-democratic politics primarily as a matter of concerning the wills of individuals, whereas environmental issues call for a politics of the common and consequently for interventionism.

Does this imply that liberal governments cannot tackle the environmental challenge? My argument is that if liberalism limits itself to a policy of neutrality and an aggregate of autonomous decisions, then it is likely to fail in this matter. Ecology implies state intervention, justified by a consideration of the common good, and hence abandoning neutrality as a justification for the liberal state and its policies. A sense of community is needed because in the environmental era social and environmental responsibilities should play a much more important role than self-interested profit-making motivation. If we do not wish to retreat to totalitarian regimes, we must take the opportunity that liberalism as a philosophy provided (as I argued in the first part of this article) and look for a liberalism that is more social: the politics of the aggregate of autonomous decisions and an economics of individual preferences are of little benefit. The politics of the common, however, that at the same time does not arbitrarily restrict liberties, can be found in the other, perhaps so far more neglected, tradition of liberalism accompanied by a strong welfare state. This conception of liberalism could, perhaps, allow for a justification of environmental policies in terms of liberal terminology, and is therefore much better suited to the environmental era.[60]

Notes

1. See Kristin Shrader-Frechette, "Science, Democracy, and Public Policy," *Critical Review* 6 (1993): 255–65.
2. Robert Goodin, "International Ethics and the Environmental Crisis," *Ethics and International Affairs* 4 (1990): 81–105, and "The High Road is Green," *Environmental Politics* 1 (1992): 1–8; Steven Yearley, *The Green Case* (London: Harper Collins Academic, 1991); A. de-Shalit, "Environmental Policies and Justice Between Generations," *European Journal of Political Research* 21 (1992): 307–16.
3. For the debate between individualistic liberalism and communitarianism, see M. Sandel, ed., *Liberalism and Its Critics* (Oxford: Blackwell, 1984); and S. Avineri and A. de-Shalit, eds., *Communitarianism and Individualism* (Oxford: Oxford University Press, 1992).
4. Throughout this essay I give a very broad and flexible interpretation of liberalism, both in a historical sense (I refer to the liberalisms of the eighteenth, nineteenth, and twentieth centuries) and in an analytical sense.
5. John Rawls, *A Theory of Justice* (Oxford: Oxford University Press, 1973), p. 512. See also M. S. Pritchard and W. L. Wade, "Justice and the Treatment of Animals: A Critique of Rawls," *Environmental Ethics* 3 (1981): 55–62; and Walter Achterberg, "Can Liberal Democracy Survive the Environmental Crisis?" in A. Dobson and P. Lucarde, eds., *The Politics of Nature* (London: Routledge, 1993), pp. 81–101.
6. Brendt Singer, "An Extension of Rawls' Theory of Justice to Environmental Ethics," *Environmental Ethics* 10 (1988): 217–32.
7. R. Taylor, "The Environmental Implications of Liberalism," *Critical Review* 6 (1993): 265–83.
8. The latter is Ronald Dworkin's formula. See his "Reverse Discrimination" in his *Taking Rights Seriously* (London: Duckworth, 1981), and his "Liberalism" in his *A Matter of Principle* (Oxford: Clarendon Press, 1986). But see also John Locke ("This equality of men by Nature . . . [is] so evident in itself, and beyond all question . . . ," in his *Second Treatise of Governments*, chap. 2, 5), the French first declaration of the rights of man and of citizens ("Men are born, and always continue, free and equal in respect to their rights"), the American Declaration of Independence, Thomas Paine's Rights of Man, and Alexis de Tocqueville's *Democracy in America*

(since everyone is equal, no one is entitled to be a despotic ruler: part 2, chap. 10).
9. This is typical of ecofeminist philosophy. See A. Collard and J. Contrucci, *Rape of the Wild* (London: The Women's Press, 1988). But also of environmental philosophy in general: see R. Routely, "Against the Inevitability of Human Chauvinism," in K. Goodpaster, ed., *Ethics and the Problems of the 21st Century* (Notre Dame: Notre Dame University Press, 1979).
10. Aldo Leopold, *A Sand County Almanac* (Oxford: Oxford University Press, 1987 [1949]).
11. Christopher Stone, "Should Trees Have Standing?" *Southern California Law Review* 45 (1972), and *Earth and Other Ethics* (New York: Harper and Row, 1987).
12. Roderick Nash, *The Rights of Nature* (Madison: University of Wisconsin Press, 1989), p. 200. See also Robert Paehlke, *Environmentalism and the Future of Progressive Politics* (New Haven: Yale University Press, 1989), pp. 8–9.
13. Paul Taylor, "Frankena on Environmental Ethics," *Monist* 64 (1981): 217, and *Respect for Nature* (Princeton: Princeton University Press, 1986), pp. 56–57.
14. J. Horton and S. Mendus, *Aspects of Toleration* (London: Methuen, 1985).
15. For example, Alastair Gunn, "Traditional Ethics and the Moral Status of Animals," *Environmental Ethics* 5 (1983): 133–54.
16. Anna Brammwell traces the origins of environmentalism to romantic philosophy and right-wing political theories. But she fails to distinguish between antirational, antimodern ruralism, and modern, scientific-based, and democratic Green environmentalism. See Anna Bramwell, *Ecology in the Twentieth Century* (New Haven: Yale University Press, 1991).
17. Michael Freeden, *The New Liberalism* (Oxford: Clarendon Press, 1978); and Ronald Dworkin, "Liberalism," in his *A Matter of Principle* (Oxford: Clarendon Press, 1986), pp. 187–88.
18. Three good examples of Green publications that warn of the dangers to individuals' rights are the books by Brian Price, *The Friends of the Earth Guide to Pollution* (London: Friends of the Earth, 1983); John May, *The Greenpeace Book of the Nuclear Age* (London: Victor Gollancz Ltd., 1989); and Victor Ferencz and John Keyes, *PlanetHood: The Keys to Your Survival and Prosperity* (Oregon: Vision Books, 1988). A famous Green act was Greenpeace's attempt to stop French nuclear tests in the Pacific, which ended with the blowing up of the Greenpeace boat by members of French Intelligence in 1985.
19. Jonathon Porritt, *The Coming of the Greens* (London: Fontana, 1988), pp. 71–76.
20. Jonathon Porritt, *Seeing Green* (Oxford: Basil Blackwell, 1984), p. 115.
21. Consider, for instance, the damage being caused by all nations to the ozone layer, the problem of acid rain, or the contamination of the Mediterranean Sea. For a further discussion of the international aspects see G. Porter and J. Brown, *Global Environmental Change* (Boulder: Westview Press, 1991); R. E. Benedick, *Ozone Diplomacy* (Cambridge: Harvard University Press, 1992); the Brundtland report, Brundtland Gro Harlem et al.; The World Commission on Environment and Development staff, *Our Common Future* (Oxford: Oxford University Press, 1987); and M. Grubb et al., *The Earth Summit Agreements* (London: Earthscan, 1993).
22. William Ophuls, *Ecology and the Politics of Scarcity* (San Francisco: W. H. Freeman, 1977), p. 145.
23. Robert Paehlke, "Democracy, Bureaucracy, and Environmentalism," *Environmental Ethics* 10 (1988): 293.

24. William Ophuls, *Ecology*, p. 227.
25. Ibid., p. 159. To be fair, Ophuls denied that he had wanted rule by elite: see his comment in William Ophuls, "On Hoffert and the Scarcity of Politics," *Environmental Ethics* 8 (1986): 287–88.
26. Carolyn Merchant, *Radical Ecology* (London: Routledge, 1992), pp. 159–63, describes such cases and such tendencies among the "big ten" environmental organizations in the U.S., and Andrew Dobson, *Green Political Thought* (London: Harper Collins Academic, 1990), discusses the differences between the public and private faces of Green argumentation.
27. I discuss the manifestations in Europe of the latter in my article, in Bob Brecher, ed., *Liberalism and the New Europe* (London: Avebury Press, 1993), pp. 149–62.
28. Jeremy Seabrook, *The Myth of the Market* (London: Green Books, 1990), pp. 13–14 and chap. 7.
29. David Moberg, "Environment and Markets," *Dissent* (1991): 511.
30. Robyn Eckersley, "Green versus Ecosocialist Economic Programmes: the Market Rules OK?" *Political Studies* 40 (1992): 319.
31. It is interesting to note that John Stuart Mill already thought this could be a solution to air pollution. See *Principles of Political Economy* (New York: The Coloral Press, 1990), p. 7.
32. Moberg, "Environment and Markets," p. 512.
33. For a more complicated distinction see Andrew Brennan, "Moral Pluralism and the Environment," *Environmental Values* 1 (1992): 15–33.
34. Robert Goodin, *Green Political Theory* (Oxford: Polity, 1992), p. 25.
35. See Robyn Eckersley, "Green versus Ecosocialists," p. 319. Some market advocates suggested that future generations' interests are, in fact, represented in any social arrangement that is decided upon now. See David Gauthier, *Morals By Agreement* (Oxford, Clarendon Press, 1986), p. 299. I argue elsewhere that bargaining with the not yet born is impossible, not only literally but also theoretically. See my *Why Posterity Matters* (London: Routledge, 1994).
36. Robyn Eckersley, "Green versus Ecosocialists," p. 318.
37. Steidlmeier, for example, argues that pollution permits have a legitimacy as "a second-best solution." See "The Morality of Pollution Permits," *Environmental Ethics* 15 (1993): 133–50.
38. See the statement issued by the Centre for Science and Environment (New Delhi) on global environmental democracy as a reaction to the "northern agenda," in *Alternatives* 17 (1992): 271–79. Also see J. Moberg, "Who Rules the Market," *Political Studies* 40 (1992): 337.
39. Robert Goodin, *Green Political Theory*, p. 7, fn. 30.
40. See also Brian Barry, "The Continuing Relevance of Socialism," in Brian Barry, *Liberty and Justice* (Oxford: Clarendon Press, 1991); and Robyn Eckersley, "Free Market Environmentalism: Friend or Foe?" *Environmental Politics* 2 (1993): 1–20.
41. This is related to the question of whether only use value should count, or whether option value (that is, whether I can use and enjoy this environmental good in the future) counts as well. See Alan Randall, "Human Preferences, Economics, and the Preservation of Species," in B. Norton, ed., *The Preservation of Species* (Princeton: Princeton University Press, 1986).
42. David Gauthier, *Morals by Agreement*, p. 341. See also John Rawls, *Political Liberalism* (New York: Columbia University Press, 1993); Bruce Ackerman, *Social*

Justice in the Liberal State (New Haven: Yale University Press, 1980); and Will Kymlicka, "Liberal Individualism and Liberal Neutrality," *Ethics* 99 (1989): 883–905. But see Joseph Raz, "Liberalism, Autonomy and the Politics of Neutral Concern," *Midwest Studies in Philosophy* 7 (1980): 89–120.

43. This rationale is put forward in Ronald Dworkin, "Liberalism," p. 202.

44. Compare R. M. Hare, "Ethical Theory and Utilitarianism," in A. Sen and B. Williams, eds., *Utilitarianism and Beyond* (Cambridge: Cambridge University Press, 1982).

45. Ronald Beiner, "Liberalism," in his *What's the Matter With Liberalism?* (Berkeley: University of California Press, 1992); and William Galston, *Liberal Purposes* (Cambridge: Cambridge University Press, 1991).

46. See, for example, Kenneth Arrow, *Social Choice and Individual Values* (New York: Wiley, 1962).

47. See also Charles Taylor, *Hegel* (Cambridge: Cambridge University Press, 1975), pp. 540–43.

48. On this see John Elster, "The Market and the Forum: Three Varieties of Political Theory," in J. Elster and A. Hylland, eds., *The Foundations of Social Choice Theory* (Cambridge: Cambridge University Press, 1986).

49. See David Miller's account of energy policies to the two models of politics in his "Deliberative Democracy and Social Choice," *Political Studies* 40 (1992): 54–66.

50. Jonathon Porritt, *Seeing Green*, p. 116.

51. T. Schrecker, "Resisting Environmental Regulations," in R. Paehlke and D. Torgerson, eds., *Managing Leviathan* (London: Belhaven Press, 1991), pp. 165–99; and H. Feiveson, F. Sinden, and R. Socolow, eds., *Boundaries of Analysis* (Cambridge, MA: Balliger Pub., 1976).

52. Garret Hardin, "The Tragedy of the Commons," *Science* 162 (1968): 1243–48.

53. William Ophuls, *Ecology and the Politics of Scarcity*, p. 186; and Mark Sagoff, *The Economy of the Earth* (Cambridge: Cambridge University Press, 1988), p. 27.

54. Mark Sagoff, "Ethics and Economics in Environmental Law," in T. Regan, ed., *Earthbound* (Prospect Heights, Illinois: Waveland Press, 1984), pp. 172–73.

55. For example, Robert Goodin, *Green Political Theory*, p. 51.

56. Arne Naess, *Ecology, Community, and Lifestyle* (Cambridge: Cambridge University Press, 1989); Holmes Rolston III, *Philosophy Gone Wild* (Buffalo: Prometheus Books, 1989).

57. Indeed, it seems that the state "has generally been the only institution with the necessary resources to provide environmental policies." K. Walker, "The State in Environmental Management," *Political Studies* 37 (1980): 25–38.

58. I elaborate on this in "Community and the Rights of Future Generations," *Journal of Applied Philosophy* 9 (1992): 105–17. See also T. Nagel, "Rawls on Justice," in N. Daniels, ed., *Reading Rawls* (Oxford: Basil Blackwell, 1975), pp. 1–16.

59. It is sometimes suggested by environmentalists that international environmental concern is also crucial for the much needed international cooperation in any valid environmental policy. See C. Spretnak and F. Capra, *Green Politics* (London: Paladin, 1985), pp. 157–81. But this forms another issue. See L. K. Caldwell, *International Environmental Policy* (Durham: Duke University Press, 1984).

60. I would like to thank David Miller and my students in the Environmental Ethics seminar for their comments.

6

Be-wildering Order:
On Finding a Home for Domestication and the Domesticated Other

David Macauley

> Each creature ... does what is necessary in its own behalf, and is domestic in its own *domus* or home.
> —Wendell Berry

> It is mankind, alone among all living beings, to which the term 'domesticated' is perhaps strictly appropriate.
> —Theophrastus

> The very design of imagination is to domesticate us in another, a celestial nature.
> —Ralph Waldo Emerson

I. Dilemmas, Definitions, and Challenges of Domestication

In his celebratory "Walking" essay, Thoreau remarked, "I love to see the domestic animals reassert their native rights,—any evidence that they have not wholly lost their original wild habits and vigor; as when my neighbor's cow breaks out of her pasture early in the Spring and boldly swims the river, a cold, gray tide, twenty-five or thirty rods wide, swollen by melted snow. It is the buffalo crossing the Mississippi."[1] In Thoreau's view, such liberating actions confer dignity on the domestic species and reveal that a latent instinctual seed is preserved secretly within the skins of most creatures. Indeed, Thoreau's literary sentiments and philosophical preferences for the wild over the tame might be borne out in part by an examination of domestication, which has contributed not only to the diminishment and endangerment of the natural world but also aided in the emerging definition—and some might add—subsequent disenchantment of human society.

Domestication implies a bringing-in or accustoming to the house (*domus*), but it is an activity and word that shares a root meaning with *domination*, suggesting as well a suffocating control and mastery over this domain. Domestication, too, stands in general contrast to the *wild*, a term that is cognate with *will*, so that it can be said of a wild animal, plant or human—unlike its domestic counterpart—that it (he or she) is likely self-willed, uncontrolled, even autonomous.[2] Despite these differences, the attitude that great numbers of people retain toward domestication is a mixed and ambivalent one, revealing complex and ambiguous dimensions to a practice many historians believe has been the single most significant "intervention" that humans have made into nature and one of the most important events in our own psychological, biological, and social history. Thus, it is valuable to look in some detail at this phenomenon for the ecological and political light it may cast upon human and nonhuman animals and their often bewildering relations in a variety of natural and social environments.[3]

One of the great values of ecology—as a critical and reconstructive enterprise—is that it enables us not only to behold the familiar order of things in new, unprecedented ways, but also to reconsider the idea of order itself in its philosophical and political dimensions. That is, critical ecology permits us to challenge and dismantle stark oppositions such as nature/culture, wilderness/civilization, animal/human, even left/right (in the political sense), and to think and act in contextual, relational, embodied and historical manners. It is my contention that a consideration of the notion and practice of domestication (along with its "others") offers a particularly rich subject for exploring and developing a new kind of "wild order," an opportunity for creatively *be-wildering* the domain of the human household without unduly obscuring or denigrating its value. It also provides a portal for viewing the way in which we have increasingly disturbed existing ecological communities, infantilized nonhuman animals, and simplified our own psyches and culture. But a critical encounter with domestication might suggest new possibilities for social and political analysis or change, too.

What follows, then (as I hope it will be read), is less a broadside to eliminate domestication altogether—an unrealistic suggestion at present—or a indictment of it *in toto* and more a call and first attempt to investigate a few of its myriad forms and complexities as part of a broader search for both a more ecological and democratic society and a more viable social theory. It is an argument that we need to reconfigure the walls of the human house and household—and its notions of order, power, and space—so as to be more receptive to the encompassing natural world. We also need to see the manner in which doors are open already to this possibility. Ultimately, it is not very instructive or constructive to view domestication (e.g., the birth and rise of agriculture) simply as an ecological "original sin," as a first and irreversible event that separated us forever from a close relation to a pristine natural world. Some theorists and activists subscribe explicitly or tacitly to this view when they erect overly sharp and precipitous divides between nature and culture, the wild and the domestic, or engage in the genetic fallacy or the fallacy of misplaced concreteness.[4] Despite the power, seduction, and partial truth of this image, domestication can and often does involve both separation and sympathy, both isolation and inclusion, and both "dominance and affection," to use Yi Fu Tuan's phrase.[5] It has radically altered the focus of human attention from the surrounding world to a more isolated, less environmentally and communally aware self, but in so doing it has sometimes inadvertently opened up new worlds, ones to which we must remain alive—if only to better resist additional steps toward the attenuation of perceptual experience, ecological complexity, and political stability.

Nietzsche, for example, argued that bad conscience (*schlectes Gewissen*) originated with the enclosure of humans within the walls of society, a form in effect of self-domestication—and, in his view, the most significant change we have yet experienced. He held further that this change was neither gradual nor voluntary—it was not an "organic adaptation"—but rather a disjunctive and disastrous event that severed us from our animal past. This event, which made us heavy (like sea creatures coming ashore) and transformed us into a human "torture chamber," was instituted and maintained through acts of violence, especially by the state. At the same time, Nietzsche recognized that despite our resultant and ongoing "homesickness for the wild," we were made interesting, even dangerous, by such actions. "The existence of an animal soul turned against itself, taking sides against itself, was something so new, profound, unheard of, enigmatic, contradictory, *and pregnant with a future* that the aspect of the earth was essentially altered."[6] Nietzsche thereby opts not for nostalgia—a word deriving from *nostos* (return home) and *algos* (pain), and implying agony, longing, and homesickness for a lost world—but instead counsels us to "remain faithful to the earth" and to strive to "naturalize humanity."[7] Most of human culture—we must remind ourselves—has lived in semipermanent houses of one kind or another for the last 15,000 years. Even if there is no going back, the possibility for creative transformation, historical critique, and critical recovery of perduring ideas and experiences still exists.

An ecological perspective on domestication challenges social and political theory to understand the "civilizing process" as frequently *weeding out the wild* to the detriment both of human society and the trans- or extrahuman world. The task, it seems, is thus to encourage a cultural "enlargement of mind" so that mind embraces (and is enveloped by) the natural order, nonhuman animals, and a more inclusive sense of the *oikos* (home). The cultural fact of domestication, in turn, also offers a challenge to ecological philosophy to think notions of place, home, and community in social and political terms, as well as in the vernacular of the natural or the personal. The "wild," in short, might be bound intimately with the "good" and the "sacred" as Thoreau and Gary Snyder have discovered, but it is certainly a more limited (and problematic) heuristic notion when one speaks of human polity and domesticated society. Animal liberationists and environmentalists—who usually focus on very different kinds of beings to the exclusion of others—could be challenged, too, to engage in dialogue with each other and to discuss their respective normative objections and alternatives to domestication.[8] Such undertakings might further reveal that the realms of the wild, the undomesticated, and animal "otherness" are not always so clearly (or purely) prepolitical or postpolitical—as many deep ecologists, for example, believe them to be—but partially suffused or imbricated with forms of power, projection, and images of human society.

Domestication concerns walls and what they include or exclude. But domestication is not only a biological and historical process. It is also an apt metaphor and trope when applied to other aspects of human culture. Like "the wild," it can be an elusive and elastic term, part natural fact and part social construct. To this degree, it is sometimes better described as "artifactual." Extending Foucault's suggestions, domestication might be understood more generally in terms of a process of making-the-same, of taming and interiorizing the other, of systematically rendering the foreign domestic, and of reigning in (even by excluding) what is perceived as threatening, chaotic, or unruly—as a defensive or utilitarian undertaking.

A number of prominent theorists and writers in diverse fields have touched upon or explored the dilemmas of domestication either directly or indirectly. Lévi-Strauss, for example, has discussed the distinction between the domesticated (*domestiquée*) and the wild (savage, *sauvage*) through a classification in which the former realm is associated with the modern era, science, engineering, abstract thinking, concepts, and history; while the latter realm is connected with the Neolithic age, myth, bricolage, intuition and imagination, signs, and atemporality.[9] Anthropologist Jack Goody has critiqued this approach, in turn, for its dichotomies and spoken of the subject in terms of the "domestication of the savage mind."[10] Intellectual and social historian Hayden White has considered the myth of the wild man (the figure who resists domesticating influences) in premodern society, exploring its function and development in Western thought and literature.[11] Zoologists, biologists, and agricultural historians such as Frederic Zeuner, Carl Sauer, V. G. Childe, J. Clutton-Brock, and Eric Isaac have debated the origins and effects of animal and plant domestication. Human domestication also seems to bear some similarities to cultural historian Norbert Elias's notion of the "civilizing process" and philosopher Herbert Marcuse's use of "pacification," though it differs in many respects as well.[12]

Despite these contributions, the discourse on domestication has rarely been couched in both ecological and political contexts, especially ones that include humans with nonhumans while being directed toward contemporary social relevance. In short, domestication and "the others" who serve as its objects—organic cultures, animals, and plants, among others—are often either left without a true home (theoretical or real) or else forcibly cordoned and housed in dominating and "unnatural" edifices (e.g., reservations, zoos, factories, artificial parks). Thus, a host of issues are germane to a social, political, and ecological investigation of the subject. Some of the most salient questions that need be raised include:

- How do notions of order and power in social and political thought take their bearings from a view of domestication?

- Are our flights from Earth into outer space and the heightened tendency to flee from the colonized life-world into the exotic, the primitive, the wilderness, or the erotic legitimate responses to an overly domesticated planet?

- Can one speak of the domestication of other cultures (e.g., Native Americans) or other species as analogous to imperialism, colonialism, or biocide?

- What significance does the inclusion (or exclusion) of animals and animality within (or from) the walls of the human community have for understanding freedom, nature, human nature, and reason?

- In what ways do nomadism, physical movement, and travel (or tourism) relate, reinforce, or challenge the hegemony of the *domus* and domestication?

- And what is the importance of domestication for evolutionary theory in terms of whether it proceeds largely through cooperation, competition, or chance variation?

To these important concerns, we can add the following:

- What are the psychophysical effects of domestication on animals, humans, and the environment?

- Do new technologies domesticate or liberate us, constrain or enable us?

- How does domestication influence or restrict the possibilities of social change or revolution?
- What are the social repercussions of the incarceration of wilderness, the reigning in of wildness, and the systematic control over sexuality under domestication?
- What are the normative dimensions of domestication? Should it be encouraged or discouraged? Of what does domestication deprive us? What does it contribute? How is environmental ethics to make sense of the distinction between wild and domestic animals?
- Finally, is domestication a metaphor, process, condition, fiction, framework, or fact when applied to humans? Is it reversible?

Given space limitations, I examine below just a few of these areas. In my view, domestication is a robust and ripe domain for critical social theory and ecological philosophy to engage. Therefore, much room exists for further debate and dialogue. In the forthcoming sections, I consider the subjects of animals and animality, order, the house, and power as they relate to the issue at hand. It is necessary, first of all, to look at some of the historical, empirical, and biological aspects of domestication before turning to more theoretical, normative, and political concerns. I conclude my discussion by suggesting a few possible ways of responding to the many challenges of domestication.

II. Other Animals and the Animal as Other

> The very beginning of Genesis tells us that God created man in order to give him dominion over fish and fowl and all creatures. Of course, Genesis was written by a man, not a horse. There is no certainty that God actually did grant man dominion over other creatures. What seems more likely, in fact, is that man invented God to sanctify the dominion he had usurped for himself over the cow and the horse.
> —Milan Kundera, *The Unbearable Lightness of Being*

Nature and History of Domestication

As a process, domestication of nonhumans generally involves, first, spatial confinement of an animal or animals; second, control over their reproduction through, for example, forced breeding or castration (which makes the male more docile); and third, eventual alteration of physical and genetic characteristics. This process usually entails the isolation of a small group of animals from others of their kind so as to prevent free mating. Sexual isolation is then followed by artificial selection to alter and "improve" given features or capacities of the animal, such as shape, color, strength, speed or kinds of milk, wool, meat, or eggs. The aim, in short, is to hold stable certain desirable mutations of a given species so that they can be put to later use for human benefit. Typically, the juvenile behavioral and anatomical traits are retained in order for an animal or species to be manipulated or made submissive more easily.

Though debate still exists, animal domestication is generally thought to have begun about 10,000 years ago in a range of locations like the Far East, Mesoamerica, and the Middle East, and to have developed more fully as humans made the transition from hunter-gatherer life to Neolithic agriculture. It took place about the same time as did the introduction of plants such as beans, onions, and squash into emerging settlements; the two processes undoubtedly reinforced one another since they both required a semisedentary existence, the regularities of domestic life, and the routinization of labor. Contrary to what many people believe, however, the first steps toward animal domestication were probably *not* due to a desire to increase food supplies. The egg-laying "productivity" of fowl, the wool covering of sheep, and the milking of cattle were *consequences* of their captive breeding, not reasons for their domestication.[13]

Cattle were likely one of the first of the great herd animals to be domesticated, probably in western Asia by sedentary farmers rather than by nomadic hunters. In fact, all present-day domestic cattle seem to derive from the wild urus, an ancient, dangerous, and fierce ancestral relation. The urus may have been selected because of their large curved horns, which resembled the crescent of the moon and which, in turn, were connected with the mythic symbolism of the Mother Goddess. (Hathor, the Egyptian goddess of the moon, for example, was a cow.) They were probably first caught and held in corrals, later to be put to death in a sacrificial ceremony. As they became more inbred or outbred (breeding between stocks or individuals who are relatively unrelated), they developed features that differed from their wild parents: longer legs, smaller heads, and straighter backs. The appearance of a more tractable and infantlike stock allowed these animals to be used for a variety of purposes beyond ritualistic means, including plowing, pulling wagons, and later as food.[14]

The close bond with our proverbial "best friend," along with the unique dilemmas of domestication, are recognizable in the remains of a late Paleolithic burial site in Northern Israel, which provides the oldest physical evidence of this process. In a 12,000-year-old tomb, two skeletons were discovered: an elderly human and a small domestic dog about five months in age. What is remarkable about these remnants is that the small pup is being held or gently embraced by the left arm of the human figure, indicating decidedly a sign of affection. As is commonly believed, the dog, or *Canis lupus familiaris*—perhaps the first domestic animal—likely descended from the wolf, or *Canis lupus,* since both share the same chromosome count and are able to interbreed with ease.

The first human-wolf contact developed around 10,000–12,000 years ago in places as varied as present-day Kurdistan and Idaho, when wolves roamed near to and connected themselves with human encampments. As the distance between the two species gradually closed, humans realized the value of these animals (who may have been raised initially for companionship or to be eaten) for their ability to drive out birds and other small mammals from the brush during a hunt and to warn against intruders with their heightened olfactory and aural senses. The dog's great plasticity in both appearance and breeding capacity (which expressed itself subsequently) is attributable to the wide diversity of genetic material inherited from numerous kinds of wolves. Unfortunately, wolves have not been held in as high esteem as dogs. The wolf, especially, became a symbol in the United States of an "untamed" nature that early colonists assiduously tried to subjugate, and the animal was hunted or exterminated with a vengeance for more than a century under bounty laws passed in most states.[15]

Following shortly after the domestication of wolves were sheep and goats, and then later a host of other animals. Sheep—who derive originally from West Asia—are in fact among the most completely domesticated of herd animals, since they depend almost entirely upon humans for survival. Their intelligence, for which they are known as a wild species, is diminished greatly in their domestic counterparts. The horse, though not as affected by changes brought about by domestication, was probably the first animal upon whom riding techniques were foisted. Domestication, which first occurred in Eastern Europe and western Asia, began in the third century B.C. when this strong and graceful animal was used for labor similar to that imposed upon cattle and camels, who hauled goods and carried people. By the second century B.C., most of the animals who now interact with or live within our society had been brought under the yoke of the human hand and the roof of the human house. Other animals, such as the turkey, would be domesticated much later when the old and new worlds met. Currently, attempts are still being made in Africa to domesticate both the gazelle and the eland, a powerful antelope that produces a prodigious amount of milk, despite the failures of the Egyptians to do so many thousands of years ago. In Australia, experiments are also being conducted on transforming the wild dingo into a domestic sheep dog.

In more *theoretical* terms, the order of domestication has been roughly: first, scavengers (e.g., the dog and jackal); second, seasonal migration animals who allowed for nomadic life (e.g., reindeer, goat, and sheep); third, those animals who fit in with settled life (e.g., cattle or honeybees); and fourth, mammals used for transportation (e.g., the horse and camel). A number of animals such as the falcon, cheetah, and leopard have been captured and *tamed* in different historical periods to fulfill desires for status and royalty, but they were not fully *domesticated* since their reproductive processes were not brought entirely under control.[16]

Theories of Domestication

Numerous theories exist as to the origin of animal domestication, including those that point variously to the pressures of population increases, sudden climate changes that forced animals and humans together, the role of religion, evolutionary chance or spontaneity, and the transition from a hunter-gatherer society to an agricultural one, among other factors. Such theories of domestication have great diagnostic significance for thinking about the sources, forms, and expressions of ecological and political problems with which they are associated. Thus, they should be considered as part of a more general search for responses to these difficulties as well. James Serpell speculates, for example, that prehistoric pet-keeping may have led to animal husbandry and livestock raising. He considers the possibility that early hunter-gatherers may have brought back young sheep and goats to the camp as playthings for children when the adult animals had been killed for food.[17] A shortcoming to this theory is that it does not explain why or how their status changed from companion animals to herd animals and how humans were able to take such a step that distanced us psychologically from such animals and the environment.

Theories that emphasize the role of religion (broadly construed) find greater evidential support and point to the fact that the first gods who appeared in animal form were not those whom humans hunted; instead they were more likely members of a semidomestic herd such as the goat, ram, ibex, and antelope—friends, in effect, of the tribe. In this view, domestication and divinity appear at about the same time historically. Full domestication resulted because such animals had to be protected from more wild predators, and so they could not be left simply to the attentive eyes and ears of a shepherding dog. As with the ram, the yak and bison came to enjoy a dual status—part sacred, part profane—since they were also symbols of fertility representing the sun or one of the elements such as earth, air, fire, or water.

Economic uses of animals thus would have been secondary or incidental to the early uses of animals for sacrificial or sacred purposes. This view is reinforced by the fact that some cultures still use domestic animals for ritualistic magic—such as poultry for divination—rather than as food, or prohibit the consumption of certain creatures—pigs, for instance—by erecting religious taboos against such practices. In India, the cow is still considered sacred, and even the urine and excrement of the animal are sometimes preserved. When domestic animals with religious significance were consumed, it was often believed that one was partaking of the body and blood of a god, or minimally a sacrificial substitute. A religious motivation can be found for the domestication of pigeons and doves, who are symbols of the Holy Ghost and universal peace, respectively; cats, who were mummified as epiphanies of the goddess Bast in Egypt and linked with witchcraft in Germanic folklore; and domestic fowl, who lay eggs (a common symbol of death and rebirth) or were used in cockfights for a ritual reenactment of divine warfare.

As agricultural historian Carl Sauer has noted: "One might say that animals were chosen for domestication that were not easy to take, which were not common, and which were difficult to make gentle—the wild mountain goats and sheep that avoid the vicinity of man, the formidable wild cattle and buffalo."[18] When this fact is considered, *early* domestication can be seen more readily as a ceremonial than a practical phenomenon, one that required a tremendous severance from the past in psychological and intellectual terms. It has, of course, moved far beyond these oft-forgotten stages into an industry and agri-*business* generally removed from its agri-*cultural* roots. In brief, domesticated animals have been transformed increasingly into *biomachines* who (or which) convert grass into milk or grain into meat. It would be very significant if this theory of domestication proves to be coherent and true, because it would lend support to the view that the ecological crisis has some religious roots, as has been surmised by some historians and philosophers.[19]

More recently, a mutualistic theory of domestication has been advanced by some theorists and writers. Drawing on the work of selected biologists and historians, Ned Budiansky, for example, suggests that since the ice age, it has been advantageous for some wild animals to choose entrance into the human world as a route for their own survival. He argues that domestication was, in effect, an "act of nature" rather than a "crime" against it, a natural process instead of a Promethean step on the part of humans. This view points to the failure of man *qua* domesticator to tame and raise moose, bear, and elk, among other animals, despite attempts to do so. It claims that those animals who have "chosen" us—"human commensals" such as house mice—or at least cooperated in their own domestication have had greater reproductive success by living with humans, as in the case of dogs and cats, who number nearly 120 million in the United States alone.

This perspective provides a helpful counterbalance to the idea that humans have always and everywhere been able to control the animal world without limitation. It questions some of the anthropocentric presumptions of a narrow humanism and tends to raise the status of other animals relative to our own self-proclaimed dominance. Instead of emphasizing struggle and competition, it stresses cooperation, reciprocal advantage, and symbiosis. It locates humans as well within the natural world, as part of the nature rather than apart from it, and views us as partners (or perhaps stewards) in an evolutionary process. To a degree, it is compatible with the evolutionary theory of Peter Kropotkin and other writers who call attention to the benefits of mutual aid and assistance in facilitating species survival and well-being. As Kropotkin has argued, "Mutual aid is as much a law of animal life as mutual struggle, but that, as a factor of evolution, it most probably has a far greater importance, inasmuch as it favours the development of such habits and characters as insure the maintenance and further development of the species, together with the greatest amount of welfare and enjoyment of life for the individual, with the least waste of energy."[20]

While a mutualistic view has many merits, it may fail to account fully for the uniquely social and cultural dimensions to domestication, such as religion, relying heavily on a controversial interpretation of biological data. The apparent biological paradox that Budiansky finds—"the only way to produce an animal with the desirable traits is through captive breeding, yet the only way they could have been captively bred is if they had the desirable traits to start with"[21]—is resolvable in part when such factors are considered and when it is acknowledged that domestication may have been a gradual process where *some* desirable traits—cultural associations or physical characteristics—were already present. Certain kinds of domestication may have been advanced by instincts or associations such as those that lead humans to respond to their own infants and that can be aroused by young mammals of the same size. Dog pups and piglets, for example, have been nursed in numerous primitive societies by women.[22]

The nature (or plausibility) of these and other theories has not been—and cannot be—examined in any detail here. Rather, the point I wish to underscore is that normative philosophy must begin to consider and take seriously the biological, anthropological, and historical theories of domestication if it is to adequately address ecopolitical questions in their complexity. The way that we view the process through which humans and other animals have been "oikonomized" affects our perspectives on being at home not only in-the-world but on-the-earth.

Domestication and Domination

The signs of animal domestication are testaments to the dilemmas and dangers it presents, ones of which we should be aware. These marks often include tooth decay or disappearance, general reduction in animal size, and specific diminishment in the size of horns, tails and mouths. The wild horse, for example, measured 52 inches on average, while early domestic types tended to be at most 48 inches. Shortening of the face and jaws is particularly common. Native intelligence also generally declines, since the brain of domestic animals tends to be approximately 20 percent lighter in weight than that of the comparable wild species, while the nervous and endocrinal systems are altered significantly. In particular, the parts of the brain that relate to hearing, scent, and sight are most affected.[23]

Sedentary or shackled existence and a lack of adequate exercise combine with an abundance of nutritionally poor food, leading frequently to obesity. The distinct coloration of many wild animals, too, becomes exceptional in domestic varieties. As to the reproductive cycle, domestic animals, unlike wild ones, will mate year round rather than during distinct seasons, and they are able to produce offspring at a much earlier age. Other common juvenile characteristics that are encouraged—especially in the case of "pets"—include curled tails, skin folds, short hair, and hanging ears. The sum effect of this process is, in short, to foster a form of infantilism and dependence and the favoring of pathological conditions.

Paul Shepard has pointed out that

> the temperament and personality of domestic animals are not only more placid than their wild counterparts, but also more flaccid—that is, there is somehow less definition. Of course there is nothing placid about an angry bull or a mean watchdog, but their mothers were tractable, and once an organism has been stripped of its wildness it can be freaked in any direction the breeder wishes. It may be made fierce without being truly wild. The latter implies an ecological niche from which the domesticated animal has been removed. Niches are hard taskmasters. Escape from them is not freedom but loss of direction.[24]

John Muir noted the difference between wild mountain sheep and the more common domestic variety, whom he characterized as "hoofed locusts" who were only "half alive." The contrast between the wild and domestic types of an animal can also be observed in the pronounced differences between the domestic and wild turkey. In the wild, turkeys are colorful, streamlined, and adept animals who can fly over 50 miles per hour, run nearly 20 miles per hour, and take large 4-foot strides. They have keen eyesight, protect themselves well in the woods and sometimes live for more than fifteen years. The domestic turkey, however, is clumsy and dependent, a creature who, in captivity, is no longer even able to reproduce without artificial insemination techniques due to the massive changes that have been induced through breeding. Their native intelligence—along with much of their native habitat—has been destroyed to the point where they sometimes actually drown by staring up into the sky with their mouths open when it is raining.[25]

The distinct human contribution to this process should not be completely downplayed or ignored by pointing out that *some* animals may choose us as much as we select them or by claiming that it is actually good for the animals, an argument common in the eighteenth century when it was held that domestication "civilized" animals and increased their numbers. As South African writer Laurens van der Post has observed:

> One has only to look in the eyes of, for instance, the animals he [man] has domesticated, to see that the compensations he offers in return for services rendered are not enough. For those eyes, when they are not on their guard and focused in the service of his bidding, like those of the dogs that follow at his heels, the horses munching in the stables and the cows in his meadows, amaze and confound one with the sadness glowing at the far end of the long look that goes back to their remote beginning. For human eyes that are still open to these things, it is a sadness that emanates from a nostalgia for a time when they were not enslaved but were free to be their immediate, instinctive selves. For ears that can hear, this nostalgia is there even in their voices. . . . In all these there is expression both of a persistent, incurable sickness for the wilderness that was their garden in the beginning, and reproach to powerful men who have malformed a natural kinship.[26]

Finally, one must remember that domestication is, by and large, an irreversible process, although some individual animals have reverted to a *feral* state, meaning that they have returned to the wild. Such recoveries of aboriginal instincts and environs have happened with goats in the Galapagos Islands, reindeer in France, mustang horses in the United States, and occasionally with cats and dogs, or pigeons in the countryside.

III. Dangers in the Domus: The House In and Out of Order

> People, in the course of the civilizing process, seek to suppress in themselves every characteristic that they feel to be 'animal.'
> —Norbert Elias, *The Civilizing Process*

> Suddenly it becomes possible that there are just *others*, that we ourselves are an "other" among others.
> —Paul Ricoeur

Masters themselves often become servants, either to what they have assumed dominance over or to a process in which they are reciprocally enslaved, captives in effect of the dynamics of dependence. This is a theme explored in some of the literature of high technology or novels such as Mary Shelley's *Frankenstein*, George Orwell's *Animal Farm*, and William Golding's *Lord of the Flies*. In this regard, the domestication of many animals appears to have had additional deleterious effects upon human society and the environment. There is some evidence to suggest that it provided a model for the enslavement of blacks and a training ground for discrimination against women. Historian Michael Peters, for instance, claims that *animal* domestication provided the necessary social preconditions and material foundations for *human* slavery in early societies, making the latter psychologically possible.[27] Similarly, Elizabeth Fisher maintains that the sexual domination of women, which she finds present in all societies, has been modeled after the domestication of animals. In brief, as early animal-keeping societies evolved, many of them began to view women as another kind of livestock to be enlisted for utilitarian ends.[28] Sauer has also argued that animal domestication initiated a division of labor within early societies, whereby men took over not only the agricultural operations of plowing and herding but also of slaughtering, butchering, and castrating, while women were led into house and garden work.[29]

The connection between attitudes toward the domestication of animals and the control of humans is also apparent in the treatment of mental patients and prisoners, who were often thought to be wild and dangerous animals. The generally enlightened eighteenth-century thinker Dr. Benjamin Rush, for example, thought that the insane could be domesticated or "tamed" by being completely deprived of food, offering as evidence for his view the fact that wild elephants in India were brought under control by denying them sustenance. He argued, too, that methods for breaking horses could be applied to violent mental patients and asserted with odd logic that in domesticating, "We multiply life, sensation and enjoyment."[30] In this regard, it is noteworthy that the earliest zoos—where wild animals could be observed under domestic conditions—provided the model for the first asylums and that humans were included in the menageries of Montezuma as "pets."[31]

The effects of the control exerted through domestication were responsible as well for many marked and often harmful changes in the New England landscape, including a multitude of fences, the disappearance of wolves, the network of rural roads, and the erosion of beaches.[32] In contrast to the Native American understanding of *belongings*, which granted possession of an animal only at the moment of death, early colonists began to *own* animals while grazing or roaming. This new conception of property led to conflicts and competition for grazing lands that, in turn, drove settlements and towns apart. More generally, domestication no doubt helped to expedite the displacement and replacement of older practices or beliefs such as communal sharing, usufruct, and a vision of land as sacred.

Domesticated livestock contributed to the deterioration of many old growth forests and the destruction of native plant species through their rampant overgrazing. Such grazing has been shown by field biologists to result in a decrease in native perennial grasses and in the density of vegetation. It also has adverse effects upon soil (lost through erosion at a rate of two billion tons each year in this country), which then reinforces the negative effect upon plants. Grazing by domesticated animals leads to decreased aeration of soils, increased runoff, and water accumulation on the surface of land.[33] The accumulated effect of this process is *desertification*, the loss of the earth's topsoil, which disappears each year by an area equal to more than two times the size of Belgium. Coupled with these changes is *deforestation*, which has destroyed Botswana (a last remaining wilderness area in Africa) and half of the rain forests of Central America and which is caused primarily by cattle raised for the hamburger market in the United States.

Moreover, with the rise of biotechnology and the manipulation of DNA, we are perhaps witnessing, too, the *domestication of the gene*, what Neil Evernden calls "the final assault on the wildness of life." In his view, such an ability "exterminates wildness at the source and places all life within the domain of human willing. Nature is domesticated in body, in concept, and, finally, one must say, in spirit."[34] Whereas domestication generally involves the slow alteration of genetic make-up, biotechnology now threatens with alacrity the image of humanity as well as species integrity. Biologist René Dubos has speculated that "There may emerge by selection a stock of human beings suited genetically to accept as a matter of course a regimented and sheltered way of life in a teeming and polluted world, from which all wildness and fantasy of nature will have disappeared. The domesticated farm animal and the laboratory rodent on a controlled regimen in a controlled environment will then become true models for the study of man."[35]

In this vein, Murray Bookchin, has argued that "everyday life has steadily acquired almost bovine characteristics.... The price we pay for this repellent reduction of humans to domesticated, shepherded and unthinking beings is costly beyond imagination." His work has shown that our loss of community itself has been akin to a kind of domestication that is bereft of meaning, purpose, and direction much like a wild animal's loss of niche. "Like our cattle, poultry, pets, and even crops," he writes, "we too have lost our wildness in a 'pacified' world that is overly administered and highly rationalized."[36] Bookchin makes a case for the idea that our own mental structures have been simplified and degraded like the genetic constitution of domesticated animals. In this sense, domestication can encourage a dangerous dependency on the part of humans—the self-domesticated animal—as well as nonhuman creatures. It is to this subject of the human house that we now turn before considering Foucault's relevance for locating a theoretical home for domestication.

IV. Housing Power

Dwelling is the basic character of being.
—Heidegger, "Building Dwelling Thinking"

Philosophy is strictly speaking a homesickness.
—Novalis

Domesticity, privacy, comfort, the concept of the home and family: these are literally, principal achievements of the Bourgeois Age
—John Lukacs, *The Bourgeois Interior*

Human domestication most basically involves living within the walls and confines of a permanent or semi-permanent house and settlement, a practice generally thought to have begun about 15,000 years ago. As anthropologist Peter Wilson has argued, "Domesticated society is founded on and dominated by the elementary and original structure, the building, which serves not just as shelter but as diagram and, more generally, as the source of metaphors of structure that make possible the social construction and reconstruction of reality."[37] In this view, the domestication of animals and plants is preceded and inspired by the domestication of human beings, a major and independent event in the evolution of hominids. Thus, sedentary, domestic life must be contrasted with the earlier primacy of nomadic existence. Spatial privacy itself also seems to be a consequence of domestication, one that comes into being with the construction of physical barriers that enhance the opportunities for concentration and creativity but at the same time complicate communication between humans and, one need add, interaction with the natural environment.[38] Such barriers provide for forms of exclusion and mediation, insulating and buffering domestic dwellers from the elements and natural sounds, sights, and smells. The walls of the human house thus literally create and structure the opposition between public and private, a distinction that comes to operate so pivotally in social and political theory.

According to Aristotle and later Hannah Arendt's interpretation of the ancient Greeks, the *oikos* (home) is a not a place of freedom. It exists for the purpose of serving the *polis* (city or civic community), which is the proper locus of citizen activity, the place for free *men*. In this sense, it is not political; it is a precondition for such a possibility, hence prepolitical. As Arendt has shown, freedom lay in the ability to leave the household—which was often ruled despotically and uncontestedly—and to move, speak, and act in a community of equals.[39] The house is a realm of privacy, but it is also a world that is removed from a common life with others, in a public and political sense. It is the sphere of necessity, inequality, labor, and life processes. The household, moreover, is the domain of the hidden and the proper location for the expression of pain, love, and goodness. In this regard it is linked closely with natural functions, including birth, death, and the maintenance of life.

In Aristotle's words, the *polis* exists by contrast "for the sake of a life which is potentially the best"[40]—that is, for the "good life," which is also a kind of second life. It is not concerned strictly with daily life or mere survival. The *polis* is a sphere of shared undertakings, a place of appearance and disclosure. Only in the *polis* can excellence, courage, and civic virtue manifest themselves. It is not labor—with its traits of endless circularity, repetition, and circadian rhythms—that belongs uniquely to the *polis*, but work, which facilitates the emergence of permanence, rare deeds, autonomy, and immortality. In this sense, freedom is won by transcending (or escaping) the *oikos* as a natural sphere of necessity rather than by opening up, extending, or incorporating this realm into the political community[41] or a broader natural community, endeavors that could prove to be more ecological in spirit and fact. In short, viewed from this classically Greek—and specifically Aristotelian—perspective, one is fully human only when one is *not* at home.

There are grounds for believing as well that the ancient Greeks may have actually feared a wild, untamed nature, which was thought of as a threatening realm beyond order, or at least beyond human order. As Arendt notes, ancient and medieval towns were commonly separated from natural surroundings by distinct walls. She argues that such walls and the wall of law that protected them provided for the possibility of a public realm and political community but also for private property, allowing people to enclose or retreat into a place of their own.[42] But if walls enclose and allow for public disclosure, they also exclude both natural and foreign elements and prevent, or at least hinder, the natural movement of others and "otherness." To this extent, the *privacy* they foster is often a *de-privation* for a natural or human community. In the deep forest or the unruly waters, too, there was little place for human reason to prevail for the Greeks. In addition to the political positioning of the *polis* over the *oikos*, there presided a dualism that gave conceptions of cosmos preference over chaos, notions of order over disorder, and ideas of a controllable society over an unbridled nature.[43] The importance of these dualities is not simply that they existed in the minds of philosophers but that they were played out in political ways that allowed *men* to conceive of the idea of dominating the environment and those who were thought of or represented as being wild, chaotic or threatening—notably, women, slaves, animals, and "barbarian" foreigners. Thus, the actual and figurative walls of the human house and community tended to augment the divisions between culture and nature and to reinforce the distinction between the human house *in order* and the natural world outside of or *out of order*.

Moreover, as suggested by a Platonic line of criticism, the individual household also represents a potential threat to other households and the realization of an ideal community in that private pursuits and private property can endanger the interests of others and hinder the realization of a collective good. When the purpose of the *oikos* is expressed solely in terms of the *polis*, as in Aristotle's *Politics*, the relations between households and between the household and community are regularly overlooked or ignored. Applying these points historically and to our contemporary situation, we might be able to better see how individual and separate households (especially nuclear families) can pose a real threat to the environment as well. Encouraging atomistic social practices (e.g., the private automobile) in turn tends to promote consumerism (hence, productionism), further waste of energy, excess garbage and pollution, clearing of additional wilderness areas in order to build single-unit housing, and perhaps even overpopulation.[44] In brief, the *economy* of individual, semiautonomous households may be harmful to the broader *ecology* of natural and social communities, especially given that in the modern world (unlike in the ancient Greek world) the recent emergence of a *social* realm has blurred, confused, and even undermined the clear distinction between the private and public spheres (and changed their nature in the process).[45] Given this development along the attendant decline of the family and the withering away of the public realm (but not the state), we are faced not simply with the personal task of keeping our own individual households "in order" but also with the need to attend to "collective housekeeping" including what lies beyond the traditionally understood domestic walls.

Since the time of both their first appearance in human settlements and their place in ancient Greek society, houses have also increasingly assumed a political dimension. Architecture itself has taken on dimensions of power. As physical structures, houses (like other buildings) control space, define property and territory, influence or restrict movement, impose physical limitations, create psychological demands, and so on. As Wilson has argued, "The evolution of domestic society is typified by the striving toward the perfection of architecture, the geometry that is its abstract counterpart and foundation, the arts and sciences which spin off from that geometry—music, sculpture, painting, mechanics—and the technologies that combine geometry and architecture, such as artillery, siege works, ballistics, plowing, formal gardening, and the theatre."[46] These subjects lead us to the work of Michel Foucault, who considered themes related to order, power, discipline, surveillance, and technology, especially as they bear upon the construction of space and knowledge. In Foucault's writings, we can perhaps discover or creatively forge a radical theoretical framework for understanding the domestication of nonhumans as well as the workings of power in the extended human household.

IV. Finding a Theoretical Home: Foucault's An-architecture

> While once the wilderness was humanity's home and protection, capital's domestication process has instilled in people a fear of the natural. "Raw" is their word for incomplete, harsh, vulgar. "Jungle" is their metaphor for a place of mindless aggression and violence. "Animal instincts" are those emotions which are suppressed and negated. Areas which have no immediate productive value are wastelands, yet these are usually teeming with life compared to the mechanized urban landscape. All that is wild has become a threat to humanity's new goals.
> —Jacques Camatte, *Against Domestication*

> Animality has escaped domestication by human symbols and values; and it is animality that reveals the dark rage, the sterile madness that lie in men's hearts.
> —Michel Foucault, *Madness and Civilization*

Foucault begins one of his most important works, *Les Mots et les choses*,[47] with a meditation upon a passage from Borges that arouses in him a subversive and productive laughter. On the one hand, this response disassembles the known order of thought, displaces or destroys the familiar terrain of knowledge, and erodes or erases the universal *taxinomia*. On the other hand, it gives rise, in turn, to his archaeology of the human sciences. It is this epistemological order, which is always anchored in or bound with political power, that allows us, as Foucault claims, to "tame the wild" panoply of existing things—to domesticate it in effect—but that is threatened itself by thinking that breaks down the perennial distinctions between the same and the other, identity and difference, similitude and alterity. The passage in question concerns an ancient Chinese encyclopedia in which animals are separated into those:

> (a) belonging to the Emperor, (b) embalmed, (c) tame, (d) sucking pigs, (e) sirens, (f) fabulous, (g) stray dogs, (h) included in the present classification, (i) frenzied, (j) innumerable, (k) drawn with a very fine camel hair brush, (l) *et cetera*, (m) having just broken the water pitcher, (n) that from a long way off look like flies.[48]

In this list of creatures, which defies our customary categories, cuts across present classifications, and mocks extant conventions, we seemingly encounter a foreign and feral system of thought, a bewildered order, or a chaotic cosmos—a kind of *chaosmos*, to use James Joyce's term. Nevertheless, as Foucault observes, this bestiary is still delimited; its capacity for contagion is localized; it defines and discriminates—between real and unreal, for example. What simultaneously transfixes and disorients the mind is the alphabetic series that links embalmed animals with tame ones, or frenzied creatures with those who have just broken the water pitcher.

Where *Les Mots et les choses* treats the order and ordering of the same (detailing the *epistème* of an epoch and marking the discursive discontinuities and mutations between epochs), Foucault's *Madness and Civilization* and *The Birth of the Clinic* consider the history of the Other—the marginal, excluded, liminal, or wild. *Madness and Civilization* in particular discusses such themes in relation not only to insanity but also at times to animality. When these works are coupled with the author's study of penal institutions, *Discipline and Punish*, and read as part of a larger trajectory, Foucault's analyses can be applied more generally to help us better understand the nature of domestication as a process that fundamentally concerns power, order and disorder, exclusion, normalizing functions, and disciplinary technologies.

Ultimately, if the discourse of domestication is to provide insights for ecological and political theory, it needs to find a theoretical home in which it can be understood and critiqued. Foucault's opus is appropriate for this enterprise—first, because it is explicitly concerned with the construction and control of social space and knowledge, especially with their effects upon the human body and also potentially—one could argue through creative extension—upon animal bodies and the wilderness. Second, Foucault's writing offers a tool box of strategic concepts and ideas for dismantling entrenched power relations and binary oppositions such as those created through the systematic control of sexuality, the reigning in or exclusion of "wildness" and "animality," and the disciplinary practices associated with domestication. In this way, his work can disrupt the dominant and domesticating "order of things." It can help undermine the ruling *arche* (or *archai*) and provide an antidote to the housing of power in the edifices of scientific knowledge or the physical structures of architecture. Stated in other terms, it might offer, in effect, a kind of epistemological *an-architecture*, or a tactical set of relocatable Archimedian points from which to raze foundations and undertake a Foucauldian "archeology."[49]

Foucault's work has been taken up and developed in terms of a critique of a variety of institutions and practices, including the university, sexual relations, employment procedures, the law, medicine, and the workplace.[50] Indeed, it invites the "specific intellectual" (Foucault's term) to deploy an arsenal of theoretical cantilevers, pulleys, skeleton keys, and chisels so as to expose, critique, and resist forces of domination and power. In this regard, Foucault's own studies are situated at the points where power "invests itself in institutions, becomes embodied in techniques, and equips itself with instruments and eventually even violent means of material intervention."[51] Given such a position and disposition toward power, there is reason to believe such critique can be extended to an understanding of the various forms of domestication.

Foucault's work on "dividing practices," for example, has elucidated how institutional walls in prisons, asylums, poorhouses, and hospitals have been structured to selectively exclude and objectify marginal figures so as to exercise, maintain, or increase power over them. Similarly and relatedly, his work on "scientific classification" has illustrated the manner in which humans have been turned into objectified subjects for disciplines of knowledge. But these analyses could also be applied to nonhumans and to institutions such as zoos, vivisection, factory farming, and agribusiness, or to the "incarceration" of wilderness areas, game preserves, or reservations (including of native peoples).[52] In these institutions and areas, animals are routinely and systematically observed by a scientific gaze (e.g., in research laboratories); bodily disciplined or punished through specific technologies (e.g., milking, taming, and riding); made docile and passive through spatial ordering (e.g., caging); and "normalized" through statistical measurements, conceptual hierarchies, and regulative procedures (e.g., breeding).[53]

In one of Foucault's most memorable discussions of disciplinary technology in the eighteenth century, he writes of Jeremy Bentham's design for the "automatic functioning of power" in the Panopticon. The scheme for this haunting mechanism of utilitarian rationality organizes space so as to create a visual order that permits both inmates and guards (or anyone for that matter) to be watched and controlled anonymously, continuously, and methodically through small cells or "theaters" viewed from a central tower. Foucault himself speculates that this project may have been inspired by a mengerie at Versailles, in which cages containing different animal species were arranged into an octagonal pavilion so as to be observed through windows from the king's salon at the center. "The Panopticon is a royal menagerie; the animal is replaced by man, individual distribution by specific grouping and king by the machinery of furtive power."[54] Thus, the effects of power on animal and human bodies run virtually parallel, reinforcing one another in the process. Animal breeding facilities, zoos, and research laboratories employ very comparable forms of surveillance and disciplinary techniques in controlling animal movement for the purposes of reproduction, public observation, and research. Meat-packing plants, for example, first developed and instituted conveyer-line production, influencing the later use of this technology by the automobile industry, even if the method employed was based upon a *disassembly* of body parts rather than an *assembly* of mechanical parts.[55]

Foucault's study of madness and insanity can likewise be related closely to understanding the subject and subjugation of animality, since nonhuman creatures have often been treated as the paradigmatic embodiments of a lack of rationality, against which the court of reason understands itself. In *Madness and Civilization*, Foucault explores how conceptions of insanity and animality were deeply imbricated from the sixteenth to eighteenth centuries so as to define and defend selected conceptions of reason and in order to relegate the "the madman" to a liminal position in human society—that is, at the interior of the exterior or the exterior of the interior. For example, Foucault shows that madness became a spectacle, "a thing to look at: no longer a monster inside oneself, but an animal with strange mechanisms, a bestiality from which man had long since been suppressed."[56] He notes that penitentiaries of "moral order" all appeared to follow the visible motto of one in Mainz, Germany, that read: "If wild beasts can be broken to yoke, it must not be despaired of correcting the man who has strayed."[57] And he claims that the image of madness that haunted the halls of hospitals during one period "borrowed its face from the mask of the beast."[58] Such a model governed the asylums, fitting them with cagelike structures and the appearance of a menagerie.

However, if the consequences for inmates and those believed to be insane were often harsh, "brutal," and pernicious, such representations of animality no doubt reciprocally reinforced the domination of real animals as well. Foucault seems to have been sensitive to this fact, although he does not dwell upon it at any length. He observes:

> From the start, Western culture has not considered it evident that animals participate in the plenitude of nature, in its wisdom and its order: this idea was a late one and long remained on the surface of culture; perhaps it has not yet penetrated very deeply into the subterranean regions of the imagination. In fact, on close examination, it becomes evident that the animal belongs rather to an anti-nature, to a negativity that threatens order and by its frenzy endangers the positive wisdom of nature. . . . Why should the fact that Western man has lived for two thousand years necessarily mean that he has recognized the possibility of an order common to reason and to animality?[59]

Foucault's answer to this question and a central point of his analysis focus upon the fact that the West's understanding of "rational animal" has been the yardstick of the way in which the freedom of reason has constituted its opposite term. Put differently, human order has unfortunately depended on the exclusion of "wild" animality.

If one were to consider Foucault's discussion of the way in which the insane have been spatially marginalized and excluded from society and then apply this analysis to understanding the treatment of actual, living animals (rather than an abstract animality), one might discover interesting relationships, perhaps even close historical analogies and parallels. Just as "fools" and "madmen"—who were once part of everyday street life in the Middle Ages—were increasingly removed to isolated asylums, the historical locations and relocations of animal slaughtering within population centers exhibit the same tendency on the part of society and industry to increasingly distance people from such activity or facilities, and to isolate (marginalize or centralize) nonhumans in the process. In early Rome, for example, slaughterhouses were situated pell-mell about the city, and the stench from offal and blood of slaughtered animals circulated freely about the streets. By Nero's time, however, they were moved to a central location in a large market structure. In Paris, *abattoirs* were originally located on main avenues, and gutters flowed with blood to the Seine while noisy animals obstructed traffic. Then, beginning in 1818, the city constructed model slaughterhouses. Similarly, in New York, "food" animals wandered the streets as late as the 1840s, when the city passed an ordinance outlawing this practice. In the United States today, large slaughter- and packing-houses, which are generally located in secluded rural or urban areas, have almost completely driven local butchers out of business, reducing greatly any regular contact that people might have with many animals as well as the process of confining, controlling, or killing them.[60]

It would be extremely valuable to explore in more detail and depth the way in which Foucault's works are specifically applicable, analogous, or relevant to domestication, but this undertaking is beyond the scope of the present paper. From the foregoing cursory discussion, however, it should be apparent that such an analysis can and needs to be undertaken. Instead, we now turn to consider very summarily and finally a few possible options for responding to domesticating influences and to opening up the human house and mind to embrace order outside itself and, particularly, the outside order.

VI. Opening the Oikos and Undoing Domestication

What of architectural beauty I now see, I know has gradually grown from within outward, out of the necessities and character of the indweller.
—Henry David Thoreau, *Walden*

It is time people themselves rebelled against being confined in box-constructions, in the same way as hens and rabbits are confined in cage-constructions that are equally foreign to their nature.
—Hundertwasser, "Mould Manifesto"

Above all, do not lose your desire to walk: every day I walk myself into a state of well-being and walk away from every illness; I have walked myself into my best thoughts.... Thus if one just keeps on walking, everything will be all right.
—Søren Kierkegaard, letter to Jette

In very loose, abstract terms, a possible response to an overly domesticated world might involve finding new ways to bring the outside in and to turn the inside outward. We could, for example, learn to develop a greater attentiveness to the value of margins, animality, and feral or nomadic realms of being. We could encourage the development of more open houses, free *physical* movement, and *social* or *political* movements of freedom, along with environmentally sensitive activities such as walking, gardening, camping, animal observation, and the like without at the same time succumbing to a romanticization of the liminal, instinctual, wild, or nostalgic. I touch upon a few of these areas very briefly below, but am unable to treat any one of them at length.

Marginal Values

Domestication depends upon the erection and maintenance of both theoretical and physical walls—stable references and structures that exclude many ideas or concepts on the one hand and most animals and natural elements on the other hand. Rather than trying to "undo" or escape domestication so directly or completely by fleeing into the wild, the exotic, or the erotic or by leaving the Earth entirely in search of a home away from Home, we might first explore existing border zones and boundary situations. In short, we need to participate in, protect, value, and enlarge places at the margins of distinctly human communities, areas that frequently border, bound, or fall between the artifactual and natural extremes: beaches, swamps, foot paths, bogs, overgrown lots, river banks and levees, gardens, meadows, orchards, rooftops and the like. Here, wild animals and plants are often emboldened to creep, peep, or pop up in order to investigate us, to find a temporary shelter or food source, or perhaps to present themselves for us to behold. Such feral, transitional regions are often overlooked in ecological discussions because they cannot be so neatly classified as natural or artificial, wild or civilized. They are situated quite literally at the edges of culture and on the virgules of thought, between extremes. Alternatively, marginal concepts (and places) such as "landscape," "earthworks," "dreamtime," "bioregions," and "the commons" are valuable for opening up the *oikos*, enlarging the cultural mind and be-wildering human order because they create a middle term (and territory) between earth and world as they are classically understood.[61]

Becoming Animal, Being at Home

To the consternating question of how we can begin to resolve the dichotomy between the civilized (or domesticated) and the wild, poet Gary Snyder responds with another question, "Do you really believe you are an animal?"[62] In other words, as individuals and as a culture we might start to acknowledge the degree to which our bodies are already wild. As Snyder remarks, "The involuntary quick turn of the head at a shout, the vertigo at looking off a precipice, the heart-in-the-throat in a moment of danger, the catch of the breath, the quiet moments relaxing, staring, reflecting—all universal responses of this mammal body."[63] In this sense, being at home implies becoming fully at ease in our animal bodies. It means as well seeing language (song, dance, poetrys, even philosophy) as evolving with and within a natural place and through the many contributions of nature, including a chorus of animal voices, the elements, and the earth itself.

Viewing *some* forms of domestication in interdependent, coparticipatory, even symbiotic terms might also help us to see animals as our cousins, brothers, and sisters or kin. As Mary Midgely has argued, some animals may have been brought into the human house not simply due to fear, but by forming bonds of kinship with humans, who tamed them through understanding our social signals. She concludes that they could accomplish this feat because they were social beings as well.[64] To view domestication always and only as a one-way process of domination is to simplify the complex relations not only between humans and nonhumans but also between domestic animals themselves. This is especially the case with cats, dogs, and other companion animals, who often seek out and enjoy our friendship and the company of other creatures. We can see this complicated process at work more metaphorically in the bestiaries and mythologies of human imagination, which are heavily peopled with animal forms and figures. On a final related note, we might also heed the insights of Deleuze and Guattari, who direct us toward nomadic possibilities and multiplicities inherent in "becoming animal" (as they understand it). In their view, "becoming animal" is a way of breaking with state, oedipal, and domesticating powers and emancipating ourselves at the same time.[65] In the end, if we can bring once-wild species into the *domus* of humans, we might broaden our understanding of place and home to include what lies outside it, without reducing real differences or violently assimilating otherness.

Physical Movement: Walking A-Way and Returning Home

If wilderness is one realm of contrast to the domestic, then travel is another antipode, in the sense that it is transition away from a permanent domicile. Departing from the house on a journey, sojourn, or trip is perhaps the most basic and common method of side-stepping self-domestication. And in fact it is a meritorious path in many ways because it introduces us to and into a larger order and world. At the same time, just as one must distinguish the many forms and types of domestication, it is necessary for us to reflect upon the multiple senses of movement and to consider which kinds are more or less liberatory, ecological, and provisional. For in traveling, one might continue to carry mentally the house one left or even a new, restricting physical shell or structure (e.g., an automobile). In this regard, the practices of nomadic cultures are particularly relevant to understanding domestication, as is the more "pedestrian" everyday activity of walking—which tends to actively promote self-awareness, sensitivity to surroundings, and respect for the earth.[66] The aborigines of Australia, for example, sing the "songlines" of the land as they walk the outback, orienting themselves through stories and invisible musical maps of the land through which they and their words pass. In this way, they are remarkably placed at home in the wild while being physically in transit.[67] Tourism is another matter, one that can be explored in terms of theory (e.g., "alienated leisure"), domestication, and the home. As Georges Van Den Abbeele has argued:

> The tourist theorizes because he is already en route and caught up in a chaotic, fragmented universe that needs to be domesticated. The very concept of 'the voyage' is this domestication in that it demarcates one's traveling like the Aristotelian plot into a beginning, a middle and an end. In the case of the tourist, the beginning and the end are the same place, 'home'. It is in relation to this home or domus then that everything which falls into the middle can be 'domesticated'.[68]

From this vantage point, the activity of domestication is doomed to fail, however, because the interpretation of the tourist always falls behind the venturing; it is forever trying to catch up and capture the escaping excess of experience.[69] With movement away from the *domus*, there arises the final and inevitable dilemma of returning home, a place whose meaning may have been altered in the process of journeying. This is once again both our point of origin and our destination.[70]

Social Movements: Contesting Power

According to Jacques Camatte, what prevents real social change or revolution is the "domestication of humanity," which occurs when capital forms itself as a human community. This domestication makes people passive and is further accentuated by what Camatte describes as the "escape of capital," the fact that economic processes are out of human control. Such escape expresses itself as overpopulation, pollution, and the depletion of "natural resources" as well as a monetary crisis. In Camatte's view, the domestication of human beings involves an acceptance of the development of capital as it is theorized by Marxism, which is "the arch-defender of the growth of productive forces."[71]

To the extent to which a capitalist economy rapidly accelerates domestication and the destruction of our given wilderness home, as Camatte theorizes, it cannot be effectively resisted simply at the personal level through physical movement or merely by including animals and "otherness" in a broader understanding of the *oikos*, for example, as suggested above. Rather, what is required are local and global political struggles against commodity relations, antidemocratic forces, and unecological technologies and developments. In this regard, radical ecological movements can help challenge the centripetal, homogenizing, and domesticating effects of capitalism by forming new communities of resistance and empowerment. But, as Camatte warns, they must not reproduce old relations of power or tolerate repressive consciousness. Rather, revolution must refuse the old terrain of struggle and engender new conceptions of space and time that allow for a reconciliation with the natural world. It remains to be seen in what way the movements of social ecology, socialist ecology, deep ecology, bioregionalism, and ecological feminism can address the dilemmas of domestication. They have undoubtedly made some inroads to change, but there is obviously far to go.

VII. Conclusion: Bewildering Order

> To preserve wild animals implies generally the creation of a forest for them to dwell in or resort to. So it is with man.
> —Henry David Thoreau, "Walking"

> Perhaps one should not talk (or write) too much about the wild world: it may be that it embarrasses other animals to have attention called to them.
> —Gary Snyder, *The Practice of the Wild*

For many species, including that of homo sapiens, domestication can be a kind of protracted extinction, a slow, crippling death of birth itself. In this sense, it belongs to a sluggish economy of violence, which unfortunately seems to proceed faster and faster each day with encouragements from capitalist social relations and modern technology. As biologist Lewis Thomas has warned, "We are beginning to treat the earth as a sort of domesticated house pet, living in an environment invented by us, part kitchen garden, part park, part zoo." But, he adds, "It is an idea we must rid ourselves of soon. . . . We are not separate beings. We are a living part of earth's life."[72] In Saint Exupery's famous story, *The Little Prince*, the fox remarks at one point: "One only understands the things that one tames. . . . Men have no more time to understand anything. They buy things already made at the shops. But there is no shop anywhere where one can buy friendship, and so men have no friends any more. If you want a friend, tame me."[73] In this passage, however, the French verb *apprivoiser* is rendered falsely as "tame." The word, instead, implies not mastery of one over another but a shared experience based upon appreciation and mutual understanding.

With a more full and empathetic knowledge of our animal brethren, the environing world, and the manner in which they, it, and we have been changed through domestication, perhaps we can move in such a direction—for it is physical *movement* and social *movements* that can help resist such complacency and control. Being at home in-the-world and on-the-earth means believing in, acting as, and even becoming animals (again) while accepting the profound presence of a be-wildered order beyond and within the four walls of the human house.[74]

Notes

1. Henry David Thoreau, "Walking" in Brooks Atkinson, ed., *Walden and Other Writings by Henry David Thoreau* (New York: The Modern Library, 1937), p. 621.
2. See *Compact Oxford English Dictionary* (Oxford: Clarendon Press, 1989), p. 465.
3. The discussion that follows applies mainly to human and nonhuman animals but by extension also to plants and—insofar as the analogies hold—to wilderness areas and other cultures.
4. See, for example, Jared Diamond, "The Worst Mistake in the History of the Human Race," *Discover* (May 1987): 64–66; and John Zerzan, *Elements of Refusal* (Seattle, WA: Left Bank Books, 1988).
5. Yi Fu Tuan, *Dominance and Affection* (New Haven: Yale University Press, 1984).
6. Friedrich Nietzsche, *On the Geneaology of Morals*, trans. Walter Kaufmann (New York: Vintage Books, 1967), p. 85.
7. Nietzsche, *Thus Spoke Zarathustra*, trans. Walter Kaufmann (New York: Penguin, 1966), p. 13.
8. For perhaps the first beginnings of dialogue on this subject between the two groups, see J. Baird Callicott, "Animal Liberation: A Triangular Affair," *Environmental Ethics* 2 (1980): 311–38; and Edward Johnson, "Animal Liberation Versus the Land Ethics," *Environmental Ethics* 3 (1981): 265–73.
9. Claude Lévi-Strauss, *The Savage Mind* (Chicago: University of Chicago Press, 1966).
10. Jack Goody, *The Domestication of the Savage Mind* (London: Cambridge University Press, 1977).
11. Hayden White, "The Forms of Wildness: Archaeology of an Idea," in *Tropics of Discourse: Essays in Cultural Criticism* (Baltimore: Johns Hopkins University Press, 1973), pp. 150–82.
12. See Norbert Elias, *The Civilizing Process: The History of Manners*, trans. Edmund Jephcott (New York: Urizen Books, 1978); and Herbert Marcuse, *One Dimensional Man* (Boston: Beacon Press, 1964).
13. My discussion of the history and nature of animal and plant domestication relies upon a variety of sources. See, generally, Juliet Clutton-Brock, *Domesticated Animals from Early Times* (Austin: University of Texas Press, 1981); Ian Mason, ed., *The Evolution of Domesticated Animals* (London: Longman, 1984); Peter J. Ucko and G. W. Dimbleby, eds., *The Domestication and Exploitation of Plants and Animals* (Chicago: Aldine Publishing, 1969); Frederick E. Zeuner, *A History of Domes-*

ticated Animals (New York: Harper and Row, 1963); Kathleen Szasz, *Petishism: Pets and Their People in the Western World* (New York: Holt, Rinehart and Winston, 1969); Keith Thomas, *Man and the Natural World* (New York: Pantheon Books, 1983); James Serpell, *In the Company of Animals: A Study of Human-Animal Relationships* (New York: Basil Blackwell, 1986); Eduard Hahn, *Die Haustierre and ihre Beziehungen zur Wirtschaft des Menschen* (Leipzig: Duncker and Humblot, 1896); J. J. Barloy, *Man and Animal: 100 Yeras of Friendship*, Trans Henry Fox (New York: Gordon and Cremoneri, 1978); Erich Isaac, *The Geography of Domestication* (Englewood Cliffs, NJ: Prentice-Hall, 1970).

14. Eric Isaac, "On the Domestication of Cattle," in Paul Shepard and Daniel Mckinley, eds., *The Subversive Science: Essays Toward an Ecology of Man* (Boston: Houghton Mifflin Co., 1969).

15. For relevant discussions, see Barry Lopez, *Of Wolves and Men* (New York: Scribner's, 1978); and Diane Antonio, "Of Wolves and Women," in Carol J. Adams and Josephine Donovan, eds., *Animals and Women: Feminist Theoretical Explorations* (Durham, NC: Duke University Press, 1995), pp. 213–30.

16. Regarding attempts to domesticate insects, the only steadfast and successful efforts have involved the bee, who has been sought for her wax, honey, and propolis and, to a lesser extent, for ritual religious needs in the Middle East. Bee hives were also hurled at the enemy during the Thirty Years' War and wasps were used by the Vietnamese against U.S. and Saigon forces.

17. Serpell, *In the Company of Animals*.

18. Carl Sauer, Quoted in Marston Bates, *The Forest and the Sea* (New York: Random House, 1960), p. 239.

19. See, for example, Lynn White, "The Historical Roots of Our Ecologic Crisis," *Science* 155 (1967): 1203–1207.

20. Peter Kropotkin, *Mutual Aid* (Boston: Extending Horizons Books, n.d.), p. 6. The work was originally published in 1902. See also David Macauley, "Evolution and Revolution: The Ecological Anarchism of Kropotkin and Bookchin," in Andrew Light, ed., *Anarchism, Nature and Society: Critical Perspectives on Social Ecology* (New York: Guilford Press, in press).

21. Stephen Budiansky, *The Covenant of the Wild: Why Animals Chose Domestication* (New York: William Morrow and Co., 1992), p. 24.

22. To my knowledge, there are no known instances of mammals who have been domesticated by other nonhuman mammals. Some ant species, however, steal the larvae and pupae of neighboring colonies, later forcing the fully grown captive ants to work as slaves. These slave-ants, though, are not then bred, even if some ants do seem to "rear" greenflies for the sweet fluid they secrete in stablelike holes in the ground.

23. See, for example, Ucko and Dimbleby, *The Domestication and Exploitation of Plants and Animals*.

24. Paul Shepard, *The Tender Carnivore and the Sacred Game* (New York: Charles Scribner's Sons, 1973), p. 10. See also his *Nature and Madness* (San Francisco: Sierra Club Books, 1982), pp. 19–46.

25. See David Macauley, "From the Forest to Factory: Turkeys and the Turkey Industry," *The Animals' Agenda* (November 1987).

26. Laurens van der Post, quoted in *The Animals' Agenda* (March 1989), p. 4.

27. Michael Peters, "Nature and Culture," in Stanley and Rosalind Godlovitch and John Harris, eds., *Animals, Men and Morals* (New York: Grove Press, 1971). See

also David Macauley, "Animal Enslavement: Are There Any Insights from Earlier Abolitionsists," *Lamakatsi* 1 (Spring 1987).
28. Elizabeth Fisher, *Woman's Creation* (New York: McGraw-Hill, 1979).
29. See Carl Sauer, *Agriculture Origins and Dispersals* (New York: American Geographical Society, 1952).
30. Benjamin Rush, quoted in Yi Fu Tuan, *Dominance and Affection*, pp. 83–84.
31. See Michel Foucault, *Discipline and Punish: The Birth of the Prison*, trans. Alan Sheridan (New York: Vintage Books, 1979).
32. William Cronin, *Changes in the Land: Indians, Colonists, and the Ecology of New England* (New York: Hill and Wang, 1983).
33. See Katey Palmer, "Return of the Natives," *Earth First!* (December 21, 1988); p. 26.
34. Neil Evernden, *The Social Creation of Nature* (Baltimore: Johns Hopkins University Press, 1992), p. 120.
35. René Dubos, *Man Adapting* (New Haven: Yale University Press, 1965).
36. Murray Bookchin, *The Modern Crisis* (Montreal: Black Rose Books, 1987), pp. 6–7; and Bookchin, *The Ecology of Freedom* (Palo Alto, CA: Chesire Books, 1982), p. 279.
37. Peter J. Wilson, *The Domestication of the Human Species* (New Haven and London: Yale University Press, 1988), p. 153.
38. Ibid., p. 173.
39. Hannah Arendt, *The Human Condition* (Chicago: University of Chicago Press, 1958). For a consideration of Arendt's relevance for ecological politics, see David Macauley, "Hannah Arendt and the Politics of Place: Earth Alienation to *Oikos*" in Macauley, *Minding Nature: The Philosophers of Ecology* (New York: Guilford Press, 1996).
40. Aristotle, *Politics* trans, Benjamin Jewett in Richard McKeon, ed., *The Basic Works of Aristotle* (New York: Random House, 1941), p. 1288. (1328b35) translation altered.
41. It must be acknowledged, however, that in Aristotle's view the *polis* (hence politics) is also natural—even if not strictly a sphere of nature (*physis*) or natural functions—because it is rooted in human nature and because earlier forms to which it is related (i.e., the family and village) are natural.
42. One might think here of Socrates who stayed within the walls of the city. See Plato's *Phaedrus*, for example, where Socrates for once ventures outside the walls of the city with his companion but remarks that he cannot learn from tree and "the open country."
43. See Bookchin, *The Ecology of Freedom*, p. 107.
44. For a critique of the nuclear family as unecological and a consideration of alternatives to it, see Joan Roelofs, "Charles Fourier: Proto-Red-Green," in Macauley, *Minding Nature*, pp. 43–58. Arendt argues, too, that our waste economy is a mark of the triumph of *animal laborans* and a consumer society. For a specific critique of the private automobile, see Julia Meaton and David Morrice, "The Ethics and Politics of Private Automobile Use," *Environmental Ethics* 18 (Spring 1996): 39–54. I would like to underscore that I am not endorsing a Platonic perspective on politics or nature; indeed, I find Aristotelian approaches to these areas more suggestive and fruitful.
45. According to Arendt, the social realm is marked by conformity, normalization, and administration. It protects private accumulation but not property, and it is

marked by the rule of noone but not no rule. The social realm encourages an "unnatural growth of the natural," promotes flight from the public world, the replacement of action by behavior, and facilitates the rise of mass society and loneliness. See Arendt, *The Human Condition*, pp. 38–49.

46. Wilson, *The Domestication of the Human Species*, p. 155.
47. The English edition is entitled *The Order of Things*, trans. Richard Howard (New York: Vintage Books, 1973).
48. Foucault, *The Order of Things*, p. xv.
49. Foucault suggests something like this idea in a 1975 interview when he says: "Writing interests me only insofar as it enlists itself into the reality of a contest, as an instrument of tactics, of illumination. I would like my books to be, as it were, lancets, or Molotov cocktails, or minefields; I would like them to self-destruct after use, like fireworks." Quoted in Allan Megill, *Prophets of Extremity: Nietzsche, Heidegger, Foucault, Derrida* (Berkeley: University of California Press, 1985), p. 243. Another plausible, competing (if problematic) framework in which to explore and critique domestication is an Aristotelian one—specifically in terms of Aristotle's teleology and metaphysical biology. I plan to investigate this possibility elsewhere.
50. See, for example, John Caputo and Mark Yount, eds., *Foucault and the Critique of Institutions* (University Park: Pennsylvania State University Press, 1993).
51. Michel Foucault, *Power/Knowledge: Selected Interviews and Other Writings, 1972–1977*, ed. Colin Gordon, trans. Colin Gordon et al. (New York: Random House, 1991), p. 96.
52. Thomas Birch notes this possibility in passing for wilderness reservations in "The Incarceration of Wildness: Wilderness Areas as Prisons," *Environmental Ethics* 12 (Spring 1990): 9. Bob Mullan and Garry Marvin begin to examine the relevance of such a framework for understanding zoos in their *Zoo Culture* (London: Weidenfeld and Nicolson, 1987), pp. 33–44.
53. See, for example, Jim Mason and Peter Singer, *Animal Factories* (New York: Crown Publishers, 1980).
54. Foucault, *Discipline and Punish*, p. 203.
55. For further elaborations of the related exercise of power on human and nonhuman animals, see David Macauley, "Animals, Ecology and Anarchism," *Lomakatsi* 1–2 (serialized in two parts from Spring 1987–Summer 1988).
56. Michel Foucault, *Madness and Civilization: A History of Insanity in the Age of Reason*, trans. Richard Howard (New York: Vintage Books, 1965), p. 70.
57. Ibid., p. 63.
58. Ibid., p. 72.
59. Ibid., p. 77.
60. See Macauley, "Animals, Ecology and Anarchism."
61. For an elequent defense of margins, especially as they relate to farming, see Wendell Berry, *The Unsettling of America: Culture and Agriculture* (New York: Avon Books, 1977), pp. 171–223.
62. Gary Snyder, *The Practice of the Wild* (San Francisco: North Point Press, 1990), p. 15.
63. Ibid., p. 16.
64. Mary Midgley, *Animals and Why They Matter* (Athens: University of Georgia Press, 1983), p. 112.
65. See Gilles Deleuze and Felix Guattari, *A Thousand Plateaus*, trans. Brian Massumi (Minneapolis: University of Minnesota Press, 1987), pp. 232–309.

66. See David Macauley, "A Few Foot Notes on Walking," *Trumpeter* 10 (1) (Winter 1993): 14–16.
67. See Bruce Chatwin, *The Songlines* (New York: Viking Penguin, 1987).
68. Georges Van den Abbeele, quoted in Meaghan Morris, "At Henry Parkes Motel," in John Frow and Meaghan Morris, eds., *Australian Cultural Studies: A Reader* (Urbana: University of Illinois Press, 1993), p. 251.
69. For relevant texts on tourism and travel, see Georges Van Den Abbeele, *Travel as Metaphor: From Montaigne to Rousseau* (Minneapolis: University of Minnesota Press, 1992); Georges Van Den Abbeele, "Sightseers: The Tourist as Theorist," *Diacritics* 10 (December 1980); and Dean MacCannell, *The Tourist: A New Theory of the Leisure Class* (New York: Schocken, 1976).
70. Another viable and valuable response to the dilemmas of domestication would be to focus upon improving and altering the nature of our homes as architectural structures so as to make them more livable, open, and liberating. As Wendell Berry and others have argued, "The modern house is so destructive . . . because it is a generalization, a product of factory and fashion, an everyplace or a noplace." *The Unsettling of America: Culture and Agriculture* (New York: Avon Books, 1978), p. 52. Along with the very relevant subjects of reinhabiting places and experiments in living arrangements (e.g., communes), this matter lies outside the scope of the present essay.
71. Jacques Camatte, *Against Domestication* (Detroit: Black and Red Press, 1975), p. 5. See also his *The Wandering of Humanity* (Detroit: Black and Red Press, 1975).
72. Lewis Thomas, *Lives of a Cell* (New York: Bantam Books, 1975)
73. Quoted in Rene Dubos, *So Human an Animal* (New York: Charles Scribner's Sons, 1968), p. 220.
74. I would like to thank Roger S. Gottlieb for his encouragement, patience, and support with this paper and to acknowledge the helpful comments and questions of the copyeditor.

Part 2

Environmental Theory and Moral Questions

7

A Sleepless Ethicist and Some of His Acquaintances,
Including the Monoculturalist, the Poetic Naturalist, the Very Famous Biologist, the Happy Scientist, and a Few Heroic Antimodernists

Roger S. Gottlieb

To paraphrase one of my teachers[1]: how easy it is to speak of our love for, and obligations to, the Earth. We might well wax eloquent, put up our feet in comfort, and even drink some organic herbal tea as we do so. Yet if there is one among us who listens deeply to the claims of the suffering Earth, and also feels the call of an ethic in which human beings and nonhuman nature are both objects of concern, that person might well wake in the predawn light and face, with fear and trembling, some troubling questions.

The Demands of Humility and the Imperatives of Politics

Peering through his bleary eyes at human history, our troubled nonsleeper tries first to figure out what is so troubling about his present ethical situation, why it is so different from other moral demands he has faced. He begins by reflecting that for most people, most of the time, there *is* an impulse toward morality. This acknowledgment does not deny the violence and casual oppressions that permeate history. Rather, these dismal occurrences merely show that there are, too often, countervailing pressures that are stronger than the impulse to goodness; or that socially constructed viewpoints sometimes make moral truth inaccessible. Further, he knows that in all cultures of which we have knowledge, there has been a call to morality. From the biblical prophets to the ethics of the Buddha, from bourgeois to socialist revolutions, from the Civil Rights movement to Stonewall, our past echoes with reminders to heed the outcry of the oppressed and to remember or discover our better selves.

Our Ethicist also knows that this outcry changes over time. Criticisms of patriarchy, for instance, constitute comparatively recent moral demands. And it is not hard to trace the shifting historical conditions that make such criticisms possible. Related developments in economic structure and political ideology (women entering the workforce and Cold War–fueled proclamations of social freedom and equality of opportunity) create the historical possibilities of a women's liberation movement and the corresponding moral demands on men that are embodied in that movement.

While the recently emerged demands of feminism may be rigorous, men can, our Ethicist believes, respond to them without contradiction. A well-intentioned, good-hearted man *can* surrender male privilege in his personal relationships—say, by making himself emotionally vulnerable, taking responsibility for his own actions, and doing his share of the dishes. He can refuse to collaborate in the oppression of women, publicly support their political equality and personal safety, and challenge male-biased cultural values. Having done so, he can then speak with a kind of moral integrity about the evils of sexism.

Now perhaps the principle reason our troubled Ethicist is awake at this ungodly hour is that he senses how different are the demands of the environmental crisis from those of, say, feminism. Industrial capitalism, imperialism, Third-World militarism, patriarchy, and consumerism have assaulted all of nature (including people)—and imposed moral demands that seem impossible to fulfill. These challenges threaten the Ethicist's sleep and any justified confidence in his society's ethical standing. It is both the civilization as a whole and the most concrete details of his own moral life that are at stake.

What is happening, our Ethicist wonders, to an essential presupposition of personal and collective morality: *that we are able to teach virtue to our children*? We claim to be bearers of values which we hope to impart to them. We claim, directly or tacitly, that we are worthy moral models. We expect children, in turn, to believe that we know what we are talking about. Family life, schools, religions, the legal system, and even political parties depend on this presupposition.

For our Ethicist, this presupposition is rendered deeply suspect by the shadows of the toxic waste dumps, elimination of other species, and cavalier nuclear testing. The Ethicist wonders how any of us can teach morality to our children, or pretend to ethical competence, when breakfast-cereal boxes list endangered and extinguished species, and kids can ask, as did his seven-year-old daughter: "Daddy, can a time come when there are no more trees?"

It is not so easy to maintain a view of humanity as the "lords of creation" when our social practices are poisoning that creation. What people have done and continue to do casts doubt on our collective ability to serve as exemplars for the next generation. Engaging in moral teaching now seems to require that we turn a blind eye to reality and engage in a kind of denial—one that may be transparent to those we wish to teach. Our Ethicist cogitates: "Don't our children sense what is absent in our moral talk; and doesn't that further lessen their already limited respect for our integrity and efficacy?" He thinks of the bitter, cynical, escapist, passive, or technoaddicted young people he knows, and wonders how much of their condition stems from the simple fact that they cannot forgive the moral avoidances and failures of the adult world. And he asks himself: "How can we profess love for our children and simultaneously destroy the ozone layer that protects them from cancer-causing sunlight? How can we teach them to love their neighbors when energy use in the First World may cause a global warming that will raise the sea level and virtually wipe out island nations? Having tampered with the climate and the protective features of the atmosphere, having dangerously overloaded our planet's carrying capacity, how can we pretend to be experts in morality?" Strangely, since so many people practice or desire environmentally damaging mass consumption, whole populations are now implicated in their own destruction. What then happens to confidence in the rational progress of humanity—or to our very right to further develop technological civilization?

Sleep recedes even further, as the Ethicist reflects that even those of us who are not identified with or who resist the dominant mainstream governments, religions, corporations, or political parties can only offer our children a vision of moral exclusion and impotence. If we disassociate ourselves from the collective agencies that are wreaking havoc with the earth, we must admit to our own inability to stop them. Again, we may avoid the subject and thus leave our children thinking we are hypocrites. Or we may admit the truth and confess to how little we can do to protect them or the earth. One step from despair, our Ethicist wonders what kind of legacy this is to leave to the next generation.

It is not only the cultural values and institutional imperatives of our dominant social structures that are causing the environmental crisis. As individuals, our daily lives make us accomplices to the catastrophe. For people above the poverty line in the First World, and for many in the other "worlds" as well, our everyday routines involve participation in a form of life that is bankrupting the future.

The Ethicist, who has some standing as an environmental thinker, wonders at his own situation. He knows, better than most, the horrible effects of what he is doing. But it always seems that there is comparatively little he can do on a personal level to withdraw, unless he so dramatically changes his life that (like the Unabomber?) he virtually leaves society. He tries to use his bike when he can, to recycle, and to eat organic vegetarian style. But his energy sources, food sources, medical sources, tax payments, clothes, travel, housing, and work continue to involve him in the very form of life he abhors. Thus, it is not just the abstract "society" or our dominant institutions whose moral status is threatened, but that of self-conscious individuals as well. Virtually all of us—including our Ethicist's like-minded colleagues—participate in and thus tacitly support destructive patterns of consumption and production. They may protest global warming, but drive individual cars to the protest site! They write environmental manifestos and use endless paper for rough drafts. As he prepares his passionate and sincere speeches, our Ethicist plugs his computer into the same power grid that Newt Gingrich and the Wise Use boys do.

Painfully, the moral demands of family, work, social responsibility, and care for the earth often pull our Ethicist in conflicting directions. The sheer effort to fulfill any *other* moral obligation seems to leave him helpless before his concern with the environment. Consider a friend of the Ethicist, a professor of philosophy he knows well. This man lives in Boston and for the academic year commutes to work in Worcester: one hundred miles, four times a week, for thirty-five weeks a year. An expert on the religious and political aspects of the environmental crisis, this professor nevertheless makes his own little contribution to global warming. He explains his situation by appealing to his younger daughter's serious special needs, and the fact that he can find no better school placement than her current one in Boston. Of course he might take the bus, but that would double the time of the already lengthy commute and cripple his ability to fulfill his other responsibilities: as father to both his children, husband, self-styled radical intellectual, teacher, colleague, and environmentalist. Responsibilities at which, it should be noted, he already feels painfully inadequate. But what happens to his personal credibility when he teaches Bill McKibben's disturbing study of global warming[2] in his environmental philosophy class? Who is he, he wonders, to rail against Exxon or General Motors, when he too pours unnecessary carbon dioxide into the air?

Despite these dilemmas, our sleepless Ethicist does not wish us to stop talking about our moral obligations to other people, life forms, or generations. He simply sees that, since conversations about environmental ethics and sustainable religion are pretty typically carried out by people who are not off the grid in any significant way, this discourse has become profoundly problematic. A similar moral pall is cast over all attempts to provide models for our children of virtues like taking responsibility for one's acts, restraint in consumption, or care for others. At the very least, our response to the violence and dominance embodied in ecocide, while passionate, had better be free of any taint of self-righteousness. If we are not to appear obvious hypocrites, our moral instruction to our children should be couched in terms that include direct admissions of our own failings. In this regard, our collective complicity in the environmental crisis dramatically reverses the usual constraints on moral instruction. We typically believe that the practicing murderer, thief, or profligate cannot teach others about the virtues of nonviolence, respect for others' property, or chastity. In the case of the morality of our relations to nonhuman nature (the Ethicist notes with no little irony), such a constraint would leave us with no one to teach anything at all!

Our Ethicist continues, feeling that this story is far from over. He asks himself: "Yet how can I balance this necessary humility with the equally necessary political critique?" Now turned social theorist, our Ethicist believes this latter necessity stems from the fact that while "humanity" as a whole is the agent of environmental destruction, humanity itself is divided by systematic inequalities of political and economic power. We are not evenly agents or victims of environmental aggression. For example, while poisonous pesticides are a global health problem, migrant farm workers exposed to them in the fields have a different moral relation to their toxic effects than do those who control the corporations making and distributing them. Generalized ill effects from pesticides include increased risk of cancer, immune deficiencies, respiratory illness, birth defects, reproductive problems, and on and on. These affect us all. Yet how can we pretend to talk seriously about them if we do not also acknowledge the approximately 200,000 deaths a year from direct exposure to high concentrations of pesticides, mainly among the populations of people who handle them?[3]

Thus the humility compelled by our *collective* participation in the violence of ecological destruction must coexist with a critical awareness of the politically and economically sanctioned *structures of inequality and domination* that contribute to the environmental crisis and its savage effects on human beings. Not all of us have equal power to decide to continue manufacturing CFCs, or to gut public transportation, or to export banned chemicals to Third-World countries. Economic, political, and military elites, usually unconstrained by the rest of the population, shape the world's social life. Ecocide, our Ethicist senses, is their stock in trade.

Our Ethicist wonders: Can human violence against all forms of life be reduced if social relations do not become more just? The outcome of many different conservation programs has taught him, for instance, that if we are to save the wildlife in Africa, we need to care as much for the people as we do for the animals.[4] Similarly, when he studied the clear-cut forests and silted up coral reefs of the Philippines, he saw that concern over "nature" was at the same time concern over truly democratic access *to* natural resources and *over* decisions concerning the path of economic development.[5] In particular, it is often impossible to preserve a fragile ecosystem when the native people who live there are dispossessed and dislocated. The goal of sustainable development, it seems to him, can't be separated from maintaining cultural diversity; and preserving biodiversity is inextricably linked to the preservation of indigenous peoples.[6]

More generally, there is no population problem or consumption problem, per se. Rather, there are problems with the unequal distribution of wealth, power, privilege, and control—control over natural resources, reproductive capacities, the development of technology, and the implementation of social policy. How many children people have, and how and how much they consume, are consequences of more fundamental social structures. The peasant family's economic need for more children, the oppressed woman's lack of access to birth control and education, the psychic desperation driving First-World folks to shop 'til they drop—these causes of overpopulation and overconsumption are themselves the effects of injustice and domination.

The Ethicist wonders: Is there a *simple* way to say what is wrong with the way we live? He is of course familiar with a number of different ways of putting it. People talk of anthropomorphism, exploitation, commodification, and consumerism. Of capitalism and militarism. Great Thinkers— much greater than our insomniac moralist—have described the modern psychic individualism that keys identity to the acquisition of objects and the expenditure of climate-altering energy.[7] Patriarchal masculinity—otherwise known as boys will be boys—demands the domination of women and other living things.

But there is another term that has captured him: "monoculture." Here economic modernization (*mal*development some call it[8]) breaks the traditional, organic (in several senses) connections of communities with the land; and replaces those relations with hypertechnologized agriculture based on a single commodified crop. An evolved natural-social system of local knowledge, natural patterns of sustainability, and multiple uses is assaulted by chemicalized and imported inputs, alien seeds, reductionist science, state planning, and global corporate integration. Differentiated but normally interdependent local communities become social interest groups—or fanatically warring sects—defined by their relation to the centralized state and international commodity exchange rather than by their relations to each other.[9]

A monocultural blight emerges in the cultural sphere as well. An entire globe suffers from attention deficit disorder as differences of tastes, culture, and identity formation get flattened out into a series of rapidly changing but endlessly repetitive basketball shoes, pop CDs, and nylon jackets.

It is these various monocultures—the endless unnatural sameness dependent on inputs from elsewhere, and the crazed expectation that wastes can be endlessly shipped "away"—that is the hallmark of an ecocidal age.

Our Ethicist senses the profound evil of these monocultures. He knows that if he is to maintain his self-respect, his moral humility had better not get in the way of his naming them for what they are and his supporting struggles against them.

The Depths of Nature

Looking deeply into the abyss of the consequences of the way he has lived, our predawn Ethicist suspects that the only way out of these dilemmas of ethical powerlessness is the realization that certain kinds of moral demands cannot be fulfilled by individuals. Only fundamental shifts in collective social practices will allow a truly moral relation to other people and the environment. He begins to see morality as social not only in the familiar sense of having to do with benefits for a group but also in the sense that the subject or agent of morality is the collectivity self-organizing itself. He has seen this approach to moral life expressed by those who see whole modes of production as morally lacking. To say that capitalism is based on exploitation, for example, is to ask for institutional rather than purely personal moral changes. It is not to challenge the ethics of a particular businessman, but that of "business" per se.

As he considers the moral quandaries and the social character of the environmental crisis, the Ethicist sees a vast cultural system of anthropocentric biases in which human beings claim themselves as separate from and superior to the rest of nature. This arrogant model has deep historical origins and also permeates the individualism of modernity. But anthropocentrism has also, our Ethicist knows, been rendered suspect by the environmental crisis itself.

Ethical life is shifting, he believes, because global *ecological* interdependence now implies both personal and collective *moral* interdependence. The most mundane activities of our daily lives can affect the well-being of people throughout the world. This moral truth reawakens our awareness of the spiritual interdependence taught by great prophets—which are now rendered material by the far-flung effects of production and consumption. We might have laughed up our sleeves when Jesus told us to love our neighbors, when Isaiah clamored for justice, when Gandhi called for universal compassion, or when Marx hoped that each would give "according to his ability" and only take what was needed. Yet when other people's smokestacks kill our forests, or our use of gasoline threatens someone else's agriculture, the smiles get wiped off our faces.

Further, as Deep Ecologists have stressed, sorrow over the devastation of the natural world reveals that personal identity does not stop at our individual or social boundaries, but includes our natural surroundings.[10] The nonhuman world, subject to so much abuse by our civilization, is revealed as a part of our own selves. Our bodies, senses, emotions, and even language evolved in a nonhuman setting.[11] That is why the more the nonhuman nature is humanized, civilized, and paved over, the more lonely, frightened, and addicted to distractions we become.

For these reasons, environmental ethics are perhaps not best expressed in terms of "rights" of nature. Rather, they are rooted in the intuition that the burning of the rain forest, the pollution of a river, or the elimination of a species is a hurt not to an Other, but to Our Very Selves. The Ethicist did not need rights theory to motivate love for his daughter or his mother, his friends or his golden retriever. Similarly, his love of, and loneliness for, nonhuman nature flourishes through the depth of his kinship with it rather than through abstract principles. The environmental crisis, though a source of great pain, is thus also an incredible opportunity to realize a vision of spiritual connection.

But connect to what? And how?

Why, to nature of course . . . and . . . naturally! How else?

And what does that mean now?

A perpetually troubled person, the Ethicist cannot help but think of another acquaintance of his, a widely read and respected anthropologist and Poetic Naturalist. One of the man's books focuses on his travels to an uninhabited island off the northwest coast, to which he sails over rough Pacific seas and where he then spends time camping and exploring. The Ethicist has been particularly struck by two passages in this book. In one, his friend describes an evening on the island:

> Toward evening we start a campfire, set up driftwood benches, and look out over the shore. Diminishing surf breaks along the point and again to a small bare island across from camp. The tide has risen thirteen feet in the last six hours, submerging the maze of reefs and tide pools so the cove now lies open to the Pacific. Adrift on its own reflection, the skiff looks tiny and vulnerable. Above the silhouette of Ocean Point, snowcapped mountains rise to the fading beams of sunset, and a wilderness of coast dwindles into the distance. The only signs of human presence are this flickering fire and the light of fishing boats anchored in the far coves of Roller Bay. Through my whole adult life, I've sought to experience unaltered, unbridled nature in

wild places such as this. I've focused my work around it, chosen my home because of it, given up economic assurance to pursue it, made it a centering point of my existence.[12]

Our Ethicist, whose family ties keep *him* in the city, envies his friend's adventures and is moved by the lonely beauty evoked in the smooth prose. And yet . . . and yet . . . our Ethicist is also moved to an ironic sadness—or a sad irony—when he rereads a passage that appeared only four pages earlier:

> We haven't come unprepared. The engine is in excellent condition, and we have a smaller spare in case this one fails. We also carry plenty of fuel, and the boat is well stocked for emergencies: fire extinguisher, flares, waterproof matches, hand-held radio, the unsinkable styrofoam punt, and sealed container of emergency gear, plus sleeping bags, tools, flashlights, tent, food, and additional camping equipment. We're wearing coveralls designed for survival in cold water and have wetsuits and surfboards along.[13]

The source of our Ethicist's irony is obvious: where is "unaltered nature" to be found when nature can only be approached with so many trappings of civilization? And how can his sincere, articulate, but perhaps not always self-conscious friend not see the contradictions in those two passages?

But the Ethicist's sadness signals a deeper and perhaps more compassionate response. For he sees that we have become like descendants of refugees who cannot approach their ancestral culture without simplified translations into the new, alien, but by now "native" tongue; or without a guidebook to the major holidays and central myths of the now barely remembered tradition. Even the most ardent of nature seekers is helpless without a vast array of plastic, metal, and goretex. And even the most devoted emerge from and return to the same old source of environmental destruction: our society.

How can we comprehend these paradoxes, the Ethicist wonders. On the one hand, we are so desperately lonely for something Other than ourselves that we call "nature." Yet it seems we can only approach, comprehend and ultimately know this Other in human terms. As a Very Famous Biologist and intrepid rain-forest researcher put it:

> If I ever seriously thought of confronting nature without the conveniences of civilization, reality soon regained my whole attention. The living sea is full of miniature horrors designed to reduce visiting biologists to their constituent amino acids in quick time . . . evolution has devised a hundred ways to macerate livers and turn blood into a para-

site's broth. So the romantic voyager swallows chloraquin, gratefully accepts gamma globulin shots, sleeps under mosquito netting.... He hopes that enough fuel was put into the Land Rover that morning, and he hurries back to camp in time for a hot meal at dusk.[14]

Perhaps, our Ethicist muses, our plight is but an extension of the first technological uses of sticks and stones. Perhaps it is another facet of McKibben's melancholy notion that as we alter the global climate and atmosphere, "nature" has ceased to exist. This reflection leaves the Ethicist sad and not a little frightened.

And, also, somewhat confused. Like his friend the Poetic Naturalist, the Ethicist too would like to encounter "unbridled, unaltered nature." But where would he go to do so? For the profound alteration of which McKibben spoke is true not only in the global sense, but in countless local settings. For instance, the "wild" lands of the American national parks—supposedly dedicated to preservation of wilderness at its most pristine—have already been significantly altered by the importation of exotic species of plants and animals. In the Sierra Mountains of California, the native herbaceous species were replaced by annual grasses to support intensive cattle and sheep grazing in the nineteenth century. In Hawaii, nonnative mosquitoes, bunchgrass, pigs, and mongooses brought in by white settlers have decimated countless local species of insects, animals, birds, and plants, and even constitute a large percentage of the parks' biomass. Quite often the newcomers, though "natural," have escaped their "natural" enemies and proliferate with little restraint. The result can approach a spontaneous, a "natural," monoculture.[15]

Our poor Ethicist's head begins to swirl when he further remembers that it is not simply the recent white settlers who have impinged on "unaltered" nature. As much as he would like to lay the blame solely on technological civilization, he has learned that when the original Polynesian settlers came to Hawaii over a thousand years ago, they brought with them twenty-five species of plants as well as dogs, pigs, and rats. As a result, "The large native Hawaiian birds appear to have been hunted to extinction by the new arrivals."[16] In another example, he has found that large areas of the American forests were managed and altered by controlled burnings initiated by Native Americans. And that at times Native Americans would stampede herds of buffalo over cliffs and take only the tongues and the humps.[17]

What then could it mean to return forests to their "natural" condition? After decades of stomping on forest fires, should we try to reintroduce the fire practices of the aboriginal culture? But those practices themselves were not constant, but evolving. And what happens when, by imitating the natives' controlled burns, we begin to threaten the air quality of the surrounding settled areas? Conversely (and now the Ethicist's head not only swirls but begins to hurt!) should we try to return the forest to what it had been before the "recent" (biologically speaking) incursion of Native Americans? But the balance of fire-resistant and fire-dependent species has already changed, and neither imposing fires nor leaving things alone can get us "back." When the Wild and Rebellious Nature Lover insists (of Yosemite) that what we must do is "keep it like it was," just what are we supposed to do?[18]

In the end, all we can say is that park visitors do see "nature"—meaning an alternative to, and an escape from, urban life. But much of this nature has been shaped by centuries of human action. And maintenance of even that now requires aggressive levels of human intervention. "Nature" has become our charge and our care.[19]

The Ethicist remembers a trip of his own to Sanibel, a small island off Florida's west coast. To escape the ubiquitous hotels and condos, upscale little malls, and elegant yet casual restaurants, he had taken a stroll through a section of the local wildlife refuge. Birdcalls delighted his ears, and he saw white ibyx, red-billed moorhens, and softly chirping black cummingtons. A water snake rippled through a pond, and an alligator lazily floated in the estuary. At the close of his restful stroll, the Ethicist found a sign describing the history of the area: to his dismay he found that not only had the entire island been altered by major dredging and waterway construction to eliminate mosquitoes, but a nonnative decorative shrub that had taken over large areas had been attacked with herbicides and bulldozers so that the native marsh grasses could return. This bit of nature, so lovely an oasis amid the desert of civilized holiday development, was in its own way not much more "natural" than the grass and ornamental flowers of the nearby golf course.

The moral dilemmas multiply. The Very Famous Biologist complains that cutting down the rain forest to grow food for peasants is like burning a Rembrandt to cook dinner.[20] Of course the Ethicist cries for the rain forest as much as the next man (or woman). But he wonders (perhaps unfairly, but his sleeplessness is now making him a bit testy) when was the last time that the Very Famous Biologist was hungry at dinner time? And he muses that if one is hungry enough, burning the Rembrandt might be the appropriate thing to do. What else, he wonders, is anything for?

The Ethicist also knows of the work—work he admires deeply—of a Very Distinguished Writer on Ecology.[21] This man celebrates the joys of hunting: the mystical communion with the prey, the sense of nature as gift, the cultivation of perceptions and focused awareness demanded by the pursuit of the game. The Ethicist wonders, this time surely unfairly, if this advocate of the hunt would enjoy being the prey of someone who wanted to celebrate the mystery of hunting, and finding, and then eating the Very Famous Writer on Ecology.

In all this confusion, and despite his growing irritability, the Ethicist begins to chart a course. He knows that often people want to honor something they call "nature." What they mean by this varies, but amid a family of related ideas he has seen the notions of: untouched by people, spontaneous, unplanned, different from mind or history, possessing an internal self-regulated balance, being an integrated system with self-generated continuity; being real, which means being *what* it is no matter what people may think or say about it. In turn, the advocates and defenders of "nature" distinguish it from the artificial and man-made, from the planned or consciously evaluated, and especially from Sudden Changes brought on by Unsound Social Practices. They criticize (quite justly, the Ethicist believes) the entire modern mind-set that would subject the beauties of the physical world to the domination, exploitation, and endless control of technological (and patriarchal) civilization. And they criticize (justly again) the extremities of a deluded postmodern thinking that claims that the idea of "nature" and the claims of science are only culturally relative human products.

Our Ethicist too would like to honor and defend "nature"—if only he could find it! He has already reflected on the difficulties of connecting with "unbridled, unaltered" nature. He knows that the untouched is by now hard to find; and in any case our access to it depends on just those technological powers that are destroying it. Really, it seems to him, claims about the love of "nature" often mean little in any general sense, though they may well carry a very deep meaning about a particular encounter. When a Poetic Naturalist tells us to "love" or "respect" nature, s/he is often reflecting on a personal and very powerful experience. One such Poet looked into the eyes of a dying wolf that he himself had shot—and saw there a "fierce green fire" that forever altered his sense of humanity's relative importance in the scheme of things. Another saw a polluted river and felt loss that was immense without being narrowly "self"-interested.[22] The Ethicist himself remembers entering one of the last remaining groves of Redwood trees and weeping, thinking of nothing so much as survivors of the Holocaust.

Yet the emotional power of these encounters—love for this entity, grief for its death or pollution—tells us nothing of some entity called Nature. What we learn—and this is of the highest importance—is something about our particular relations to this particular river, wolf, or redwood grove. The appeal to nature is at best a summary of myriad such particular encounters; at worst it is an evasion of the contradictions of which the Ethicist is too much aware: contradictions about our own seemingly inescapable use of and control over what surrounds us.

The Ethicist is also well aware of the contradictions in the existence (though not, of course, in the writings) of the Heroic Antimodernists and Ecofeminists. Finding origins at times in Descartes, or Frances Bacon, or Plato, or the sky-gods of monotheism, these thinkers have impressive philosophical criticisms of the modern western addiction to power over and control of the physical world. The Ethicist thinks in particular of a Very Great German Philosopher, loved as an inspiring critic of modernist madness by many and detested as a fascist obscurantist by others, who repeatedly criticized the inability of western thought and practice to "let being be."[23] Yet despite his profound verbal rejections of the calculating, manipulating spirit of everyone from Plato to Marx to modern engineers, didn't this philosopher want his morning coffee like everyone else? Or, more importantly, didn't he want the printing presses to work—at least when they printed *his* books?

The Ecofeminists (both women and men, and the Ethicist counts himself as one) typically root the environmental crisis in a culturally male attitude that seeks to control rather than connect, to dominate nature rather than celebrate our intimate relations with it. Ecofeminist research has shown how men sought escape from the feminine and the fleshy and then fashioned that greatest of all cultural calamities: the modern, male, hyper-individuated, hyper-controlling, hypermathematizing, and hypermeasuring ego.[24] The Ethicist appreciates the depth of the Ecofeminist criticism, and would himself wish for a culture in which the drive for control were replaced by the love of empathy. Yet he must admit that he and his fellow Ecofeminists want to know how much gas is in the car when they drive to the library; want to be able to calculate their paychecks; and especially want to eliminate the bacteria that cause ear infections in their children. In our daily lives we, no less than Descartes or the CEO of Dow Chemical, try to control our surroundings.

Thus the antiscientific stand of the Antimodernists does not gives us the key to nature. Can we perhaps get it from science, which will show us—all postmodern frippery to the contrary notwithstanding—the True Reality of the Real World of the Natural? The Ethicist thinks here of a deep argument that tries to do just this. In this argument a Happy Scientist rejects the fashionable notion that scientific claims are really just social products. He asserts that "the taxonomies of aboriginal societies are virtually always the same in structure as those of modern, scientific cultures . . . and . . . aboriginal taxonomies typically recognize the same entities as species as do modern taxonomists." When, for instance, a leading ornithologist traveled to the remote mountains of New Guinea, he found the native peoples identified the same 136 species of local birds that he did. It may well be, the argument goes on, that it is not culture that determines our understanding of nature but rather two other factors. First, "the human sensory/perceptual apparatus" enables us to see the world in a certain way and therefore to classify and analyze it accordingly. Second, it is not the generalities of culture, but the particularities of personal needs that enable us to know deeply. A modern zoologist earns his living by knowing his animals; a "primitive" hunter earns his the same way.[25]

This argument intrigues the Ethicist. He admires its willingness to engage with brilliant and ironic claims that "everything is made up as we go along" by citing specific contexts in which such claims seem pretty implausible. However, just as the Poetic Naturalist carries unnatural baggage with him, and the Ecofeminist and the Antimodernist philosophers desire control despite themselves, surely the Happy Scientist has a few problems in finding nature. To begin with, the Ethicist observes, the scientist does not study nature, describe nature, or know any more about nature than anyone else. Rather, the scientist studies and describes and publishes papers about stars and ants, rain forests and dinoflagellates. The word "nature" appears in no scientific theories, allows for no explanations or predictions, does no theoretical work. When a respected conservation biologist suggests that nature is "real" and not just a human construct (is really "out there"), the Ethicist notes sadly, he misses the point. For to say that science is telling us things about nature in addition to things about ants or molecules is already to add to what the scientist has said, and to add in a way that has none of the authority of the scientist's claims about particular birds, planets or chemical processes. To use "nature" in this way is to cast the conceptual net to meet certain social needs and pressures: to draw a line and thereby distinguish Nature from History, or Society, or Illusion or Ideology, or Mind. The aboriginals do the same when they speak of the Great Spirit or the Earth Mother. And all these forms of poetry—or philosophy, or religion, or myth—have their place and their use. But they are not of the same kind of knowledge that we get from either theorems in chemistry or traditional lore about herbs. Our claims about nature (including all of his own, the Ethicist hastens to add)—that it is the source of blessing, that it expresses the love of the Great Mother, that it is a social product, or that it is "real"—are of a different order, and serve different social functions, from the concrete studies we have made of the world around us. And this holds whether those concrete studies derive from the most nontechnological of hunter-gatherers or the latest hot flash from MIT.

And the situation is further complicated by the fact that even concrete science, *especially* in a modern society in which knowledge is institutionally connected to economics and politics, may in fact lead us far from what we need to know. For example: while scientists and natives may indeed classify birds the same way, there are far too many cases when scientists are at odds with the locals on the how to manage plants, forests, and water. In the monocultural cases of the Green Revolution, the technological elite had no idea that its import of nonnative eucalyptus trees would destroy the local ecological balance in the forest; or that large dams in Indonesia would ruin the water conservation practices there. But the local, nonscientific people did know. And resisted. And were dismissed as ignoramuses.[26] The sad truth is, then, that while the knowledge that science seeks may be rooted in our bodily structure and our social interests, *we can never be sure in general when those same social interests don't distort what we think we know.* The zoologists classifying birds did well; the agronomists of the Green Revolution failed miserably. Perhaps this is so because those agronomists were too deeply embedded in the effort to turn local ecosystems into sources of commodities for the global market. How different things might have been if their goal had been to preserve complex local diversity of humans and nonhumans alike.[27]

The Ethicist realizes he can put this another way. Science cannot tell us where nature is, not only because science does not study nature, but because *we* cannot ever be sure where *science* is. Just because the door of someone's office has the word "science" on it does not prove that this is not another case where the manipulative interests of imperialism are masquerading as the latest word in technological expertise. In each case, our Ethicist realizes (feeling tired already at the thought of how much work this commits him to) the concrete details of the concrete situation have to be looked at in detail. Which is it: the birds of Indonesia or the destroyed forests of the Green Revolution? The brilliance of radiotelescopes or the horrors of thalidomide?[28]

The Ethicist, coming to a turning point, is thus clear that the Happy Scientist cannot give us nature. And neither can the Poetic Naturalist's immediacy of vision or the Antimodernist, Ecofeminist distaste for "control."

And here is where the Ethicist, tired but seeing a faint glimmer of peace, thinks he has finally understood something. We cannot handle the problem of the environmental crisis by invoking nature, he thinks, at least not with nature understood as something that can be neatly summarized by the kinds of concepts invoked by the Poetic Naturalist, the Antimodernist, or the Happy Scientist.

But what we *can* do is probably more modest and certainly more demanding: to examine the complex patterns of relationships in which we live. If we want to try to transform what we are doing to our surroundings we have to attend to those relationships in all their difficult details.

Relaxing a bit, the Ethicist reminisces about a time when he taught a class on these matters. A bright, somewhat caustic, but quite scientifically knowledgeable (more knowledgeable than the Ethicist, in any case) student questioned him closely:

Why should we care about how people are changing the world? Isn't nature in flux anyway? There was a time when the earth had no atmosphere, no oceans. There will come a time when the sun will explode and we will all go along with it. Aren't those things as natural as anything else? So why make a big deal out of leaks from a nuclear plant or ozone depletion?

The Ethicist had been stunned by the simple power of the question. And he had had to think deeply before he replied.

This is a valid perspective. But can you really maintain it? Imagine (God forbid) that you have just found out that your mother has lung cancer. And one of the causes is the second hand smoke from your cigarettes. Would you say: "Oh well, the world will end in six billion years anyway. Who cares?" Or would you be struck with guilt and sorrow? And be willing to help in any way to save her? Looked at, experienced, felt, understood—the present set of ecosystems and species of our earth is like that mother. We do not seek to hasten her demise, nor would we willingly or easily choose to help change her from what we think of as healthy to one riddled with cancer. (Although we might want her to make a few minor changes: even a mother isn't perfect.) But our love for her is not based in her unchangingness, or the fact that somehow she exists without us. It is based on who she is for us—and who we are with her.

Simply and wonderfully: We are who we are only through our connections to our surroundings. We need not stand back and wave generalizations about nature.[29] We need to lean in closer and attend to how we actually live.

The Ethicist smiles for the first time. Has he gotten something right? He asks himself: How then *are* we connected to what is around us? Too late (too early?) to be comprehensive about it, he makes a short list: We are *dependent* for food, water, air, and material. We *interpenetrate* with this other, through the billions of microorganisms that live in our bodies. We feel an *empathic kinship* with many animals. We *learn* to think, to talk, to understand moral qualities and physical truths by observing and experiencing the nonhuman realm (carefully, the Ethicist avoided saying "nature" here). We are *enthralled* by the beauty of ocean or field or mountain. *Spiritually*, we may sometimes feel so exalted by our nonhuman surroundings that our isolated ego gives way to an identification with a larger, more complete, whole; sometimes nature seems to be God or the face of God or evidence of God. Our surroundings (the Ethicist almost said "nature" again) are a source of *knowledge* and an arena for the *pursuit of knowledge*. Not unlike our connections to cultural traditions, our surroundings also give us a sense of *continuity*, a sense of *participation* in what is longer lasting than our own brief lives. As we might treasure our heritage as Jews or Frenchmen, so we can value our connection to a lasting river or forest.

As our surroundings include people as well as animals and plants and large environmental settings, these natural items (the Ethicist just couldn't help himself there) are also *mediators* in our relations with other people. We affect future generations and people in distant lands by the way we treat nature (the Ethicist just gives up, at last).

Finally, our surroundings hold images that *reveal our own worth and character*. When we look at a well-tilled garden, a clean river, a leaking toxic landfill, or the return of wolves to Yellowstone, we see a good part of the meaning of our lives.

Rolling along, forgetting for a moment all the difficulties bound up in each of these relationships, the Ethicist sees that it is not an absolute called Nature that thus calls to us but a series of relationships, each with its own internal goods. To see what we must do, we must simply (though not easily) *attend to the demands of each relationship*. If we depend on our surroundings, do we want to poison them? Or weaken (e.g., through antibiotics) the essential internal organisms in our own bodies? Can we, without deep loss, torture or poison or eliminate the species we love or that give us solace and beauty and spiritual inspiration? Don't we need to examine the effects of our actions on others? And don't we want to think twice about hurting people through our actions toward nature? Killing strangers or family through pollution? Killing descendants through pollution? Don't we feel bad about ourselves when we look at the world and see what a mess we've made of it![30]

In short, attention must be paid. When we consider our surroundings, do we want them to be more "real" in the sense of being more impressive, more deeply connected to other settings, more providing of information?[31] If so, then we had better be careful—take a close look at the consequences—when we replace living trees with plastic ones, or warm-blooded dolphins by Disney, animatronic Flippers. To call the originals more "natural" may just be to say that as aesthetic objects and subjects of empathy, as sources of information about the larger world, and as supports of our own existence, they have a very different character from their replacements.

Of course, the Ethicist goes on, knowing even here he must be honest: sometimes we will opt for comfort and safety over information or spiritual value; we will screen out the mosquitoes, replace weeds with tomato plants, kill the bacteria in the cut finger, and try to keep the sharks out of the swimming area. Yet in neither case have we needed to invoke nature as a generality or to pretend that the dolphins and forests were never touched by people. We only had to attend to what is actually going on.

The Ethicist knows (for he has been taught by Modern Wisewomen[32]) that to understand these relationships we will have to pay close attention to our emotional responses to them. How do we feel when we experience what is being done *in* our relationships? Is there not a wisdom, a teaching, in the answer to that question? The Wisewomen were the ones who taught him that grief for the slaughtered species is evidence (information, even a "fact") about the depth of our connections to them.

(Here, our Ethicist muses, we must be careful. For he suspects that emotional responses *by themselves* tell us little except the most crucial of teachings: that attention must be paid. Emotions, like evidence of the senses, like claims of scientists, like arguments by philosophers, carry no certainty but only suggest a center of concern. While he has always been deeply critical of that disorder of feelings that consists of not having any, he also knows there is the danger that our immediate emotional responses might lead us to weep for the Very Cuddly Panda but ignore the Slightly Creepy Bat—whose role in local ecosystems is much more pivotal.[33] Just as we must work with any particular scientific program to see whether it embodies truth or domination, we must be willing to work with our emotional lives to see what vision of life, what emotional maturity, and what ethical or spiritual wisdom they support and express.)[34]

What this all comes to is that there is, the Ethicist realizes, no shortcut way to avoid considering the particular details of any relationship. In each case we have to inquire into motivations and consequences, which means we have to have some sense of why we do what we do and at least some access to critical faculties, which allow us to challenge the validity of motivation and consequence.

From where would this come? Well, to begin with, we have to be as open as possible to the Voice of the Other: to inquire into the effects of our actions on other human communities and on the nonhuman communities. Let's listen to what, for instance, indigenous peoples say about what is happening to the places they live; let us look, quite closely, at the coral reefs dying from warming oceans; let us examine in great detail the air-pollution-stunted lungs of inner city kids—or the water birds so grossly affected by toxins in the Great Lakes that they are born with organs outside their bodies.

And we need to look quite closely at ourselves. Personally, and as communities: why are we doing what we do, and what do we get out of it? Why is this animal being killed or this species made extinct? What is at stake when we introduce a new antibiotic and help breed ever more powerful bacteria? When this particular wetland is paved over, are we meeting a need of substance or simply giving a temporary fix to a form of life that will take more and more and more until nothing is left? (The Ethicist remembers the vacuum-cleanerlike man at the bottom of the sea in *Yellow Submarine*: he sucked up everything in sight until the whole world went down his tunnellike snout. The world was only saved when he sucked up himself.)

None of these reflections, the Ethicist muses (feeling like he has accomplished only the barest of beginnings), will solve the problem. But they will help us talk about what is wrong with the way we live without getting into the trap of saying the problem is that we are changing an otherwise-unchanging nature, or that we should isolate ourseves from "nature" and nature from ourselves, or that nature and people are either totally different or indistinguishable. Instead of appealing to these problematic notions, we have the difficult and complex task of examining each relationship in turn. If we actually look at the clear-cut forests, depleted fish stocks, or burning rain forest; if we face the effects of a polluted environment on breast cancer, prostate cancer, or dropping sperm counts; well, we might make some changes. If we could not, we would find out what stood in the way: a elite of powerful people who cannot be touched—and a widespread addiction to a form of life that is producing and consuming itself to death.

Images of a small elite and widespread addiction bring the Ethicist back to the notion of monoculture. He believes that it is in the nature of addictive processes to turn Every Other Thing into The Same: all possessions sold to buy one drug, all relationships sacrificed to placate the pusher. Environmentally, the complexity of the living forest becomes the straight lines of the tree farms or the barren clear-cut. Socially, the last tribes of the rain forest are transformed into a strange species of Christians wearing nylon windbreakers and despising the old shamans. At the same time, the power elites are wonderfully interdependent with the culture of addiction. Commanding their empires of industrialism and militarism, ecocide comes naturally to them.

Is it, the Ethicists wonders, really our destiny (as some Marxists would have it) to "humanize everything"—both our own nature and that of everything else? He knows many brilliant minds have pointed out that there is only a difference of degree but humanity's first use of sticks and the genetic engineering of overweight, arthritic pigs. Still, our Ethicist wonders if such claims do not smack a little of technological sophistry. And he for one would say "no" to a world that looks exactly like us. He will not appeal to God or Goddess, to Nature or Gaia, to say why he feels this way (although he could). Rather, he will simply tell people, in great detail, *what such a world would be like*. And he will do his best to get them to listen and pay attention to the information and pay attention to how they themselves feel when they get the facts.

You see, our Ethicist might say: it is not just that we go to, and alter, and conceptualize nature, and thus that our surroundings bear the imprint of language, society, and history. It is also true that *nature comes to us*; that it sets limits to the scope of social processes; and that every social project or conceptual innovation can only unfold within the boundaries of our relationships with beings who are not us. To say there is a relationship here is to say that there is more than One Being. Since we are in a relationship, it behooves us to take a look at that which we are relating to and not collapse into a solipsism that makes everything just another boring dimension of ourselves.

Were he confronted by one of the Brilliant, Ironic Intellectuals who brilliantly and ironically argues for the textuality, malleability, and sociality of Just About Everything, our Ethicist might return to the idea of monoculture, and point out in a little more detail the consequences of the Green Revolution in India's Punjab region.[35] There, with the best of intentions (rural development, preventing agrarian communist movements, and finding a market for fertilizer companies), engineered seeds, chemical fertilizer, pesticides, and vastly increased water needs were imposed. The result was toxic chemicals in the soil and water, destabilization of water usage, ruined land, and increased loss of crops to pests. And on the social side: farmers in debt and violent social movements prompted by social dislocation. Though the traditional humanization of the Punjab landscape had been a far cry from wilderness, it had followed certain basic principles of renewability, recycling, multiple uses, and the fulfillment of diverse needs. To say that we are natural beings or that there is nature may only (*only!*) be to say that we cannot flout those principles and expect not to suffer. To put it another way: While we may not always or easily know what is animal or what human, what is "natural" and what man-made, we can often. In some cases, genetic engineering may complicate matters (though do the overweight, arthritic pigs we have created really fool anyone?), but we still know that the pine tree is more "natural" than the wooden telephone pole and both more natural than the aluminum flagpole. Of course, all of them originate in the same universe, but if we pay attention to the relationships involved, we can easily see the difference between an old-growth forest, a tree farm, or a pile of metal poles. Confusing boundary cases should not cloud what we know; nor should we get so caught up in the rhetoric of the powers of science that we think it has mastered secrets that still remain far beyond its ken.[36]

To the detriment of his rest, our ethicist knows that it is not just Poetic Naturalists or Monoculturalists who do not pay the most careful attention to their connections with the nonhuman world. He must admit (if only to himself) that there are often deep flaws in his own capacity to identify or empathize with human and nonhuman Others alike. The endless busyness of daily life, the overwhelming scope of so much past and present suffering, the limited circles of care that keep people focused on family, career, nearby community: all these easily lead him (though not when he is giving his talks!) to ignore far-off rain forests, nonhuman species, or future generations.

More troubling, the Ethicist has to admit that a generalized acceptance of a less anthropocentric ethic, a sense of our "deep ecological" connections to other species and entire ecosystems, does not erase the fact that in the ordinary course of our existence human beings necessarily use and displace other life forms. Just like every other animal, we need to consume and alter our environment. We must eat, live, warm ourselves, and utilize natural resources to create social forms and cultural institutions. And when collective or personal goals are threatened, we strike back. For example, a newfound love of life will not extend to the AIDS virus; and our wariness at tampering with the sacred character of nature may well be suspended when it comes to genetic engineering to cure cystic fibrosis. Ghetto rats will probably escape the purview that holds all of life as sacred, as might the black flies that cause widespread blindness in Africa.[37] The end of anthropocentric values will not eliminate the hard choices we face—choices about how much to take for ourselves and how much to leave for others; how much to exercise the control we increase day-by-day, and how much to surrender.

Can He Give Peace a Chance?

Our Ethicist can hope that an awareness of the ecocidal violence caused by our personal desires, cultural values, and social institutions would help us develop a more modest and circumspect attitude toward our enterprise of dominating the Earth. We could be more unassuming both in our self-assessment as a species and our desires for a better life—even though this self-assessment might require a difficult and contentious social transformation. We may learn that a truly "higher standard of living" cannot be achieved until we curtail our current environmental aggression; and that no amount of toys will cure our loneliness for both natural and human community. It is only by dwelling repeatedly on this last slim hope that the Ethicist manages to get a few moments of sleep before his overfilled day begins.

Yet what will happen when he tries to sleep tomorrow night?

Ethical humility, political critique, the overcoming of anthropocentrism, and the constraints of personal and social limitations—all these run through the brain of our troubled Ethicist.

Is he condemned forever to moral insomnia?

To paraphrase (this time with even greater license) the same teacher I did at the beginning: perhaps he—perhaps all of us—*should* be awake in the middle of the night, should be self-consciously self-conscious when we try to be moral teachers, to be direct and passionate in our condemnation of injustice against people and nature, and to wonder, over and over again, when people should come before the rest of creation.

Personally, our Ethicist often finds it nearly impossible to escape his own sense of failure. How little he can do! How little he can change his own life! Perhaps this self-torture stems from illusory models of male omnipotence. Possibly his brain is still addled by images of his childhood

TV cowboy heroes, heroes who ride into troubled towns and set things right before the last commercial message.

Again, it may be that these ruminations on his own failings merely obscure (usefully, in the short run at least) the depths of his own grief. Better to blame himself than to admit how hopeless it all seems.

Whatever the origins or psychic functions, it is a sad business: because our Ethicist is a decent person and deserves to be at peace with himself. ("Even a poor Ethicist," we might paraphrase, "deserves some happiness!")

Can there be another possibility for our Ethicist? Well, there is no way he can fully elude his grief for Gaia; the troubling paradoxes of his own contradictory behavior; or his rage against the perfunctory cruelties of modern societies. But alongside those painful narratives another may well be possible. Not as a replacement, but as a solace; not as ultimately *more* true, but as *equally* true.

In this narrative our self-consciousness, passion, humility, and doubt coexist with a deepened sense of communion in the mysteries of creation. Encircling rather than erasing the legitimate strenuousness and angst of the modern moral subject—whose trademark isolated ego literally feels the weight of the world on its shoulders—is a more encompassing and calming vision. As our love for other people and nature animates our passion and humility, it can also at times soothe our troubled and anxious ethical selves—if, that is, we recognize such selfhood as not bounded by our own troubled ego but as in dancing partnership with other people, with maple trees and rivers, stars and moons. The love that animates our passion is not only a burden of duty but also a force that wells up out of the depths of our utter connection with all that surrounds us. In his celebration of this profound kinship, our Ethicist can suspend his hopeless guilt and hopeless rage. He might realize that he can only do his bit and then, like an autumn leaf, drift down to the forest floor to prepare the way for other growing things. Just as the sun itself, all any of us can do is shine until the light that god or goddess gave us is darkened by the soft embrace of time—an embrace that, in the end, will claim all beings. We can choose a narrative of bitterness, or one of delight. Or, as in the most likely of scenarios, we will end up telling both. The Ethicist need not suppress his fears for tomorrow, or his guilts and regrets for yesterday. But he may also welcome the peace that comes from his own at-once joyous, sorrowful, and exuberant participation in eternity.

Notes

Thanks to Miriam Greenspan and audiences at Yale University and the Boston Theological Institute Conference "Envisioning Equity" for helpful responses to earlier versions of this essay.
1. The style here mimics Kierkegaard in *Fear and Trembling*.
2. Bill McKibben, *The End of Nature* (New York: Times Books, 1989).
3. E. Chivian et al., eds., *Critical Condition: Human Health and the Environment* (Cambridge: MIT Press, 1993).
4. Raymond Bonner, *At the Hands of Man* (New York: Knopf, 1993).
5. Robin Broad, *Plundering Paradise* (Berkeley: University of California Press, 1993).
6. For an account of a contemporary struggle, see Joe Kane, *Savages* (New York: Knopf, 1995).
7. Jurgen Habermas, *Legitimation Crisis* (Boston: Beacon Press, 1978).
8. Vandana Shiva, *Staying Alive: Women, Ecology and Development* (London: Zed Books, 1989).
9. This analysis comes from Vandana Shiva, *The Violence of the Green Revolution* (London: Zed Books, 1991), and *Monocultures of the Mind* (London: Zed Books, 1993).
10. For a survey, see Roger S. Gottlieb, "Spiritual Deep Ecology and the Left: An Attempt at Reconciliation," *Capitalism, Nature, Socialism: A Journal of Socialist Ecology* 6 (3) (September 1995).
11. David Abram, *The Spell of the Sensuous* (New York: Pantheon, 1996).
12. Richard Nelson, *The Island Within* (New York: Random House, 1989) p. 93.
13. Ibid., p. 89.
14. E. O. Wilson: *Biophilia* (Cambridge: Harvard University Press, 1984), pp. 12–13.
15. David M. Graber, "Resolute Biocentrism: The Dilemmas of Wilderness in National Parks," in Michael E. Soule and Gary Lease, eds., *Reinventing Nature? Responses to Postmodern Deconstruction* (Washington DC: Island Press, 1995).
16. Ibid., p. 130.
17. See Noel Perrin, "Forever Virgin: The American View of America," in Daniel Halpern, ed., *On Nature: Nature, Landscapes and Natural History* (San Francisco: North Point Press, 1987).
18. Edward Abbey, *The Journey Home: Some Words in Defense of the American West* (New York: Penguin, 1977), p. 145.
19. These last paragraphs are much indebted to Graber's brilliant essay.
20. Wilson, *Biophilia*.
21. Paul Shepard, in works such as *Nature and Madness* (San Francisco: Sierra Club Books, 1982).
22. References are to Aldo Leopold and J. Baird Callicott.
23. Of course, Martin Heidegger.
24. See Susan Bordo, "The Cartesian Masculinization of Thought," *Signs* 11 (3) (Spring 1986); Karen Warren, "The Power and Promise of Ecological Feminism," *Environmental Ethics* 12 (Spring 1990).
25. Michael E. Soule, "The Social Siege of Nature," in Soule and Lease, *Reinventing Nature?* pp. 150–53.

26. Shiva, *Monocultures*, p. 55.

27. For another example, consider Barry Lopez's observations (*Of Wolves and Men* [New York: Scribner's, 1978]) that wildlife biologists did all sorts of research into wolves without consulting native peoples who had been coexisting with them for centuries. The wolves as represented by scientists and by those who live with them, notes Lopez, were very different.

28. How similar this is to the political realm. Just because someone's door has the label "champion of the people" does not mean the person is not aspiring to the role of petty tyrant.

29. It is as if we said: the problem with killing that man is that you've assaulted nature! But surely we can make much stronger claims than that. For example, we can say that you have, for no justifiable reason, deprived this man of life; that he had a wife and children and friends to whom you have caused great and unnecessary unhappiness, etc.

30. Compare this list with Stephen R. Kellert's account of the nine values that nature has for us: *The Value of Life* (Washington, DC: Island Press, 1995). Also, one could easily map the different schools of environmentalism (conservationism, social ecology, etc.) onto this list by seeing which of these relationships forms their principle concern.

31. Albert Borgman, "The Reality of Nature and the Nature of Reality," in Soule and Lease, *Reinventing Nature?*

32. Joanna Macy and Miriam Greenspan.

33. See Diane Ackerman's study of bats in *The Moon by Whale Light* (New York: Vintage, 1991).

34. For the view that emotions are essentially sources of prejudice and thus have no role in arguments about animal rights, see Evelyn Pluhar, *Beyond Prejudice* (Raleigh, NC: Duke University Press, 1995). Pluhar's mistake is to think that emotions are like conclusions to arguments or like theoretical principles. They are, rather, like information or data: They must be worked up into a framework, but should not be ignored.

35. See Shiva, *The Violence of the Green Revolution*.

36. This is directed against the "cyborg" arguments of Donna Haraway.

37. Sue Hubbell, *Broadsides from the Other Orders: A Book of Bugs* (New York: Random House, 1993), pp. 74–89.

8

Imperialism and Environmentalism

Eric Katz

On the evening of December 10, 1992, the Northeast coast of the United States began to experience a violent winter storm, a Nor'easter, that was described by many officials and long-time residents as the worst storm in forty years, perhaps the worst in a century. The storm had already dumped up to five feet of snow across the states of Pennsylvania and Maryland, and it hit the New York metropolitan area with driving rain and wind gusts up to ninety miles an hour. This storm was not a hurricane, for it did not originate in the tropics, but its effects were much more acute than those of the most vicious hurricanes of this century. One reason for the severity of the effects of the storm was that it lasted for four days, through eight high-tide cycles. The shoreline communities of New Jersey, New York, Long Island, and Connecticut were devastated: seawalls collapsed; streets were flooded; bridges, tunnels, and railroads were closed; and beaches and dunes were washed out to sea. Within a mile of my summer home on Fire Island, over twenty houses were destroyed by the ocean tidal surges. In the twenty-five years that I had visited or lived on that barrier beach, I had never seen one house wash away.

I begin with this description of the so-called storm of December 1992 because it is, for me personally, the most direct representation of the power of the natural world. Whatever else nature may be, it is a complex combination of forces that exerts its various powers throughout the entire

Copyright 1995 by *Social Theory and Practice*, Vol. 21, No. 2 (Summer 1995).

physical world. In this essay I explore an idea—imperialism—as a way for understanding the relationship of power between nature and humanity. My working assumption is that the expressions, metaphors, and representations of the power of nature that we choose determine our normative stance to the natural world. When we adopt an antagonistic attitude, as I believe we do, we interpret the natural world as an evil force that must be conquered. We view nature in terms of force and domination, as an objectified "other" that must be controlled.

The use of the concept of imperialism in conjunction with both the policy of environmentalism and the forces of nature requires an explanation. In brief, I am consciously trying to expand the normal use of the concept of imperialism—indeed, one might say, to use an appropriately violent turn of phrase, that I am attempting to explode our normal concepts, especially those concerning the relationship between humanity and the natural environment. In this essay, I consider the idea of imperialism—and all that it represents concerning power, force, and domination—as a model or metaphor for understanding the human relationship with nature. Now a metaphor of imperialism and power is rather different from the benign and optimistic metaphor of "the balance of nature" so often used by environmental philosophers, scientists, activists, and policy makers. So one purpose of this philosophical examination into the power of nature is to open a dialogue about the forms of rhetoric that help to determine environmental policy. Which metaphors or models of the human/nature relationship are more appropriate? Should we view nature as a complex of aggressive forces, or as a balanced system of cooperation and harmony? Why should we choose one metaphor over another? Should we even make one exclusive choice? Why not use both metaphors —the benign and the violent—or others that may cross our minds? The metaphor we choose in a particular situation will have a fundamental impact on our ideas concerning the appropriate role of human action in the natural world.

In this essay, the terms "nature," the "natural world," the "natural environment," and "natural processes" are used as roughly equivalent to that part of the world that lies outside human activity or human culture. Although "nature" can be used in a very broad sense to mean the natural or physical laws of the universe, I am not using that definition in this argument. The broad definition of nature as physical law does not exclude any physical phenomenon from being labeled as natural—such a characterization is virtually useless in moral philosophy or political theory, where distinctions must be made between various kinds of human and natural activity. My meaning of nature is more restrictive, for it only concerns the ecosystems, bioregions, and undeveloped habitats of the Earth— and the entities and processes that constitute these physical spaces and systems. In brief, I use the term "nature" to mean the ecosystemic processes of the Earth.

It is essential for my argument that imperialism represent more than the traditional idea of national and cultural conquest. It is more than one nation or community expanding its influence onto a less dominant nation-state, culture, or community. Imperialism is a form of domination—almost any instance of domination—in which one entity uses or takes advantage of another entity for its self-aggrandizement, to increase its power, its life, its comfort. Imperialism is thus an exercise of power. An entity uses its power to alter the world, to influence other entities, so that its own interests may be satisfied. Imperialism is, as it were, the policy that seeks to satisfy interests. It is in this sense that Bertrand Russell said that all living beings are imperialists;[1] similarly, William Leiss speaks of the "imperialism of human needs."[2] In philosophical terminology, imperialism results from any successful attempt to exert power over the "other."

This sense of imperialism is connected to environmental policies because the relationship of power need not be located in the exclusive domain of human beings, human institutions, human nation-states. Imperialism expresses a relationship of power and domination, but the entities on either end of the power relationship can be either human or nonhuman. Nature—natural entities and systems—can be the "other" that is dominated by humanity, or it can be the dominating and imperialistic force that subdues some aspect of humanity. Logically, there are four basic possibilities:

1. Imperialistic humans exercise power over other humans; for example, the colonization of the Americas.

2. Imperialistic humans exercise power over nature; for example, the practice of agriculture.

3. Imperialistic nature exercises power over humans; for example, the destruction of beach houses in the December 1992 storm.

4. Imperialistic nature exercises power over other nature; for example, the process of forest succession.

One immediate objection to this logical categorization is that it is based on an abnormal use of the idea of nature. Nature may be the "object" of a power-relation, in which case humans are the imperialists dominating a nonhuman "other," but nature cannot be the subject of an imperialistic power relation. To regard nature as the subject exercising power, as in the third and fourth categories, is the error of anthropomorphism, seeing nature as a being with human qualities.

Although there is some truth to this objection, it is not fatal to my overall project or argument in this essay. As I noted above, I am attempting to revise the traditional ways in which we think about nature. I want to consider nature as the subject of an ongoing history, a subject that acts and develops in ways that are comparable—in some moral or axiological sense—to human action. As Colin Duncan has claimed, "While Nature is certainly not a person . . . it does have some of the attributes of a Hegelian subject. It can be both victim-like and agent-like."[3] This metaphor may be a form of anthropomorphism, but it is different from the fully developed anthropomorphizing of nature found, for example, in Greek mythology. I am not attributing to nature the emotions, intentions, and reasoning powers of human subjects; I am only suggesting that it is incorrect to think of nature as a mere object, a passive recipient of human activity.

Here we can also see how important it is to use the restricted meaning of nature introduced above—nature as the ecosystemic processes of the biological world. If we are to have a sense of nature as a subject or agent, acting upon humanity and human institutions, then it is imperative that we not view nature as simply the totality of physical laws of the universe. The laws of the universe do not *act on* anything, they are merely the *descriptions* of the ways in which the entities and systems of the universe operate. The idea that natural systems—or some parts of the natural world—act like imperialists, dominating other entities or systems, requires a sense of nature as a distinct system, able to stand in a relation with an "other." It is thus useful to use the terminology of Lynton Keith Caldwell, who distinguishes the "Earth of nature" from the "world of man." The difference, he claims, is obvious: "the Earth is a creation of the cosmos and is independent of man. The world is a human artifact; it is a conceptual creation of human experience and information." It so happens that in the present epoch, the human world is "geographically congruent" with the Earth of nature,[4] but this does not diminish their ontological distinctiveness. Nature is the sum total of nonhuman, biological, and evolutionary ecosystemic processes of the Earth's physical system.

But this raises a second objection: the idea that "nature" and "humanity" are opposite points in a power relationship tends to reinforce the very separation of humanity and nature that an enlightened environmental policy seeks to overcome. It is surely one goal of the environmental movement to end the common belief that humans are separate from natural processes—to instill the ecological idea that humans are an interdependent part of the natural system, requiring a well-functioning natural environment to survive. But the notion of interdependence itself requires, at least conceptually, the idea that there are separate entities that are, in fact, interrelated. To say that humans are connected to natural processes requires an idea of a distinct human presence, a distinct human ontology, that is nevertheless dependent on the "otherness" of the nonhuman natural world. It is not a mistake to claim that humans can dominate nature or that nature can dominate humanity, even though they are inseparably related. As Don E. Marietta Jr. once stated to me in conversation, the "sugar can dominate the pudding"—even in the case where the blending of distinct elements forms a harmonious unity, one element can dominate the whole.

This essay will not examine all four categories of imperialism, because they are not all of equal interest. The fourth type, nature's domination of nature, simply does not concern us, since we humans are not parties to the dispute, except indirectly. But what is intriguing about this schema is that almost all discussions of imperialism, even discussions of environmental imperialism, involve the first type, the domination of one human group by another. Environmental imperialism is generally seen as the alteration of natural systems by one human group as it colonizes and subsequently dominates or subdues another human group. Thus the European colonizers of the Americas used agriculture and a new conception of private property to modify the ecology and environmental context of the indigenous American peoples. In this first type of environmental imperialism, the effects of human activity on the natural world, although important, are indirect, secondary, or derivative. The basic form of imperialism is understood to be an expression of power between two human societies.

Although an analysis and understanding of this type of imperialism is clearly important—as the work of Cronon, Crosby, and Merchant demonstrates[5]—I believe that we should also consider the second and third types of imperialism, the direct relationship of power between humans and nature. At the end of this essay I will comment briefly on the third form of imperialism, nature's power over human civilization, but I want to consider primarily the second form of imperialism, the human domination of nature, because this is the form over which we have some control, and hence, the one for which we have some responsibility.

The human domination of nature is, of course, an old topic, even within the field of environmental philosophy: twenty-five years ago William Leiss's analysis of the influence of Francis Bacon was entitled *The Domination of Nature*.[6] Clearly, the idea that humanity has exerted a fundamental influence on natural processes has become a standard and noncontroversial datum of the environmentalist worldview. One need only read the introductory chapters of Al Gore's *Earth in the Balance* to see how the idea of human domination has permeated the mainstream consciousness of the environmental movement.[7] But to what extent is this domination a form of *imperialism*? What is the relationship of power between humanity and nature? What is the philosophical meaning—ontological and axiological—of this relationship?

Humanity has done more than influence natural processes. It has done more than "tame" wild nature for the increased comfort, wealth, and power of human beings. My central thesis is that humanity has attempted to modify and to mold natural processes for the satisfaction of human interests, to create an artificial or artifactual world that produces the most benefit for human beings. This artificial world at best resembles nature, where it is convenient. It is the basic policy of human civilization—even where that policy is unarticulated—to modify or to conquer the natural world, to subdue Nature for the furtherance of human good. This policy has been the central project of western civilization since the Enlightenment, and although it has proven in many respects to be a failure, its fundamental meaning and motivation is clear: the primary goal of western civilization, especially western science and technology, has been the control and domination of nature for the promotion of human benefit—the human imperialism over nature.

The central meaning of western civilization regarding human activity in the natural world is that human interests, the maintenance and improvement of human life, lie at the center of all value determinations. Human progress is the purpose of all human activity. Nature and natural resources are worthless, mere objects for exploitation by the dominant human species, until they are transformed, through human labor, into cultural instruments for human betterment. The analogy with traditional notions of imperialism is obvious. Humanity conquers and colonizes the natural world, with the moral justification that human good is the only determination of value, just as one nation-state might conquer another, judging its own interests to be supreme.

One chilling example of the goal of the human imperialistic domination of nature can be found in a classic essay in the field of environmental ethics, Martin Krieger's "What's Wrong with Plastic Trees?"[8] Krieger argues that if the goal of social policy is the maximization of human satisfaction and the promotion of social justice, then there is no overriding reason to preserve nature. An adequate technology, combined with a proper education through mass culture, could create an artificial plasticized world that would produce more happiness for the human population, at a lower cost, than the preservation of the natural world. Krieger writes: "There is no lack of merit in natural environments, but this merit is not canonical,"[9] and he concludes that "artificial prairies and wildernesses have been created, and there is no reason to believe that these artificial environments need be unsatisfactory for those who experience them."[10] Thus he proposes an environmental policy of "responsible interventions" to mold both natural processes and the popular human responses to them. The goal of this new interventionist environmental policy should be the maximization of human satisfaction and social justice.

Krieger's vision of a plastic artificial world designed for the maximization of human happiness is perhaps the most extreme case of human imperialism over the natural world. The terminology of "responsible interventions" even echoes the justification for the use of military force in U.S. foreign policy. Under Krieger's model, the natural world is not only subjugated to meet the demands of the dominating force (that is, humanity); it is also virtually destroyed, replaced by a more docile, cooperative system of entities. The natural world becomes a mere pacified colony of the invading forces of humanity; aspects of the natural world that further human interests are preserved, or even augmented, while the aspects that interfere with human interests are destroyed, replaced, or exiled.

Is Krieger's theory of environmental policy merely an extreme case? If it were, then we could probably ignore it, for as an extreme it would not have much influence on mainstream environmental policy or philosophy. But Krieger's imperialistic vision has surprisingly strong echoes in other theorists of environmental policy. Chris Maser, for example, is a former research scientist for the U.S. Department of Interior Bureau of Land Management, who argues in favor of a policy of "sustainable forestry" in *The Redesigned Forest*.[11] Maser is a critic of the short-term expediency of present-day forestry practices. He attacks the plantation mentality of forest management, and argues for a forestry policy that "restores" the forest as it is harvested.

I have analyzed and criticized Maser's arguments for restoration elsewhere and will not repeat the full discussion here.[12] In brief, Maser's model of the relationship between humanity and the natural world reinforces a notion of human imperialism over nature. Maser sees the natural world as a mechanism with a design that can be discovered and manipulated by human science and technology. At one point, he compares the mechanism of a forest with an automobile as he argues empirically against the substitution of an artificial forest for a natural one: "A forest cannot be 'rebuilt' and remain the same forest, but we could probably rebuild a forest similar to the original if we knew how. No one has ever done it. . . . [W]e do not have a parts catalog, or a maintenance manual."[13] The implication is that if we did have a catalog and manual, then the restoration and redesign of forests would be practically and morally acceptable.

For Maser, the ultimate motivation for forest redesign is the furtherance of long-term human interests. Maser argues that the central problem with current practices is that they are "exclusive of all other values except the production of fast-grown woodfiber."[14] What concerns Maser is the elimination of other human values and interests. "We need to learn to see the forest as the factory that produces raw materials" to meet our "common goal[:] . . . a sustainable forest for a sustainable industry for a sustainable environment for a sustainable human population."[15] By using the complex knowledge of forest ecology, restorationist foresters will redesign forests to better achieve human purposes.

Maser is an interesting case because, unlike Krieger, he is a committed environmentalist, indeed a hero to enlightened foresters and government officials dealing with practical environmental policy. But a philosophical analysis of his views clearly reveals the perspective of the human imperialism over nature. The fundamental goal for Maser is the proper use of the natural world to meet human needs—this is the basic motivation for any imperialism. His only quarrel with current forestry practices is in the methodology or the style of the domination.

What then is wrong with the human imperialism over nature? Is the consideration of this type of imperialism a normative issue? Is not the human domination of nature necessary for human survival and human progress? Or is this imperialism a kind of moral evil, perhaps analogous to the human domination of other humans?

Consider two problems or reasons why the human imperialism over nature is morally suspect. First, imperialism subverts or destroys the value of the subjugated entity. The value of the subjugated entity may be either good or bad relative to human interests—for example, a good rain forest or a bad disease organism. But regardless of whether the value is good or bad, the action of human imperialism or domination changes the pre-existing status. When we replace the natural world with an artificial human substitution, we also replace a particular complex of values. The act of replacement itself imparts a new reality and a new set of values. The new system may appear to be similar to the original natural system, but it will be, in actuality, a human artifact. It will be valuable, and indeed useful, for the human goods it furthers or promotes. If the new system were not at least thought to be valuable and useful for human interests, then we would not bother to create it in the first place—one is not an imperialist over a worthless entity. But at the same time that we acquire the benefits of the new subjugated system or entity, we will lose access to the values of the original system or entity—we will lose the value of the undisturbed natural world.

But is there an undisturbed natural world? Is there a nature that is independent of human power and human activity? The idea that human imperialism causes a loss to the value of the natural world raises important questions regarding the very meaning of nature itself. Thus, we can see that my original definition and discussion of the meaning of nature—as those ecosystemic processes independent from human culture and civilization—was clearly provisional; any detailed analysis of the meaning and value of human interaction with the natural world is forced to address the problem of the meaning of nature. One question results from the realization that the human impact on the natural world is so pervasive that nature as a separate system no longer exists. Is there a pure natural value that should be protected? A popular treatment of this theme can be found in Bill McKibben's *The End of Nature*,[16] where McKibben focuses on the fact that humanity has changed the chemical composition of the atmosphere. Since the atmosphere literally touches all of nature, once it has been changed then the physical interactions that make up the natural system have also been changed. It can be argued, then, that no independent nature exists, and a concern for the value of a pristine, untouched nature is a product of a misplaced or overly sentimental nostalgia.

McKibben also suggests another question when he writes: "When I say 'nature', I mean a certain set of human ideas about the world and our place in it."[17] Nature is not only a system of physical entities that has been subjected to pervasive human influence, it is also a set of human ideas for describing and categorizing the world. Nature is a human idea. The models by which we describe the world of nature are human constructs; our theories of nature are themselves artifacts serving human ends. This, then, is another answer to the question of what nature is—it is a human construct. We human thinkers choose the model, metaphor, or idea of nature that will best serve our purpose—just as I have chosen, in this essay, to view nature as the total set of ecosystemic processes. Thus nature is the victim of another form of human imperialism, what we can term an epistemological imperialism, for nature's very meaning is determined by human thought. The *concept* of nature, as well as its physical processes, is dominated by human thought and practice. An analysis of this domination is clearly a normative issue.

The second normative problem with the human imperialism over nature concerns the third form of imperialism, the power that nature exerts over humanity. The plain fact is that humanity has not been entirely successful in exerting its influence over nature and natural systems for the furtherance of human good. This lack of success is apparent no matter which view of nature we take to be most relevant, for humanity has not succeeded in changing the physical laws of the universe, nor has it been successful in recreating functioning natural ecosystems. True, we have been quite effective in destroying natural systems, but our efforts at replacing them with useful human-friendly systems have been nearly catastrophic. One interpretation of the current environmental crisis is that it represents nature's backlash—its counterinsurgency—against the forces of human colonization. It is here that we recognize the awesome power of nature to reassert its independence from human attempts to control it. The ecosystemic processes that constitute nature react against human activity, in part by reducing our domain and dominion.[18]

The idea of nature's reaction, its counterinsurgency, gives a context and meaning to the fact of the beach houses being swept away by ocean tides and a hundred-year storm. Nature's power over humanity reveals the depth of the human arrogance that builds homes on sand. But the same point of view applies to our use of pesticides, fertilizers, nuclear energy—the list of nature's backlash is endless. So the second reason that the human imperialism over nature is a moral problem is that it cannot answer the response of nature's domination of humanity. Our attempts to dominate nature reveal a disturbing lack of prudence, to say the least. Despite our technology and science, nature is more powerful than human civilization.

Several conceptual issues invite further discussion. First is the importance of *intentionality* in the analysis of domination. It appears that using the notion of imperialism to describe the acts of nature is problematic, whether these acts are directed toward humanity (the third type) or toward other natural entities (the fourth type). Nature's actions do not have intentionality. When the December storm washed away beach houses on the northeastern coast of the United States it was not the intention of nature to exert its power over humanity. So my notion of imperialism clearly broadens the traditional use of the concept, not only in the ideas of the proper subjects and objects of imperialism and domination, but in the basic structure of the imperialistic act. The moral analysis of domination may not require the existence of intention.

It should be noted, however, that not all intrahuman acts of domination require intentionality either. If racism and sexism are considered as forms of imperialistic domination (as I would so judge them), then there do exist instances of so-called institutional racism or sexism that are unconnected to specific or direct intentional actions. A person who is not a racist can act in a manner that promotes racism, for example, by following the procedures of an institution that unknowingly is structured to discriminate; thus some hiring and promotion requirements at a university may be racist (or sexist) in an institutional sense. The point is that domination can be unintentional.

The focus on intentionality becomes relevant when we consider our moral evaluation of actions and their consequences. Since humans have, in general, control over the activities that they instigate in their interactions with the natural world, it is easy to place blame or praise on the activities chosen. But it is senseless to blame nature for the destruction of beach houses or for a forest succession. Without the possibility of intention, the moral evaluation of the acting subject becomes impossible.

A related issue would be the entire complex of problems involving the teleological structure of natural systems. Do natural entities and systems lack intentionality while at the same time displaying purposive activity? Can purpose exist without intentionality? Is purposive activity all that is necessary for the concepts of imperialism and domination? Unfortunately, I do not have answers to these questions, which involve fundamental issues in the philosophy of biology and perhaps in the foundations of metaphysics. I raise the issues here only because I want to demonstrate the significance of the discussion over imperialism and environmentalism; I seek to provoke a far-ranging debate.

The final conceptual issue is, perhaps, the most important question raised by the preceding discussion. All four types of imperialism—but especially the second and third—reveal the necessity of the exercise of power over other entities. Humans must try to control some aspects of the natural world for their very survival; nature must respond as it asserts its independence from human design. The essential issue is then one of determining the proper scope of intervention and interaction. When are humans permitted to intervene in natural processes, and when should they refrain? When does an intervention become an attempt at domination? After all, humans must eat and grow food, must build houses, roads, cities, must cure disease. Are these all acts of imperialistic domination that are morally suspect?

The answer to this problem lies in determining the proper metaphor, the appropriate rhetoric, for understanding the human relationship to nature. When one nation-state interacts with another nation-state in a spirit and a structure of mutual benefit and partnership, then it cannot be considered an act of imperialism or domination. Not all interactions are instances of unequal power relationships; to use a biological concept, some actions are symbioses. Thus, to understand whether or not a human intervention into natural processes constitutes an act of imperialism, we must consider the model or metaphor under which the act is structured. Do we humans seek a balance with nature, a type of partnership, or a power relationship of control and domination? Are our agricultural processes, for example, organic, working with natural processes, or are they highly technological, seeking control through artificial fertilizers and pesticides?

To discover a criterion for the proper intervention in the natural world is the fundamental task in determining the real meaning of environmental policy. Only with a morally justifiable criterion will human activity within nature—whether this be political action, government environmental policy, resource management, ecological restoration, or wilderness and habitat preservation—avoid the stigma of domination. By placing this problem in the framework of imperialism and domination, I am suggesting that it is the broadest possible issue regarding human action. To be morally justified, all human activity, even that between humans, requires a standard of appropriate intervention. The determination of that standard is the central question of moral philosophy.[19]

Notes

1. Bertrand Russell, *An Outline of Philosophy* (New York: New American Library, 1974), p. 30. Cited in Holmes Rolston III, *Environmental Ethics: Duties to and Values in the Natural World* (Philadelphia: Temple University Press, 1988), p. 326.
2. William Leiss, "The Imperialism of Human Needs," *North American Review* 259 (4) (1974): 27–34.
3. Colin A. M. Duncan, "On Identifying a Sound Environmental Ethic in History: Prolegomena to Any Future Environmental History," *Environmental History Review* 15 (1991): 8. See also Eric Katz, "The Call of the Wild: The Struggle Against Domination and the 'Technological Fix' of Nature," *Environmental Ethics* 14 (1992): 265–73; and Rolston, *Environmental Ethics*.
4. Lynton Keith Caldwell, *Between Two Worlds: Science, the Environmental Movement and Policy Choice* (Cambridge: Cambridge University Press, 1990), p. 39.
5. See William Cronon, *Changes in the Land: Indians, Colonists, and the Ecology of New England* (New York: Hill & Wang, 1983); Alfred W. Crosby, *Ecological Imperialism: The Biological Expansion of Europe, 900–1900* (Cambridge: Cambridge University Press, 1986); and Carolyn Merchant, Ecological *Revolutions: Nature, Gender, and Science in New England* (Chapel Hill: University of North Carolina Press, 1989).
6. William Leiss, *The Domination of Nature* (Boston: Beacon, 1974).
7. Al Gore, Jr., *Earth in the Balance: Ecology and the Human Spirit* (Boston: Houghton Mifflin, 1992). Gore writes of the "assault of humanity on nature," of human domination of nature and the power of humanity as being equivalent to natural forces.
8. Martin Krieger, "What's Wrong with Plastic Trees?" *Science* 179 (1973): 446–55.
9. Ibid., p. 451.
10. Ibid., p. 453.
11. Chris Maser, *The Redesigned Forest* (San Pedro: R&E Miles, 1988).
12. See Eric Katz, "The Big Lie: Human Restoration of Nature," *Research in Philosophy and Technology* 12 (1992): 231–41, "Artefacts and Functions: A Note on the Value of Nature," *Environmental Values* 2 (1993): 223–32, and "The Call of the Wild."
13. Maser, pp. 88–89.
14. Ibid., p. 94.
15. Ibid., pp. 148–49.
16. Bill McKibben, *The End of Nature* (New York: Random House, 1989).
17. Ibid., p. 8.
18. Duncan, p. 9, for the notion of reducing our domain.
19. An earlier and much shorter version of this paper was read at the Eighth Annual Technological Literacy Conference of the National Association for Science, Technology, and Society, Washington, DC, January 16, 1993. I am grateful to Paul Durbin, David Rothenberg, and Chris Sellars for their comments on that earlier draft. Another version of this paper was presented at a Working Session of the Society for Philosophy and Technology Biennial Conference on the theme Technology, Nature, and Cultural Diversity, in Peñíscola, Spain, May 21, 1993. I am grateful for the comments received at that time, especially from Phil Shepard, Eduardo Sabrovsky, Andrew Light, and Laura Westra. This work is supported, in part, by the New Jersey Institute of Technology under Separately Budgeted Research Grants for the academic years 1992–1994.

9

Habermas and the Ethics of Nature

Steven Vogel

I

Central to the critique of "enlightenment" developed by Horkheimer and Adorno (and taken up by others such as Marcuse as well) in the classical period of the Frankfurt School was the claim that enlightenment was intrinsically bound up with the "disenchantment" and "domination" of a nature viewed solely as an object of instrumental manipulation. Indeed, the Frankfurt School's critique of contemporary society was offered up in a certain sense "in nature's name"—both that of the damaged inner nature of humans stuck in the fatal dialectic of enlightenment and an outer nature robbed of all qualities save those that render it amenable to human use. Thus the eloquent call for social change in these authors is as much a call for the "liberation of nature" as it is for the liberation of humans.[1] This theme of a reconciliation with nature is a familiar and evocative one, and clearly one that has influenced recent environmentalist philosophy (including deep ecology).

One of Habermas's achievements has been to raise far-reaching objections to this entire line of thought. His sustained investigation of the question of the normative foundations of social theory has revealed the irresolvable dilemmas entailed by the sort of naturalism that so entranced the earlier generation of critical theorists. Social critique cannot take place in nature's name unless those who engage in it have some prior access to an "objective" account of what nature is and what nature requires; but it is just this access that the entire epistemological tradition in which the Frankfurt School was embedded had shown to be impossible. Authors such as Marcuse and especially Adorno were constantly struggling to find a suitable way both of acknowledging the inevitable sociality and historicity of what we experience as "nature" and at the same time of claiming to have found some paradoxical mode of access (art? mimesis? negative dialectics?) to that not-yet-socialized (and so not-yet-dominated) nature with which reconciliation was to be effected. Habermas's point was that this paradox could not be resolved, and that it bore a connection to the embarrassing difficulties Critical Theory faced in justifying its own normative claims. By pointing out the internal relations linking critique and normativity to communication, he has convincingly argued that those claims can only be justified within the social realm itself and not through any utopian appeal to "what the poor earth would like to be."[2] Concepts like "domination," "liberation," and "reconciliation" all possess normative content tying them uniquely to relations between potential members of a communicative community, and so applying them to nature simply involves a category mistake.

Persuasive though this is, there yet remains an uneasy ambiguity about nature throughout Habermas's work, and in particular about its relation to the social realm. In his writings of the late 1960s, nature appears only as the object of instrumental action and is explicitly excluded from the realm of (communicative) interaction. Later on—partly in response to postempiricism—he admits a discursive element to the natural sciences that he had earlier taken as exclusively associated with the instrumental, but still speaks as though the lifeworld experiences in which we interact with objects are somehow separate from the communicative realm. And in *The Theory of Communicative Action*, nature is hardly to be found. Indeed, it is not clear where nature is left after the turn proposed in that book from the "philosophy of consciousness" to a theory of communicative action takes place. Does the paradigm shift Habermas suggests mean that (to use somewhat misleading language) now only subject-subject relations are a fit matter for philosophical examination, and that subject-object relations fall by the wayside? Or does it mean that subject-object relations too require a communicative analysis?

Now Habermas's views of nature, and his critique of the naturalism of his predecessors, have certainly been the subject of sharp criticisms from ecologically minded points of view. A number of authors, including Joel Whitebook and Henning Ottmann, have objected to the impossibility within his system of conceptualizing a reconciliation with nature or of criticizing contemporary relations to nature.[3] Indeed, they have argued, the only relation to nature he recognizes is a dominative one—the instrumental relation he associates in his earlier texts with "work," and in his later ones with Weberian purposive-rational action and the specific mode of rationalization it involves. His emphasis on language and the linguistic basis of ethics, at the same time, seems to betray an anthropocentric (or "logocentric") bias whereby nature ends up necessarily banished from the normative realm. Why is there no room, they ask, for alternative relations to nature, oriented not toward domination but perhaps toward harmony, awe, a recognition of intrinsic value, an acknowledgment of Otherness—relations that would possess from the start an ethical character?

I think that these critiques are right to question the very limited range of possible human relations to nature that Habermas admits, and right to suggest that a society where nature appears only as the passive object of instrumental action is one that ought to be criticized, and that to this extent an "ethics of nature" is both possible and necessary; but I think that they are wrong to imagine that any "alternative" relations to nature that might arise could give us any closer access to nature in its "otherness" or could "let nature be"—and wrong to think that we need (or could use) such access to develop a more appropriate "environmental ethic." I want in this essay to consider Whitebook and Ottmann's arguments, and then to indicate why I think they miss the mark. I will also criticize a more recent attempt by John Dryzek to employ a modified version of discourse ethics in support of a nonanthropocentric "ecological rationality" that again would lead to a "respect" for nature in its otherness.[4] The real problem with Habermas's account of nature, I will argue, is not that it fails to capture nature's "otherness" but rather that it does not sufficiently see that *nature too is simply part of the social or communicative realm*. In that sense it is not Other at all, I will suggest—and our misleading sense of its otherness may shed more light on the source of our environmental problems than might at first be thought.

II

The closest thing to a full-blown "theory of nature" in Habermas shows up in his early *Knowledge and Human Interests* and in the important essay on "Technology and Science as 'Ideology'" written during the same period in criticism of Marcuse.[5] In those works, Habermas developed his account of what he called the "knowledge-constitutive interests," according to which the object domains of the sciences (both natural sciences and *Geisteswissenschaften*) are seen as *constituted* subject to particular frameworks of human action. Humans engage both in "work," Habermas argued—instrumental action oriented toward manipulation of the external environment—and "interaction"—communicative action directed toward other social agents in order to coordinate behaviors. These frameworks of action are in some sense built into the structure of the species, and from them emerge the fundamental human "interests"—the "technical interest" in prediction and control of nature on the one hand, and the "communicative interest" in the achievement of mutual understanding among linguistic subjects on the other.[6] A distinct domain of objects is constituted by and through each interest: the realm of nature is that aspect of the world constituted by the technical interest and appropriately studied by the methods of the natural sciences, while the social realm is that constituted by the communicative interest and studied by the *Geisteswissenschaften*. The methods of these two kinds of science differ accordingly, Habermas argued: monological, hypothetical-deductive methods for the natural sciences, hermeneutic methods for the sciences of the human realm.[7]

This account permitted Habermas to assert the methodological uniqueness of the *Geisteswissenschaften* (and especially that of a critical social theory) over and against the claims of the positivist thesis of the unity of scientific method while at the same time acknowledging the validity of something like the positivist account of method specifically for the (obviously successful) natural sciences. Such a strategy allowed him to avoid what seemed to him two opposed dangers: the scientistic danger, on the one hand, of a view that reduces all social theory to value-free objective science and hence renders critique impossible; and the romantic danger on the other (symbolized for Habermas by Marcuse, and the earlier Frankfurt School in general) of rejecting science and technology *tout court* as part of an overall critique of contemporary culture. Marcuse had raised the possibility of something like a "New Science," an approach to nature oriented no longer toward domination but rather toward the attempt to live in harmony or partnership with it.[8] Against this Habermas argued that (methodologically speaking) the "old" natural science we have is the only one possible, associated as it is with a species-wide interest based on the biologically invariant behavioral framework of work. Marcuse's mistake was to misapply categories appropriate to interaction—domination, partnership, etc.—in a realm where they don't belong, that of nature. It is the mirror image of the scientistic mistake that tries to apply categories of prediction, control, efficiency, and so forth to a social realm in which the essentially ethical categories associated with communication are required instead.[9]

Although interesting and suggestive, this early position of Habermas's is deeply flawed, as he soon began to see; discussion of "knowledge-constitutive interests" had disappeared from his works by the mid-1970s. I and others have written elsewhere in greater detail about the inadequacies of this account as a discussion of nature and natural science; let me quickly mention two problems here.[10] One is the circularity built into what Habermas called his "quasi-transcendentalism": if all the objects of science, including natural objects, are constituted on the basis of the built-in frameworks of human action (work and interaction), what then can be said about those frameworks themselves? They presumably emerged from "natural" characteristics of the species in some sort of evolutionary process—but what science could possibly tell us about *them*? Habermas is forced to postulate a "nature in itself" that gives rise to the human interests and that must be kept distinct from the "nature for us" that is constituted subject to those interests and studied by the natural sciences. But what exactly the status of that "quasi-noumenal" nature is, and how we could ever claim to know it (which we must do in order to justify the theory of interests itself, of course), remains absolutely murky.[11]

The second problem is an embarrassment that Habermas shares with many continental thinkers of that period, especially those sympathetic to neo-Diltheyan critiques of positivism. Chastened by the unquestionable success of natural science in explaining physical phenomena and developing technologies on the basis of those explanations, those who wished to insist upon a hermeneutic, *verstehende*, critical element to social theory tended to do so by carefully and politely acknowledging positivism's correctness as an account of scientific method while arguing, as Habermas did, that the validity of such methods was however limited to the realm of nature. The trouble is that during this same period fewer and fewer philosophers of science in what until then had been the positivist tradition found themselves able to believe that those methods had been correctly described by positivism even in *that* realm. Habermas was thus left in the awkward position of assuming and even defending the validity of a certain kind of account of natural scientific method—and of its difference from the communicatively mediated methods of the *Geisteswissenschaften*—at the same time that postempiricist philosophy of science was coming to reject that very account. The new unity of science thesis threatened to be one that saw all science as hermeneutic, not as hypothetico-deductive; Habermas was unprepared for this development, which causes serious difficulties for his theory and the deep instinct for dualism that underlies it.[12]

For Habermas *needs* to keep nature and the social distinct, precisely to prevent the politicization of science he associates with Nazi physics and Lysenkoist genetics and which he sees as a disaster.[13] To introduce a social element into natural science, in the way that the hermeneutic turn necessitates, would be to raise the specter of a relativism whereby social change might be so tightly interwoven with something like Kuhnian paradigm shifts that each social order could be seen as having its "own" science, and even its own nature too. This of course is in one sense just what Marcuse was proposing, and it is just what Habermas wanted to avoid. By separating the interest underlying natural science from the one underlying social interaction, Habermas hoped to be able to detach the normative from the "technical" and so to protect objective science from any reintroduction of the sorts of "subjectivistic" concerns that he saw it as one of modernity's great achievements to have eliminated from scientific reasoning.

The cost of this inoculation of natural science from the normative, as Whitebook and Ottmann point out, is precisely the impossibility of conceptualizing nature itself as possessing moral worth, or of recognizing a moral dimension to the way in which we interact with it. Indeed, even after the theory of knowledge-constitutive interests disappears from Habermas's work this problem remains, and in fact grows more acute. In later works such as *The Theory of Communicative Action*, the underlying dualism is defended not by an appeal to interest but directly to *language*. Each speech-act, Habermas now argues, implicitly makes three validity claims—to "truth," "rightness," and "truthfulness" (or "sincerity")—and hence posits a relation of the speaker to three distinct "worlds": *the* world of objectivity, *our* world of the social, and *my* world of the subjective. Again, the objective-cum-scientific is here carefully distinguished from the intersubjective-cum-normative.[14] In the new "discourse ethics," fundamental moral principles receive their rational justification by being derived from the communicative structures necessarily presupposed in the practice of argumentation. We have obligations toward other humans precisely because our communicative interactions with them always presuppose, even if only counterfactually, the possibility of reaching a consensus in which "the only force employed is the force of the better argument."[15] But then it is only those with whom we could potentially stand in a reciprocal relation based on language-use to whom we owe such obligations; animals, other living entities, and certainly "nature" as a whole seem necessarily excluded.[16]

Since we cannot communicate with nature, it can have no moral status: that seems to be the clear, if sometimes uncomfortable, consequence of Habermas's position both early and late.[17] At best, all that the position might justify would be a form of anthropocentrism in which nature's value is indirect, stemming in the first instance only from its role as necessary material precondition for the possibility of social interaction. Nature may of course also turn out to have value because humans appreciate it, or because that value follows from other religious or aesthetic values they hold; but even so the locus from which value is bestowed for Habermas clearly remains within the human community, and thus there can be no question here of "intrinsic" value in any strong sense.[18] And in any case whatever value nature possesses remains stubbornly independent of its role in science, which for Habermas remains rigidly bound to something like the technical interest. In the last analysis, he writes, the only "theoretically fruitful attitude" toward nature is the "objectivating" one of the natural sciences—the one whose relation to normative considerations he explicitly denies.[19]

Whitebook and Ottmann both criticize Habermas on this score, rejecting his limitation of possible theoretical approaches to nature to the narrow range of the instrumental and his associated denial of any possibility of finding ethical values *in nature*, of justifying that is something like an "environmental ethics."[20] Each attempts to suggest a source for an alternative approach to nature and to show how that approach might make an environmental ethic possible. As I have already said, I share their sense that Habermas has quite arbitrarily limited the possible scope of human relations to nature, and that it is important to think about how overcoming Habermas's adamant commitment to dualism might make it possible to derive what I would prefer to call an ethics *of* (not "in") nature; but I want to argue that they look for it in exactly the wrong places and with the wrong assumptions, and that in fact it is precisely in those elements of Habermas's view that they most want to criticize that the most helpful insights for thinking about "environmental ethics" are to be found.

III

Ottmann and Whitebook employ similar strategies in their critiques, attempting to identify a kind of experience of nature that would point to the possibility of a relation to it other than the "instrumental" one of prediction and control. For Ottmann, influenced here perhaps by Heideggerian themes, that possibility is pointed to simply by our experience of nature's otherness. Even within the operation of the technical interest itself, he argues, we can discern the existence of a dimension of nature "beyond control and exploitation": in its resistance to our technical interventions, in the phenomena of environmental disruption and the revenge of nature, he suggests, we experience something which serves as a barrier to our own apparently "boundless will-to-control." This, he believes, indicates the fundamental limitedness of the nature revealed to us by the technical interest: it is and must be embedded within something larger, which technical reason alone could never encompass.[21]

Whitebook, writing more as a sympathetic critic, suggests that Habermas's identification of technical rationality as the only possible theoretical approach to nature could be countered by a critique at the level of the philosophy of science, one that would take the form of exhibiting a scientific approach to nature that does not satisfy the strictures of Habermas's account of instrumental, technically oriented science.[22] He offers biology as his example. The argument gets noticeably vague here, but the general tenor of it is clear: Whitebook interprets biology along something like neovitalist lines, and in particular in an antireductionist manner. Categories of "life" and of teleology, he suggests, may turn out to be ineliminable from biological theories. But there is no room for such categories within the sort of deductive-nomological sciences that

Habermas sees the technical interest as necessarily generating. Hence in biology, Whitebook claims, we find an instance of a "theoretically fruitful" science that nonetheless conceptualizes nature under categories other than those of strict prediction and control.[23]

Neither of these arguments seems to me very persuasive. It is first of all not really clear why Habermas's account of science need commit him to reductionism about biology, and although it may perhaps be true that his view cannot countenance concepts such as teleology (in the strong sense) or *élan vital*, Whitebook is simply *asserting* (and only, he admits, as one "speculative" and unfashionable possibility) that such concepts will turn out to be necessary in biological research—an assertion that only seems *less* plausible today, more than fifteen years after the publication of Whitebook's article. In Ottmann's argument, too, all we really get is the bald assertion that in our technical interventions in nature we do as a matter of fact encounter an aspect of it that transcends the possibility of technical control. If this is to be more than an appeal to some otherwise unanalyzed mystical intuition, it would seem to refer to the undeniable fact that some of our technical interventions fail to produce the effects we predict. But the fact that our interventions face resistance and even failure simply indicates a fact about the world we encounter *in* those interventions. To use Habermas's earlier vocabulary, it shows us something about the world constituted subject to the technical interest; it simply means that *within* the world of the technical things can be misconstrued and overlooked. It could only be said to tell us something about some *other* entity "beyond" that world with the supplement of a whole set of additional (and highly controversial) substantive ontological assumptions.

But the real problem with this kind of argument strategy is a deeper one. For what both Whitebook and Ottmann want to do in their critiques of Habermas is to find a mode of access to nature that not only goes beyond the instrumental but actually encounters nature's own "intrinsic qualities"—that is to say, those qualities it has prior to or independent of the appearance on the epistemological scene of anything like a human "interest" (in Habermas's sense) at all. Both Whitebook's appeal to teleologies and Ottmann's appeal to a nature that goes beyond the "boundless will to control" represent attempts to conceive of nature (as Ottmann puts it) not merely as a means for human purposes but as an "end-in-itself."[24] The effect of this, it is important to note, is not so much to dissolve the dualism between the social and the natural in Habermas's work as it is to intensify it: the flaw in Habermas's account turns out (a bit surprisingly) to be that even his notion of the instrumental—and indeed, the idea of "knowledge-constitutive interests" in general—still leaves too much of the taint of the human on the nature it reveals. Habermas's sharp distinction between the realms opened up by technical and the communicative interest is not dualistic *enough* for Whitebook and Ottmann: they think something like an ecological ethic requires an appeal to a nature not merely independent of the communicative interest in particular (which is Habermas's move) but rather independent of interest *in general*—that's the only way "anthropocentrism" can be avoided.[25] The appeal thus is back to what Habermas called "nature in itself," the nature that underlies the interests and generates them, but which on his account could never really be known (since all knowledge is necessarily "interested").

But I would argue rather that it is just this dualism between nature and the social that produced Habermas's problem about nature in the first place, and that Whitebook and Ottmann's attempt to radicalize the dualism only serves to exacerbate it. Habermas's original distinction between work and interaction, we have seen, had (among other things) the motivation of avoiding the dangers of relativism and of the politicization of science that seem to emerge from an account of knowledge as always "interested": the problem is avoided by asserting that the technical "interest" behind science is an ahistorical and universal interest of the species, not any particular and historically specific social interest (which would have to be communicatively mediated). But the notion of universal hardwired interests thereby posited needs in turn to be defended by some appeal to a presocial nature in itself that generates them, and it turns out that the difficulties faced by such an appeal (and in particular about how such a nature could be known) simply recapitulate the problem of relativism in another guise: if all knowledge is subject to interest, how is disinterested or "objective" knowledge of the interests themselves possible? Habermas hasn't avoided the problem, only pushed it back into the past.[26]

And Whitebook and Ottmann haven't avoided it either; instead they simply ignore it. To imagine possible access to nature in itself (to nature's otherness, for Ottmann, or to its intrinsic teleologies, for Whitebook) is to regress behind the epistemological insight Habermas took over from the earlier Frankfurt School and from Hegelian Marxism in general: the insight into the irreducible entanglement of the real and socially situated knower in the object known. Notice that there are really *two* dualisms operating in Habermas's earlier account: one between the world we come to know subject to the interests and a posited "nature in itself," and the other—which arises only *within* the former world, i.e., already subject to interest—between the social world and the world of "nature for us." "Ecological" critiques of Habermas like those of Whitebook and Ottmann typically want to dissolve the *first* of these dualisms, by insisting upon the possibility of some access to a nature independent of interest. But that dualism isn't dissolved by pretending that the epistemological problem underlying it doesn't exist. Habermas is simply *right* about the epistemological situation here: the worlds we encounter in our experience, including aesthetic and mystical experiences, are always still socially "constituted" realms, constituted subject to something like an "interest" (although I myself might prefer to say: constituted in our practices). We have no access and can have no access to anything like a presocial nature in itself; the very idea is incoherent, because all access is socially mediated. Hence one "dissolves" this dualism in the same way that Hegel dissolved the analogous Kantian one: by *dropping* the (impossible) concept of the "noumenal" or, in this case, of nature in itself. There *is* no nature "behind" or "before" the one we know, and transform, in our practices—and the nature we know, the one we inhabit, *is always already social*.

But to say that is to say that it is really the *second* dualism mentioned above that needs to be overcome. It is Habermas's separation of nature from the social, not his separation of "nature in itself" from the one we encounter in our instrumental actions, that prevents him from conceiving of the possibility of an "ethics of nature." Finding an ethics of nature is not going to involve discovering a source for ethics in some noumenal (and thus in principle inaccessible) nature, but rather precisely of acknowledging that *just because* nature itself is always already social it is also always already part of the ethical. Habermas is right to insist that the ethical and the social are intrinsically intertwined, and that it is language use and communication that connect them; and he is right to conclude from this—against all naturalisms—that therefore nothing extrasocial (like "nature in itself") can be a source of ethical insights.[27] But he's wrong to think that those insights therefore cannot extend to the nature we actually inhabit and encounter, precisely because that nature is *not* and never could be a presocial nature in itself. If I am right, then finding an ethics of nature in Habermas thus will require adding *more* sociality, *more* "interestedness," to his account of nature, not—as in Whitebook and Ottmann—less.

IV

One way *not* to do this, however, is to treat nonhuman natural entities as communicating *members* of a (suitably modified) discursive community. This (tempting) solution to the apparent problem of anthropocentrism in Habermas's account of discourse ethics has been proposed by several critics, including recently John Dryzek in an article that has received some comment.[28] A properly modified discourse ethics, Dryzek argues, might recognize moral responsibilities to nonhuman entities on the basis of what he describes as "a principle of rough equality in communicative capability."[29] Lovelock's Gaia hypothesis, for instance, implies (Dryzek writes) that "we live *in* a highly differentiated, self-regulating global system whose 'intelligence,' . . . though not *conscious*, is of a complexity equal to that of any group of humans," an intelligence that must be understood as including the property of agency; even if we do not accept the

Gaia hypothesis, furthermore, indications of agency can be found at lower levels of biological organization as well.[30] Biological research—he refers here to Keller's study of McClintock, to Jane Goodall, and to others—itself has an essentially hermeneutic element, Dryzek further argues, "involving imaginative attempts to reconstruct the actions-in-context of other thinking beings."[31] Teleological accounts of ecosystemic development and succession are mentioned in this context as well. The result of this sort of "recognition of agency in nature," he concludes, "makes the restriction of communicative rationality to purely human communities appear arbitrary."[32] The procedural standards of communicative rationality (now redubbed by Dryzek "ecological rationality") must then be extended to include specifically ecological considerations.[33]

But tempting though such a move is, it is based on a misunderstanding, or at least on an equivocation in the concept of "communication." The examples Dryzek offers here are not cases of communication in the strong sense that Habermas must employ if his argument for discourse ethics is to work. It is crucial for Habermas that language use take the form of a speaker making semantically significant *assertions* that raise potentially redeemable *validity claims*—i.e., that it take the form of *speech-acts*. He needs to be able to show that in every such act the speaker implicitly accepts the responsibility to demonstrate, if challenged by an interlocutor, why and in what sense what the act asserts is in fact the case (is true, is right, is truthful); this is what makes possible that strange Janus-faced character to language use that Habermas emphasizes, whereby at one and the same time it is contextually bound (and fallible) and yet also, because the claims it makes are implicitly universal ones, it possesses an "ideal moment of unconditionality."[34] Only this allows him to argue that any speaker has always already accepted a principle of universalization whereby the legitimacy of normative claims is tied to their discursive acceptability by all those potentially affected, and further to assert that a speaker who denied such a principle would be guilty of a performative contradiction.[35] And this "universalization" therefore extends, it is important to see, precisely to the set of possible discursive participants—which is to say, to those who can engage in the same sorts of speech-acts and hence be shown to have already accepted the same sorts of principles.

But the kind of "communication" Dryzek speaks of in the nonhuman context does not have this sort of character. *Signaling* is not *discourse*: a complex system—organic or inorganic—can in a certain sense surely "communicate" its current state, including its "needs" and "intentions," but it does not raise *validity claims*. Dryzek's examples of "communication" in nature are based on complex feedback loops or other sorts of behaviors that may indeed best be described in something like teleological terms; but that an organism functions teleologically is not sufficient to show that it is communicating in anything like the sense of a discourse ethics. The occurrence of a certain change in a teleologically organized system may indicate something about a change in its environment, but it does not *assert* anything about it, much less make a truth claim implicitly redeemable in discourse. This is exactly why the suggestion that biological research into such systems requires quasi-hermeneutic techniques is such a problematic one: in the absence of any possibility of a discursive *response* by the entities under investigation in the form of a challenge to the proffered interpretations, deep methodological problems arise as to how the illegitimate projection of anthropomorphic presuppositions could possibly be avoided.

Dryzek's argument is in any case inappropriately indirect: he uses the characteristics he mentions only to show that there is "agency in nature," and it is only on the basis of that putative agency that he suggests that something like a communicative competence can be ascribed to natural organisms as well. But in the context of Habermas's "communicative turn," with its injunction to substitute pragmatic investigations of intersubjective relations for metaphysical speculation about isolated subjects, such an argument gets things backwards: "agency" can only be determined on the basis of the potential for intersubjectivity (which means: communication) and not vice versa. In short, then, the argument Dryzek proposes is either disingenuous or self-canceling: he wants to bring nonhuman nature into the realm of communicative ethics, but can only do so at the cost of so diluting the concept of communication that its internal connection to the ethical is lost.[36]

V

Indeed, I am pessimistic about the chances of *any* such attempt to develop a "discourse ethics of nature" by treating natural entities as themselves capable of discourse. Yet to say this is not to say that nature is not social. It *is* social, I would argue, but in a very different sense—in the sense that "nature," and more precisely the environment or *Umwelt* that we inhabit, is always itself socially constituted and constructed. The world we live in—the *one* world, not divisible into ontologically distinct "social" and "natural" realms—is a world constituted in and through the socially (and linguistically) organized practices we engage in every day—including, of course, the practices of technology and of science. Never a "nature in itself," the nature we actually inhabit is one that always already shows the mark of the human: first, because we perceive and experience it, study and dream about it, in terms that are from the beginning social through and through, but second also because the objects and landscapes through which we experience it are always themselves—when closely examined—in part the product of earlier social practices.[37] *This* is what I mean when I speak of the need to criticize Habermas by showing there to be more, not less, sociality in nature.

Whitebook is certainly right to propose using the philosophy of science to question Habermas's account of natural scientific procedure, but the philosophy of science he ought to have considered is not neovitalism but instead the very postempiricism that made Habermas's notion of a monologic science of nature seem out-of-date even when he first proposed it. What Kuhnian and post-Kuhnian developments in the philosophy of science suggest, of course, is precisely that science—and so too the nature it reveals—can no longer be seen as independent of the social, of history, etc. *This* is how to counter Habermas's critique of Marcuse. Whitebook and Ottmann, like Marcuse, want to suggest that technical rationality is not the only approach possible to nature, and that an alternative science can be imagined that no longer treats nature as meaningless and valueless matter; Habermas's insistent denial of the possibility of such an alternative turns out to depend principally on his naive (pre-Kuhnian) philosophy of science. Once the postempiricist "social turn" occurs, his argument fails: there is no *single* "theoretically fruitful" approach to nature, but rather the question of what is fruitful and what is not turns out itself to be answerable in socially and historically varying ways, which means that alternative approaches are certainly imaginable. The view of nature as mere matter for instrumental manipulation criticized by Marcuse and the earlier Frankfurt School is not, as Habermas tried to argue, built into the structure of the species or of "work," but rather—as they had originally asserted—is associated with a *particular* social order and a *particular* historical epoch. A new kind of society, then, might well involve a new science, and with it a "new nature" as well.

But the trouble is that ecologically minded critiques (including Marcuse's) tend to conceptualize such a "new science" (or what Whitebook alludes to as a possible "epochal transformation" in worldview) in an equally pre-Kuhnian manner—as renouncing control, as "letting nature be," and thus as involving ultimately a more correct relation to nature "as it really is" prior to the social.[38] *This* makes no sense, for the reasons I have suggested: we have no access to nature in itself. Rather, such a "new science" or new approach to nature must instead be seen in the postempiricist context as a *social* possibility, which is to say as something that a new social order might decide to take up for its *own* (good!) reasons as the result of a discursively mediated decision, not as a discovery about the "real" character of *nature's* intrinsic needs. Instead of pretending to be a science that had finally achieved access to the truth of nature (which was, after all, what the "old science" claimed too), it would be a science that self-consciously acknowledged its own sociality and so also the sociality of the nature that it encounters/constitutes.[39]

Habermas's ecological critics would no doubt be shocked by this conclusion. They believe that the trouble with contemporary approaches to nature is that these approaches view nature as valueless, as mere matter for human manipulation, and to them this last proposal would look like more of the same. But I think that they would be mistaken. The conclusion follows, rather, from taking seriously Habermas's deeply important insight that value is always social and communicative, and seeing that it means that value in nature could only come from more—and more explicit—sociality, not less. To say this is *not* to treat nature as purely instrumental and valueless matter, but quite the opposite.

The model for the argument here goes back to Hegel, and to Marx. It is the idea that alienation occurs when an object that is really a social product appears instead as an independent power, and that freedom consists in the recognition and reappropriation by a social subject of that which it has helped to produce. What is wrong with the way nature appears to us (and to our natural science) today is that it seems to be utterly independent of us and even opposed to us. Its sociality is hidden: we see it as separate from us, as dangerous (as impossibly complex, as always taking its revenge), and fail to see that rather it is always something in which we are deeply and actively enmeshed. Thus it appears precisely as the Other, and it is this very otherness that leads to fear and to the desire to control. Only a nature viewed as so separate from us would be something we would feel the (frightened) need to "dominate" or the (equally frightened) need to "preserve." In either case it appears as alien, something to be either overcome or propitiated—not as simply part of the (social) world that we both inhabit and continuously transform.

To recognize nature's sociality, then, would not be to deny it value but rather to see it precisely as *having* value. Our environment is something for which we are responsible, something that we "constitute" not only in our worldviews or paradigms, but quite literally in our *work*—in the myriad socially and linguistically mediated practices that are constantly remaking the world, for better or (as today, mostly) for worse. Nature does not appear as "valueless" on such an account, but precisely as the opposite: it is because the world that surrounds us is something *for which we are responsible* that we ought to respect it, protect it, and attempt to make it more livable and beautiful. This is not "intrinsic" value, because it is not a value in "nature in itself," and not therefore a value in "otherness" of the sort that radically antianthropocentric environmental philosophers sometimes want to assert; but the possibility of the latter is just the sort of thing that communicative ethics has given us very good reasons to doubt. The model here might instead be a different one, like the value we find in a work of art.

Whitebook distinguishes between Enlightenment and anti-Enlightenment tendencies in Critical Theory, particularly with respect to the distinction between those who see the disembedding of humans from nature as a positive step towards autonomy and *Mündigkeit* and those for whom the disenchantment of nature was a step toward catastrophe.[40] (He is noticeably ambivalent between these two.) He takes Habermas to be a "thoroughgoing disenchanter,"[41] which is certainly right, and worries that in his work the original impulse of Critical Theory, the impulse for a reconciliation with nature, has been lost. But this distinction between disembedding and reconciliation is a misleading one, at least in terms of the argument I am offering. For as I have already suggested, it is Habermas's ecological critics (like Whitebook) who seem to want to *strengthen* the dualism between nature and the social—a dualism that I, on the other hand, have suggested needs to be *dissolved*.

The point, which again is a Hegelian one, is that the unity of nature and the social can be thought of in two ways: as a recognition of the naturalness of the social or of the sociality of nature. To insist on the latter *is* to think of reconciliation, but a reconciliation with, so to speak, the signs reversed. The problem is not to choose between reconciliation and disembedding, but rather to decide whether reconciliation is to be conceptualized as passive or as active. To imagine it, as Habermas's critics (along with Marcuse and the earlier Frankfurt School) sometimes do, as a passive harmonization of human practices and attitudes with a nature taken as independent of and prior to the social, is not to do justice to the real character of the kind of dialectical unity in question here. Humans are of course products of nature, but of a nature that in turn has itself always already been transformed by human practices and so already partakes of the social (and hence of the ethical too). Active reconciliation suggests rather a process whereby humans acknowledge their own responsibility for those transformations, and choose to engage in them now self-consciously and in full recognition of what they are doing.

To be reconciled with nature would then mean: To *make* the world that surrounds us a good one, a beautiful one, one whose structures we can discursively defend. Such an "environmental ethic" finds an ethics in the environment because it sees that environment as *already* social; thus it accepts the Marcusean critique of a technological rationality that denies all value to nature while still doing justice to the Habermasian insight that all value is tied to communication. Rather than appealing to a nature in itself which does not speak and about which nothing can be known, it would thus ground its principles in a conversation among humans about the sort of environment they wish to inhabit. We construct the world we inhabit, as I have already said, in our practices: the problem is to justify those practices, which as Habermas has shown means to justify them discursively. We get our environmental ethic, then, not from nature, but from ourselves.

Notes

1. "Nature, too, awaits the revolution!" writes Marcuse in *Counterrevolution and Revolt* (Boston: Beacon Press, 1972), p. 74.
2. This is Marcuse's translation of a phrase from Adorno's *Ästhetische Theorie*. *Counterrevolution and Revolt* , p. 66.
3. Joel Whitebook, "The Problem of Nature in Habermas," *Telos* 40 (Summer 1979): 41–69; Henning Ottmann, "Cognitive Interests and Self-Reflection," in John B. Thompson and David Held, eds., *Habermas: Critical Debates* (Cambridge: MIT Press, 1982), pp. 79–97. Whitebook and Ottmann's articles are in a sense dated, oriented as they are in part toward epistemological positions that Habermas no longer defends. Yet they continue to receive attention (especially Whitebook's) in discussions of "nature in Habermas"—and often rather uncritical attention at that. In any case, the basic structure of their critiques can in fact be carried over without much change to Habermas's more recent work (where nature tends to receive even shorter shrift than it did in the earlier writings); furthermore, as I shall argue below, both the conceptual difficulties they encounter *and* the ones they reveal in Habermas's position remain significantly illustrative for any attempt to develop a "critical theory of nature." Whitebook's essay, incidentally, has recently been reprinted (with a new introduction) in David Macauley, ed., *Minding Nature: The Philosophers of Ecology* (New York: Guilford press, 1996); references below are to the original publication in *Telos*.
4. John Dryzek, "Green Reason: Communicative Ethics for the Biosphere," *Environmental Ethics*, Vol. 12 (Fall 1990): 195–210.
5. Jürgen Habermas, *Knowledge and Human Interests*, trans. Jeremy Shapiro (Boston: Beacon Press, 1968). The essay on Marcuse appears in Habermas's *Toward a Rational Society*, trans. Jeremy Shapiro (Boston: Beacon Press, 1970).
6. I am leaving out of the account here what Habermas calls the "emancipatory interest" associated with self-reflection. This was always the most problematic of the three, and the one whose difficulties above all led him in the 1970s to revise fundamentally his whole theory.
7. See Habermas, *Knowledge and Human Interests*, pp. 191–98; and *Toward a Rational Society*, pp. 91–93.
8. See, e.g., Herbert Marcuse, *One-Dimensional Man* (Boston: Beacon Press, 1964), pp. 166–67.
9. Habermas, *Toward a Rational Society*, p.88.
10. For a more general discussion of Habermas's earlier views of nature (and of his critique of Marcuse) see my *Against Nature: The Concept of Nature in Critical Theory* (Albany: SUNY Press, 1996), chap. 5. See also (among others) Thomas McCarthy, *The Critical Theory of Jürgen Habermas* (Cambridge: MIT Press, 1978), chap. 2; C. Fred Alford, *Science and the Revenge of Nature* (Tampa: University Presses of Florida, 1985), chaps. 5–6; David Held, *Introduction to Critical Theory* (Berkeley: University of California Press, 1980), pp. 389–98; Russell Keat, *The Politics of Social Theory* (Chicago: University of Chicago Press, 1981), chap. 3; Richard Bernstein, *The Restructuring of Social and Political Theory* (Philadelphia: University of Pennsylvania Press, 1976), pp. 219–25; Rick Roderick, *Habermas and the Foundations of Critical Theory* (New York: St. Martin's Press, 1986), pp. 62–73.

11. Cf. Whitebook, "The Problem of Nature in Habermas," pp. 48–49.
12. Cf. Mary Hesse, *Revolutions and Reconstructions in the Philosophy of Science* (Bloomington: Indiana University Press, 1980), pp. 169–73.
13. See Habermas, *Knowledge and Human Interests*, p. 315.
14. Jürgen Habermas, *The Theory of Communicative Action*, vol. 1, trans. Thomas McCarthy (Boston: Beacon Press, 1984), pp. 295–310.
15. Jürgen Habermas, "Postscript to *Knowledge and Human Interests*," *Philosophy of Social Science* 3 (1973): 168. (Translation modified.)
16. See, e.g., Jürgen Habermas, *Justification and Application*, trans. Ciaran P. Cronin (Cambridge: MIT Press, 1993), pp. 105–111. But see below, note 28.
17. When pressed on the moral status of nature, his discomfort is always palpable. He wants to be able to assert it, but knows he cannot. See note 16, and Jürgen Habermas, "Reply to My Critics," in Thompson and Held, *Habermas: Critical Debates*, pp. 243–250. See also Vogel, *Against Nature*, pp. 154–60.
18. Whitebook, whose essay is notable for its ambivalence, clarifies well how a communicative ethics might be able to do most (but not all) of the work a nonanthropocentric ethics wants to do. See his "The Problem of Nature in Habermas," pp. 61–64. Habermas refers approvingly to this account in "Reply to My Critics," p. 247.
19. Habermas, "Reply to My Critics," pp. 243–44.
20. See also Thomas McCarthy's critique of Habermas along similar lines in "Reflections on Rationalization in *The Theory of Communicative Action*," in Richard Bernstein, ed., *Habermas and Modernity* (Cambridge: MIT Press, 1985), as well as Habermas's response in "Questions and Counterquestions" in the same volume.
21. Ottmann, "Cognitive Interests and Self-Reflection," pp. 87–92.
22. Whitebook, "The Problem of Nature in Habermas," p. 55.
23. Ibid., pp. 56–61.
24. Ottmann, "Cognitive Interests and Self-Reflection," p. 91.
25. A similar critique based on Habermas's later work would object to the way *the* world (of nature) is derived only on the basis of an analysis of the necessary presuppositions of "our" use of language.
26. See Vogel, *Against Nature*, pp. 120–24.
27. And these ideas run through both his early and his later work like a *leitmotiv*.
28. See Dryzek, "Green Reasons." Note that of course in a way this was Marcuse's original proposal as well, with his talk (criticized strongly by Habermas) of treating nature like "a subject in its own right—a subject with which to live in a common human universe." See Marcuse, *Counterrevolution and Revolt*, p. 60. More recently, though, even Habermas has accepted something like this as a possible account of our moral intuitions regarding animals—see *Justification and Application*, pp. 109–111. Cf. also Karl-Otto Apel, "The Ecological Crisis as a Problem for Discourse Ethics," and Gunnar Skirbekk, "The Beauty and the Beast," both in A. Øfsti, ed., *Ecology and Ethics* (Trondheim, Norway: Tapir Trykk, 1992).
29. Dryzek, "Green Reason," p. 208.
30. Ibid., p. 205.
31. Ibid., p. 206.
32. Ibid. Whitebook, who as we have seen also appeals to teleology and vitalism in general, can sometimes be understood as arguing in a similar fashion. See his reference to "pre-human nature as *incipient spirit*" in "The Problem of Nature in Habermas," p. 61.

33. Note of course that Dryzek is not arguing that nonhuman natural entities ought to be treated as equal communicative partners with humans and hence deserve something like rights. Such a move, although it would clearly have its problems, would at least preserve the moments of equality and universality in the Habermasian argument. What he *is* proposing is somewhat vaguer—a suggestion that other entities, and other sorts of communication, simply need to be taken into account. Thus, for instance, he writes that "we should be wary of highly centralized decision mechanisms . . . which could dominate, ignore, or suppress local ecological signals. . . . Diffuse feedback processes in the natural world should be matched by diffuse decision processes in human societies." "Green Reason," p. 208.

34. See Habermas, *Justification and Application*, p. 146.

35. See Jürgen Habermas, *Moral Consciousness and Communicative Action*, trans. Christian Lenhardt and Shierry Weber Nicholsen (Cambridge: MIT Press, 1990), pp. 92–93.

36. Of course this is not to deny that certain nonhuman animals may turn out to be language users in the strong sense that Habermas needs; the point here is only that arguments of this type have to be *much* more careful about what exactly that would require. Habermas's argument suggests a special moral status for *language users*, not for humans. Whether only humans are language users (in this sense) is an empirical question for linguistics, ethology, etc., not a philosophical one.

37. See my "Marx and Alienation from Nature," *Social Theory and Practice* 14 (3) (1988): 367–87, as well as *Against Nature*, pp. 35–39.

38. Whitebook, "The Problem of Nature in Habermas," p. 69.

39. Dryzek and Whitebook both try to avoid the problem of the sociality of nature by appealing to biology to ground their accounts of what nature "is." In so doing, though, they show themselves smitten by the very scientism Habermas and his predecessors in the Frankfurt School were all so firmly concerned to overcome: *science*, it seems, is supposed to tell us both what nature is and how we should treat it. When they each then go on to concede that in fact the (teleological, nonreductionistic) view of biology they champion is at best highly controversial—and not at all a mainstream one—they tip their hands, revealing their social/ethical commitments in fact to be *prior* to their choice of biological theory and not derived from it. See Whitebook, "The Problem of Nature in Habermas," pp. 57–59; and Dryzek, "Green Reason," p. 206.

40. Whitebook, "The Problem of Nature in Habermas," pp. 64–69.

41. Ibid., p. 44.

10

The Problem of Knowledge in Environmental Thought:
A Counterchallenge

Robert Kirkman

There can be little doubt that questions and issues concerning our environment pose a profound challenge to ethical, social, and political theories. Our growing understanding of the world in which we live, and of the possible ramifications of our actions, must be taken into account somehow as we go about assessing what we humans have done and what we are going to do next. But it seems to many as though conventional ways of making such decisions and assessing their results, along with conventional ways of understanding humans as social and political beings, are inadequate either to encompass or to resolve environmental problems.

My purpose, however, is not to contribute directly to the process of raising this challenge, but rather to raise a sort of therapeutic counterchallenge. This is to say that I have grave doubts concerning the manner in which ethical, social, and political theories have been rejected, revised, or replaced on environmental grounds. At issue are the intellectual bases of environmentalism in general, especially insofar as these involve factual claims about the world, many of which are either unwarranted or misleading. For the sake of brevity, I will here focus on only one set of these doubts, namely, those concerning the use of scientific (especially ecological) concepts in support of moral, social and political stances on environmental issues, particularly as those stances have been formalized in the diverse field of academic environmental philosophy.

Before I continue, I should note that the field of environmental philosophy is quite diverse, and not all of its practitioners fall easily into the pattern I am about to describe. The field encompasses Deep Ecology on the one end; the more staid and traditional environmental ethics on the other end; with ecofeminist, bioregional, geopsychological, social-ecological, and phenomenological approaches arrayed between them. Nevertheless, there is a common thread—and a common difficulty—shared widely among environmental thinkers. If I were to give this tendency a name, I would call it "the ideology of relatedness."

An ideology, in the sense in which I am using it here, is a system of principles adopted ad hoc in the service of a social or political opposition. These principles are held to be above suspicion by their adherents merely because of their *usefulness* for a particular cause; they are thus exempt from any of the usual criteria of coherence or adequacy of grounds. In general, environmental philosophy operates by attempting to identify the principles that underlie the perceived destructiveness of western civilization, and then by asserting principles that are diametrically opposed to them.

The arch-villain of environmental philosophers is, frequently, René Descartes, whose doctrines of dualism, reductionism, and mechanism are said to have alienated humans from Nature, and to have driven all of the sacredness and intrinsic value from the natural world. Nature is, for the Cartesian, only a collection of bits of dead matter that we, as the only spiritual entities, are free to manipulate to our own ends. This is said to be the root cause of our alienation from our environment, which is said to be the root cause of our technocratic arrogance, which in turn is said to be the root cause of environmental and resource problems. Even in cases where Cartesianism is not singled out, the pattern holds: there is something wrong with the way we *think* about our environment and our relationship to it, and the task of the philosopher is to root out that something and replace it with something new.

Such a frankly ideological opposition to Descartes and his ilk at the philosophical level is bound to repeat the pattern established by Romanticism in the nineteenth century: against mechanism, posit organicism; against reductionism, posit holism; against dualism, posit the unity of mind and nature; against the primacy of reason, posit the primacy of intuition or of feeling; against instrumentalism, posit intrinsic value; against alienation, posit a primordial state of harmony to which we must strive to return. In short, the imperative of relatedness-thinking is to repersonalize, resacralize, and respiritualize the world. This, at least in part, is why we are told from all quarters that we must rethink our "relationship" with our environment, or with the Earth, both as it is and as we think it ought to be, where "relationship" is understood in interpersonal terms. Until now, the story seems to go, our relationship has been a "bad" one, characterized by disrespect, disharmony, disobedience, or, at its most extreme, attempted "geocide"; with this goes the implication that there is only one clear alternative: we need to be, in some sense, kinder to and more respectful of our environment.

To be fair, this sort of reaction is understandable. Just as Romanticism, in its context, offered what seemed to be the only viable alternative to the excesses of rationalism in science and technology in the modern period, so contemporary environmental thought seizes on what seems to be the only viable alternative to those same excesses as they are manifested in the late twentieth century: the unprecedented advances of technology, the rise of technocracy, and the globalization of the impact of human activities. I stress this point in order to make it clear that I do not believe that environmental thinkers are tilting at windmills—the problems they confront are very real and very serious—nor do I question their motives. What I am doing is raising what I take to be grave doubts about the validity of the intellectual means by which they would address these problems.

It seems to me that, in taking up their ideological position, environmental thinkers are in effect setting themselves up as authorities concerning the way the world is as well as the way it ought to be, and that they may not be entitled to do so. In particular, there is a fairly consistent pattern of misunderstanding and misappropriating scientific concepts in the service of environmental thought. This is not to say that science on its own can serve as an absolute authority concerning the way the world is—I would not advocate "scientism"—but it should be granted that if environmental thinkers want to surpass or even overthrow scientific knowledge, they should at least take their rival seriously enough to understand its concepts and procedures in a sophisticated manner. In failing to do so, they run the risk of neglecting their own intellectual underpinnings, which entails the further risk of being misled: at best, they might simply assume that they know more than they do in fact; at worst, they might simply be blinded by what they want to be true. In any case, there are bound to be grave intellectual and practical consequences of such negligence.

Of course, there is much more to be said here. Most importantly, I have not yet established that environmental thinkers are as negligent as I claim; to do so, I will briefly examine four prominent examples of environmental philosophy.

In choosing my first example, I am running the risk of being accused of blasphemy, as the work in question has been regarded by many or most environmentalists as the equivalent of sacred writ. The tradition of contemporary environmental philosophy receives its first clear exposition in the work of Aldo Leopold, whose Land Ethic has had untold influence on later thinkers and activists.

Leopold states his ethic as follows: "A land ethic changes the role of *Homo sapiens* from conqueror of the land-community to plain member and citizen of it. It implies respect for his fellow-members, and also respect for the community as such." We arrive at the recognition of our membership, Leopold writes, by way of the "ethical sequence," according to which the realm of moral considerability expands outward until it encompasses not just all humans, but "soils, waters, plants and animals, or collectively: the land." This leads him to conclude that "a thing is right when it tends to preserve the integrity, stability, and beauty of the biotic community," and that it is wrong "when it tends otherwise."[1] All of this sounds plausible enough, and has been an inspiration to several generations of environmentalists. A closer look is required to identify the problem.

Consider the concept of community. In its ecological sense, a biotic community is an association of plants and animals in a given area, in which the abundance and distribution of each species is regulated by the abundance and distribution of all of the other species. Leopold adds a twist of ecosystem theory, noting the flows of energy and matter through what he calls "the land pyramid." The term "community" in this ecological context had its origins in a direct metaphor from human community, but its meaning in the context of ecology has already been refined so as to exclude most (though not yet all) anthropomorphic implications. The trees in a woodland community are not taken to be "citizens" in the full human sense, since they have neither rights nor obligations within that community; there are no political institutions among prairie grasses. It is a community only in the sense that it is an association of a number of individuals of a number of species, each of which has a role to play in the maintenance of some sort of stability and recognizable structure over time.

So what are we to make of Leopold's claim of human citizenship in such communities? There are two possible interpretations. On the one hand, he could be using the term "citizen" metaphorically, in an "ecological" sense, simply to make the point that humans participate in the flows and cycles of biotic communities. If this is the case, the analogy is limited but basically sound: the relevant similarity of "citizenship" is that of participation in and dependence on a larger, more or less integrated biogeochemical system; for the purposes of the analogy, questions of rights or obligations are set aside.

On the other hand, Leopold could be using both "community" and "citizen" in their original, human senses, with some carryover from the scientific concept of biotic community, in order to suggest that our human rights and obligations ought to be extended to the nonhuman "citizens" of what is in every other respect a human community. If this is his claim, it would seem as though he is doing nothing more than playing upon the ambiguity of "community" and so casting doubt on the validity and usefulness of his position.

The situation is still more complicated, however, in that Leopold seems to be making both of these claims at once on the basis of a reduction of human community to the terms of ecological community. This is clear in his definition of ethics. According to Leopold, an ecological ethic is "a limitation of freedom of action in the struggle for existence"; a philosophical ethic is "a differentiation of social from anti-social conduct." Leopold claims that these are "two definitions of one thing," with the common trait taken to be a striving for cooperation. As he states it, "politics and economics are advanced symbioses" in that they limit free action on the basis of a limited "ethic" of cooperation.[2]

Leopold's appropriation of the community concept out of both its ecological context and its original, human context raises serious questions about the validity of the resulting land ethic. The problem, as I see it, is that he has rendered the scope of the concept too broadly by playing the metaphor too freely. As a result, he has reduced "community" to a vague sense of "cooperation" or "relatedness" that is neither grounded in nor applicable to either ethical, social, or political theory on the one hand, or ecological research on the other. For example, Leopold seems simply to assume a clear understanding of what constitutes the "integrity, stability, and beauty of the biotic community," along with a clear set of procedures for determining whether particular human actions are helpful or harmful; it is at least possible, if not likely, that neither of these can be taken for granted.

Again, I should emphasize my belief that Leopold's motives are sound: he is seeking an alternative to the destructive practices he has witnessed. His work represents a decisive step forward in our understanding of human life in our environment; in many ways, his thought is wonderfully sophisticated, especially in his assessment of the practical problems of land use and in his recognition of the scope and limits of scientific knowledge. I fully acknowledge the importance of this project, even though I believe that the means by which Leopold carried it out are, in at least one respect, profoundly flawed. If I am correct, then it is significant that Leopold's heirs in the field of environmental philosophy seem to follow too closely in his footsteps, perpetuating and even magnifying what I take to be Leopold's error.

Perhaps the most direct lineage from Leopold is the diverse tradition of environmental ethics. For the most part, this field consists of various efforts to establish a sense of obligation or duty to the natural world, usually founded on some notion of the "intrinsic value" of certain nonhuman entities. Like Leopold, environmental ethicists tend to base their principles as much as possible on detailed knowledge of the natural world, extrapolating their principles, in part, from the concepts and theories of ecology and evolutionary biology.

To their credit, the theorists of environmental ethics are generally more circumspect about drawing ethical conclusions from scientific findings than was Leopold. For example, Holmes Rolston III, in his definitive book on the subject, is careful to distinguish between interhuman ethics and environmental ethics, denoting a more careful delineation of culture and nature. He notes that:

> In an environmental ethic, what humans want to value is not compassion, charity, rights, personality, fairness or even pleasure and the pursuit of happiness. Those values belong to interhuman ethics—in culture, not nature—and to look for them here is to make a category mistake. What humans value in nature is an ecology, a pregnant Earth, a projective and prolife system in which (considering biology alone, not culture) individuals can prosper but are also sacrificed indifferently to their pains and pleasures, individual well-being a lofty but passing role in a storied natural history .

He makes it clear that he is not extending interhuman ethics per se beyond the human realm, only the "logic of ethics," in order to establish human duties to the nonhuman world. In other words, he is not seeking to apply the ten commandments to the natural world, but to generate "a new commandment about landscapes and ecosystems."[3]

While this is in many ways a more sophisticated approach to ethics than that adopted by Leopold, Rolston's thought still parallels that of his intellectual forebear. Just as Leopold uses the term "community" in an ambiguous and overgeneralized manner in order to establish his Land Ethic, a similar kind of blurring occurs where the scientific basis of "natural value" is concerned, hinging on the ambiguity of the term "value" in an ecological context.

Taken as a whole, Rolston maintains, Nature is inventive and "projective":

> We confront a *projective nature*, one restlessly full of projects—stars, comets, planets, moons, and also rocks, crystals, rivers, canyons, seas. The life in which these astronomical and geological processes culminate is still more impressive, but it is of a piece with the whole projective system. . . . One cannot be impressed with life in isolation from its originating matrix. Nature is a fountain of life, and the whole fountain—not just the life that issues from it—is of value.

For Rolston, the fulfillment of these natural projects is intrinsically valuable, meaning that it is good independent of any human evaluation of it. The value of this system of projective nature exerts a claim on human behavior, in that it elicits an experience of value on the part of humans when it is discovered, which then entails an obligation or a duty to respect and preserve the system. "The only fully responsible behavior is to seek an appropriate relationship to the parental environment, which is projecting all this display of value."[4]

The most important problem Rolston faces is that of deriving this sense of obligation (an "ought") from observation or theory regarding the objective state of nature (an "is"). Rolston sets aside the is-ought problem as applying only in human ethics, and states outright that "in environmental ethics one's beliefs about nature, which are based upon but exceed biological and ecological science, have everything to do with beliefs about duty. The way the world *is* informs the way it *ought to be.*" It is important to note that, for Rolston, this evaluation transcends scientific description; it is not ecology but "metaecology." In making the transition from *is* to *good* to *ought*, Rolston writes, "we leave science to enter the domain of evaluation, from which ethics follows." Further, "no amount of research can verify that the right is the optimum biotic community, yet ecological description generates this valuing of nature, endorsing the systemic rightness."[5]

The question I would raise here is whether Rolston has simply chosen his view of what is in order to support his view of what ought to be. Consider that his approach to environmental ethics relies on the assumption of the truth of a given set of scientific theories and observations, even as he seeks to transcend the sciences. He assumes, for example, that ecosystems have some degree of integrity and stability, an "optimum" toward which they naturally strive, modeled on but not as tightly integrated as individual organisms. He recognizes that this might not be the case, based on the fact that there have been arguments made in ecology for an individualistic theory of biotic community. He ultimately rejects such a possibility out of hand, however; in so doing, he seems to want to establish himself as an authority in the assessment of ecological theory. It is also a possibility, according to some recent ecological theories, that there is not a single stable optimum in any ecosystem or biotic community, that ecological systems are characterized by perpetual change. If one of these theories were to gain ascendancy over that which Rolston prefers, the locus of environmental values itself might shift—or the project of environmental ethics might simply fall apart.

One of the motivations for and shapers of environmental ethics is its character as a movement of opposition, albeit a somewhat narrow one. I would argue that the theory of intrinsic natural value takes shape as a rejection of two aspects of traditional value theory, namely the principles that values are dependent on subjective (i.e., human) valuers and that nature only has value insofar as it is provides the means for human ends. This rejection seems to be motivated by the belief that traditional value theory is somehow to blame in environmental degradation, or at least that it is not adequate to the task of responding to that degradation. It also seems to be the case that the movement of opposition provides not only the motivation for environmental ethics but also the ad hoc justification for its principles. This is suggested by the manner in which Rolston carefully chooses principles and theories that, when elaborated deductively into a cosmological and ethical system, will provide him with the desired result, namely, a set of clearly defined obligations to the natural world.

Like environmental ethics, the principles of deep ecology—the third form of environmental philosophy I have chosen—are defined largely by and through opposition to what is perceived as the "dominant" worldview, except that the opposition is broadened to include not just anthropocentric value theories, but also all of western civilization, especially western science and technology. It is among the practitioners of deep ecology that may be found some of the most outspoken critics of "Cartesianism." While Rolston and other environmental ethicists make tacit assumptions about the way the world works in order to bolster their ethical project, it is the *primary* project of Deep Ecology to "articulate a comprehensive religious and philosophical worldview," a "new ecological philosophy for our time" on the basis of some sort of "ecological consciousness."[6] I have already outlined the pattern of this form of opposition, in which holism is substituted for reductionism, and so on.

Deep Ecologists' adherence to holism, with its corollaries of organicism and monism, is expressed in the first of two basic principles as "self-realization." At the core of this principle is the conscious act of "identification" of the self with the other: I cannot become fully myself, the principle states, unless I "identify" myself with the broader Self of "organic wholeness." Just as I come to understand another person by recognizing the ways in which that person is like me, so I am to recognize the sameness, the identity, between myself and nonhuman beings, between myself and the biosphere. This is a melding of the spiritual and the biological into what is called "spiritual/biological personhood."[7]

Deep Ecologists Devall and Sessions establish their principles, including that of self-realization, as self-evident "*ultimate norms* or intuitions which are themselves not derivable from other principles or intuitions," and which "cannot be validated, of course, by the methodology of modern science."[8] Arne Naess, the founder of the Deep Ecology movement, states the relationship of Deep Ecology to ecology as follows:

> In so far as ecology movements deserve our attention, they are *ecophilosophical* rather than ecological. Ecology is a *limited* science which makes *use* of scientific methods. Philosophy is the most general forum of debate on fundamentals, descriptive as well as prescriptive. . . . By an *ecosophy* I mean a philosophy of ecological harmony or equilibrium.[9]

This "philosophy of ecological harmony" involves "hypotheses concerning the state of affairs in our universe," hypotheses that are posited and, by being cast in terms of a "debate on fundamentals," exempted from further scientific scrutiny. This is the basis on which Naess can claim that, in nature, cooperation—the "ability to coexist and cooperate"—is more important than competition—the "ability to kill, exploit and suppress."[10]

From the point of view of Deep Ecology as a movement of opposition, the science that is called "ecology" comes down on both sides of the ideological divide as both an ally and a foe. As a science of *relationships*, it is seen as having contributed to our understanding of the natural world by its "rediscovery" of organic wholeness,[11] or at least by its discovery of larger wholes with which we can identify or that we are bound to respect. Ecology is a source of metaphors, which seem to be selected according to their intellectual fit with the ultimate norms of the movement.

As a *science* of relationships, however, ecology is seen as participating in the Cartesian tradition of reductive and exploitative science. Concerning the use of models in ecosystem ecology, Devall and Sessions write:

> The sensuousness of the natural world is left out of their purely formal, computerized mathematical abstractions. Much of scientific ecological theory is based on cybernetic systems theory—a contribution of the Cartesian seventeenth-century view of the universe as a machine—and should be held suspect for that reason.[12]

Ecology is denigrated as the reduction of wholeness to chemical and mathematical terms and thus as the justification for further technological exploitation and degradation of the environment.[13]

The new worldview is thus established on the basis of science, but also in spite of science; it seeks no further support from that direction. Even though the materials for the construction of the deep ecological worldview have been around for some time (Devall and Sessions credit numerous sources), this holistic biocentrism is taken to be a sound metaphysical principle simply because it stands directly opposed to what are taken to be established or dominant metaphysical principles. As the ideology of an opposition movement, its coherence and its ability to stand on its own merits outside the narrow context of that opposition are never brought into question.

Deep Ecology takes its place among the most radical elements of the environmental movement; also in the radical camp is a theory that has long been at odds with the principles of Deep Ecology, namely, social ecology as defined and defended by Murray Bookchin. Bookchin is broadly critical of other environmental thought as perpetuating the image of humans as opposed to nature, or of humans as needing to be reduced to biological terms. Deep Ecology is a favorite target, especially its doctrine of "biocentrism." Bookchin nurtures an abiding contempt for this sort of "spirituality" in environmentalism: "'Deep ecology' was spawned among well-to-do people who have been raised on a spiritual diet of Eastern cults mixed with Hollywood and Disneyland fantasies."[14]

By contrast, social ecology is a radical political movement that seeks to establish an "ecological" society, based on the insight that the domination of nature by humans follows from the domination of humans by humans. Thus, Bookchin concludes, a nonhierarchical, decentralized, democratic society would be inherently "ecological." This vision of the future is shared widely among radical environmentalists and has, ironically, been adopted by Deep Ecologists to inform their vision of the future of humanity. Bookchin sets himself apart in his emphasis on the social aspect of that vision. "Ecology alone, firmly rooted in *social* criticism and a vision of *social* reconstruction, can provide us with the means for remaking society in a way that will benefit nature *and* humanity."[15] In developing this position, Bookchin encounters many of the same difficulties as Deep Ecologists, especially where the sciences are concerned: he relies upon scientific concepts, but keeps himself at a safe distance from the process by which those concepts have been developed.

Social ecology relies on the assumption that both society and nature were harmonious before humans began to dominate each other and the world around them. In other words, domination and hierarchy are relatively recent human inventions that exist neither in the state of Nature nor in the earliest human societies. In this belief, Bookchin is clearly indebted to Rousseau's image of the noble savage, the idea that humans lived in freedom and harmony before the coming of civilization. To support his claim that domination and exploitation do not exist in the natural world, he relies on a selective ethology. In *Remaking Society*, he looks to the behavior of animals toward one another, but it is notable that his examples are intraspecific, seemingly ignoring the possibility that predation is a kind of "domination" or "exploitation." Further, he rejects even the appearance of hierarchy in primate societies, the "high status" of alpha males for example, because they do not rule "through institutional forms of violence as social elites do."[16] Perhaps not, but this is hardly a guarantee of "harmony" in the natural world.

Bookchin offers an equally idiosyncratic reading of evolutionary theory. He takes it as a "presupposition" or an "intuition" that natural evolution is driven by an "inherent striving" or "nisus" toward greater differentiation, complexity, freedom, and subjectivity.[17] At one point, he states that nature is nothing more than this evolutionary striving. "This drama is the study of a nature rendered more and more aware of itself, a nature that slowly acquires new powers of subjectivity, and one that gives rise to a remarkable primate life-form, called human beings, that have the power to choose, alter, and reconstruct their environment." Moreover, evolution is "participatory":

> A brown hare that mutates into a white one and sees a snow-covered terrain in which to camouflage itself is *acting* on behalf of its own survival, not simply "adapting" in order to survive. It is not merely being "selected" by its environment; it is selecting its own environment, and making a *choice* that expresses a small measure of subjectivity and judgment.[18]

The support he offers for this claim is simply that it is an "intuition" based on a paleolontological record that he takes to be unambiguous in its significance: there is an observable trend toward more complexity, therefore there must be an inherent striving in nature toward more complexity. Furthermore, humans have emerged rather late in this history, and they are conscious beings, so he concludes that there must be an inherent striving in nature toward subjectivity and freedom. But, the question may still be asked, why should we take the leap from the description of the fossil record to the almost normative claim that nature tends toward freedom? Surely there are other possibilities.

In general, Bookchin presents his theories of nature as though they were established scientific fact, although the evidence he offers is selective and sparse. While he ultimately appeals to intuition, it seems likely that the true justification for most if not all of Bookchin's claims about the state of the world is ad hoc: his take on ethology, evolution, and anthropology help him to define, establish, and defend a radical social agenda. It is on this basis that he contends that "every social evolution, in fact, is virtually an extension of natural evolution into a distinctly human realm."[19]

In the end, it seems as though Bookchin shares a serious dilemma with Deep Ecology: he relies on the sciences, presupposes their concepts and theories; but he also must reject them on ideological grounds. He does not state this explicitly, but it seems evident that the sciences on which he relies are very much the products of modern, technological, hierarchical, capitalist culture. It is also a distinct possibility for Bookchin that the sciences themselves are "infected," so to speak, by whatever structures of oppression are built into the society that produces them.

Let these four examples suffice as evidence that the pattern I have identified in environmental thought does indeed exist. As I have noted, there are countless other examples, and among them are no doubt important exceptions. Nevertheless, I have deliberately selected prominent and influential figures to serve as examples, in order to suggest that the problem is pervasive, and that there is need for a re-examination of the epistemological bases of environmental thought at every level.

One way of characterizing environmental problems is as a set of variations on the more general problem of how we are to maintain our values—our lives, our security, our hopes, even our own aesthetic appreciation for and fascination by our natural surroundings—within the limitations and risks imposed upon us by our surroundings. Ever more frequently, we are confronted with the task of protecting that which we value against the consequences of our own actions. From such a perspective, there seems to be a profound connection between environmental problems and the problem of knowledge. It becomes vitally important that we determine the extent and limits of our knowledge of our surroundings, that we critically examine our expectations of ourselves and of the world that surrounds us, when we find ourselves pressed to decide what is required—personally, socially, and politically—to protect that which we value. If what we claim to know about the world is based on assumptions, then we must examine those assumptions to avoid being led astray.

In the western tradition, speculative philosophy has made the boldest claims for certainty, whereby philosophers are to adjudicate what is the case and what is not. It is little wonder, then, that environmental philosophers often move into a speculative mode when articulating the philosophical support for their claims about the way things are, frequently appealing to intuition or to fundamental principles that transcend the sciences. Such speculative systems rely, minimally, on the assumption that, by intuition or by reason, the human mind can discover the ultimate truth of things; for this to be the case, mind and nature must share some fundamental link that renders the world intelligible to us. At the very least, this article of philosophical faith—the "intelligibility thesis"—must be subjected to sustained scrutiny. I suspect that such scrutiny must result in the rejection of the intelligibility thesis and the acknowledgment that the world is, in its depths, opaque to us; I further suspect that environmental philosophy, as I have characterized it, does not even attain the highest standards of speculative philosophy.

I will not attempt to substantiate my suspicions here, but I would point out some features of environmental thinking in particular that are revealing. It is suggestive, though not conclusive, that the fundamental principles of environmental philosophers seem not to be self-evident so much as ideologically necessary; they are principles that, if accepted, lead naturally to the sort of social and political conclusions their advocates desire. This might not be such a bad thing, but if someone were to disagree with those principles, for whatever reason, there could be no further rational or intuitive appeal on the basis of which to resolve the dispute: one either does or does not believe. As a result, environmental philosophers might easily find themselves in the position of preaching to the choir.

If my suspicion is correct that the world is ultimately opaque to us, however, then the sciences will be no more capable of grasping the ultimate truth of things than can speculation. Nevertheless, as a matter of historical observation, science has surpassed philosophy as the accepted authority in matters of knowledge; it is only fitting that it, too, should come under scrutiny. What emerges from such scrutiny is a picture of science as a means of attaining knowledge that is simply of a different order from speculative philosophy: far from being the parodied "Cartesianism" that has been derided by environmental thinkers, and equally far from being the source of concepts appropriated indiscriminately for cosmological or ethical speculation, the sciences are a complex and sophisticated endeavor that must be understood as such.

This is not the place to articulate a full-fledged philosophy of science, so I will offer instead an outline or overview of the sciences that I have found useful. Perhaps the most important point to note is that the results of scientific investigation are never final or certain; theories are maintained not because of any claim to absolute "truth" but because of their explanatory and predictive powers, their ability to illuminate certain aspects of our experience. In other words, scientific theory offers us, not without bias or ambiguity or conflict, some small glimpse of what we can expect of various aspects of the world around us; what it clearly does not offer us is an intuitive grasp of the whole. Even so, the world is complex and frequently surprising, and it often overturns our expectations. For this reason, scientific investigation is by its very nature an open-ended process, where even long-standing theories may be swept away, or at least altered or refined, in light of further tests, debates, and discoveries. Strangely, at their best, the strengths of the sciences lie in their very tentativeness: they remain flexible in the face of a world that confounds expectation, and so can move on to theories that are ever more refined and comprehensive.

It is helpful to think of scientific concepts as specialized metaphors, tentative efforts to grasp the complexity of our surrounding world in terms the intellect can accept and understand; as our expectations change, so the metaphors are adapted to fit new circumstances. This process of adaptation is, in part, why scientific concepts cannot be appropriated directly from their scientific context and applied to other realms of human knowledge; the result of such a transposition is a blurring and loss of meaning, as in the case of Leopold's use of the community concept.

There is clearly much more to be said on all of these matters, but I did not set out to resolve at a single stroke every problem that might arise. Rather, I have simply offered a therapeutic counterchallenge to environmental thought: those who would raise the "environmental challenge" must be exceedingly careful when making claims about the world, especially when drawing from the sciences to do so. I should further clarify that my purpose in raising epistemological doubts about environmental philosophy has not been exclusively negative; there is no point in dwelling on what I have construed as the errors of the past and of the present, nor is there any point in tearing down the genuine accomplishments of environmental thinkers. If there is any justification for examining errors and misunderstandings, it is to make it possible to steer a course around them. In effect, then, I have taken only the first few steps down a long and difficult path.

At the end of this path, as far as I can see, is a very different sort of envrionmental thinking, one characterized by epistemological modesty (not to say skepticism) based on an ongoing critical assessment of the bases of factual and normative claims about ourselves and our natural environment. In light of such an assessment, caution and circumspection would reign supreme in intellectual and practical matters alike; when nothing is truly certain, it is best to tread lightly. Still more intriguing is the prospect of an epistemological pluralism, an openness to multiple perspectives and interpretations of the world that might serve to illuminate limitations and possibilities we might otherwise have ignored. It is my belief and my hope that such a tradition of environmental thought would be far more coherent and effective than one patterned after the ideology of relatedness; it would provide us with a far more sturdy and flexible framework in which to reconsider our lives, our practices, and our institutions.

Notes

1. Aldo Leopold, *A Sand County Almanac, and Sketches Here and There* (Oxford: Oxford University Press, 1949), pp. 202–204, 224.
2. Ibid., p. 202.
3. Holmes Rolston III, *Environmental Ethics: Duties to and Values in the Natural World* (Philadelphia: Temple University Press, 1988), pp. 225, 161, 228.
4. Ibid., pp. 197, 198.
5. Ibid., p. 231.
6. Bill Devall and George Sessions, *Deep Ecology: Living as if Nature Mattered* (Salt Lake City: Gibbs Smith, Publisher, 1985), pp. ix, 65.
7. Ibid., p. 67.
8. Ibid., p. 66.
9. Arne Naess, "The Shallow and the Deep, Long-Range Ecology Movement. A Summary," *Inquiry* 16 (1973): 99.
10. Ibid., p. 96.
11. Devall and Sessions, *Deep Ecology*, p. 85.
12. Ibid., p. 89.
13. Naess, Devall, and Sessions hint broadly at this connection between science and domination. For a more sustained criticism from an ecofeminist perspective, see Carolyn Merchant, *The Death of Nature: Women, Ecology and the Scientific Revolution* (New York: Harper and Row, 1980).
14. Murray Bookchin, *Remaking Society: Pathways to a Green Future* (Boston: South End Press, 1990), p. 10.
15. Ibid., p. 13.
16. Ibid., p. 33.
17. Ibid., pp. 42, 200. For a more detailed discussion of Bookchin's theory of evolution, see his *The Philosophy of Social Ecology: Essays in Dialectical Naturalism* (Montreal: Black Rose Books, 1990).
18. Bookchin, *Remaking Society*, pp. 41, 37.
19. Ibid., p. 25.

II

Feeding People versus Saving Nature?

Holmes Rolston, III

When we must choose between feeding the hungry and conserving nature, people ought to come first. A bumper sticker reads: Hungry loggers eat spotted owls. That pinpoints an ethical issue, pure and simple, and often one where the humanist protagonist, taking high moral ground, intends to put the environmentalist on the defensive. You wouldn't let the Ethiopians starve to save some butterfly, would you?

"Human beings are at the centre of concerns for sustainable development." So the *Rio Declaration* begins. Once this was to be an *Earth Charter*, but the developing nations were more interested in getting the needs of their poor met. The developed nations are wealthy enough to be concerned about saving nature. The developing nations want the anthropocentrism, loud and clear. These humans, they add, "are entitled to a healthy and productive life in harmony with nature," but there too they seem as concerned with their entitlements as with any care for nature.[1] Can we fault them for it?

We have to be circumspect. To isolate so simple a tradeoff as hungry people versus nature is perhaps artificial. If we are too far abstracted from the complex circumstances of decision, we may not be facing any serious operational issue. When we simplify the question, it may become, minus its many qualifications, a different question. The gestalt configures the question, and the same question reconfigured can be different. So we must analyze the general matrix and then confront the more particular people-versus-nature issue.

Humans win? Nature loses? After analysis, sometimes it turns out that humans are not really winning if they are sacrificing the nature that is their life support system. Humans win by conserving nature—and these winners include the poor and the hungry. "In order to achieve sustainable development, environmental protection shall constitute an integral part of the development process and cannot be considered in isolation from it."[2] After all, food has to be produced by growing it in some reasonably healthy natural system, and the clean water that the poor need is also good for fauna and flora. Extractive reserves give people an incentive to conserve. Tourism can often benefit the local poor and the wildlife, as well as tourists. One ought to seek win-win solutions wherever one can. Pragmatically, these are often the only kind likely to succeed.

Yet just as obviously, there are times when nature is sacrificed for human development; most development is of this kind. By no means is all warranted, but that which gets people fed seems basic and urgent. Then nature should lose and people win. Or are there are times when at least some humans should lose and some nature should win? We are here interested in these latter occasions. Can we ever say that we should save nature rather than feed people?

1. Feed People First? Do We? Ought We?

"Feed people first!" That has a ring of righteousness. The *Rio Declaration* insists, "All States and all people shall cooperate in the essential task of eradicating poverty as an indispensable requirement."[3] In the biblical parable of the great judgment, the righteous have ministered to the needy, and Jesus welcomes them to their reward. "I was hungry and you gave me food, I was thirsty and you gave me drink." Those who refused to help are damned (Matt. 28:31–46). The vision of heaven is that "they shall hunger no more, neither thirst any more" (Rev. 7:16), and Jesus teaches his disciples to pray that this will of God be done on Earth, as it is in heaven. "Give us this day our daily bread" (Matt. 5:11). These are such basic values, if there is to be any ethics at all, surely food comes first.

Or does it? If giving others their daily bread were always the first concern, the Christians would never have built an organ or a sanctuary with a stained-glass window, but rather always given all to the poor. There is also the biblical story of the woman who washed Jesus' feet with expensive ointment. When the disciples complained that it should have been sold and given to the poor, Jesus replied, "You always have the poor with you" (Matt. 26:11). While the poor are a continuing concern, with whom Jesus demonstrated ample solidarity, there are other commendable values in human life, "beautiful things," in Jesus' phrase. The poor are always there, and if we did nothing else of value until there were no more poor, we would do nothing else of value at all.

Eradicating poverty is an indispensable requirement! Yes, but set these ideals beside the plain fact that we all daily prefer other values. Every time we buy a Christmas gift for a wife or husband, or go to a symphony concert, or give a college education to a child, or drive a new car home, or turn on the air conditioner, we spend money that might have helped to eradicate poverty. We mostly choose to do things we value more than feeding the hungry.

An ethicist may reply, yes, that is the fact of the matter. But no normative ought follows from the description of this behavior. We ought not to behave so. But such widespread behavior, engaged in almost universally by persons who regard themselves as being ethical, including readers of this article, is strong evidence that we in fact not only have these norms but also think we ought to have them. To be sure, we also think that charity is appropriate, and we censure those who are wholly insensitive to the plight of others. But we place decisions here on a scale of degree, and we do not feel guilty about all these other values we pursue, while yet some people somewhere on earth are starving.

If one were to always feed the hungry first, doing nothing else until no one in the world were hungry, this would paralyze civilization. People would not have invented writing, or smelted iron, or written music, or invented airplanes. Plato would not have written his dialogues, or Aquinas the *Summa Theologica*; Edison would not have discovered the electric light bulb or Einstein the theory of relativity. We both do and ought to devote ourselves to various worthy causes, yet persons in our own communities and elsewhere go hungry.

A few of these activities redound subsequently to help the poor, but the possible feedback to alleviating poverty cannot be the sole justification of advancing these multiple cultural values. Let us remember this when we ask whether saving natural values might sometimes take precedence. Our moral systems in fact do not teach us to feed the poor first. The Ten Commandments do not say that; the Golden Rule does not; Kant did not say that; nor does the utilitarian greatest good for the greatest number imply that. Eradicating poverty may be indispensable but not always prior to all other cultural values. It may not always be prior to conserving natural values either.

2. Choosing for People To Die

But food is absolutely vital. "Thou shalt not kill" is one of the commandments. Next to the evil of taking life is taking the sustenance for life. Is not saving nature—thereby preventing hunting, harvesting, or development by those who need the produce of that land to put food in their mouths— almost like killing? Surely one ought not to choose for someone else to die, an innocent who is only trying to eat; everyone has a right to life. To fence out the hungry is choosing that people will die. That can't be right.

Or can it? In broader social policy we make many decisions that cause people to die. When in 1988 we increased the national speed limit on rural Interstate highways from 55 to 65 miles per hour, we chose for 400 persons to die each year.[4] We decide against hiring more police, though if we did some murders would be avoided. The city council spends that money on a new art museum, or to give the schoolteachers a raise. Congress decides not to pass a national health care program that would subsidize medical insurance for some now uninsured who cannot otherwise afford it; and some such persons will, as a result, fail to get timely medical care and die of preventable diseases.

We may decide to leave existing air pollution standards in place because it is expensive for industry to install new scrubbers, even though there is statistical evidence that a certain number of persons will contract diseases and die prematurely. All money budgeted for the National Endowment for the Humanities, and almost all that budgeted for the National Science Foundation, could be spent to prevent the deaths of babies that die from malnutrition. We do not know exactly who will die, but we know that some will; we often have reasonable estimates how many. The situation would be similar, should we choose to save nature rather than to feed people.

U.S. soldiers go abroad to stabilize an African nation from which starving refugees are fleeing, and we feel good about it. All those unfortunate people cannot come here, but at least we can go there and help. All this masks, however, how we really choose to fight others rather than to feed them. The developed countries spend as much on military power in a year as the poorest two billion people on Earth earn in total income. The developed countries in 1990 provided $56 billion in economic aid to the poorer countries, but they also sold $36 billion worth of arms to them. At a cost of less than half their military expenditures, the developing countries could provide a package of basic health care services and clinical care that would save 10 million lives a year. World military spending in 1992 exceeded $600 billion. U.S. military spending accounted for nearly half this amount, yet in the United States one person in seven lives below the poverty line and over 37 million people lack any form of health care coverage.[5] These are choices that cause people to die, both abroad and at home.

But such spending, a moralist critic will object, is wrong. This only reports what people do decide, not what they ought to decide. Yes, but few are going to argue that we ought to spend nothing on military defense until all the poor are fed, clothed, housed. We believe that many of the values achieved in the United States, which place us among the wealthier nations, are worth protecting, even while others starve. Europeans and others will give similar arguments. Say if you like that this only puts our self-interest over theirs, but in fact we all do act to protect what we value, even if this decision results in death for those beyond our borders. That seems to mean that a majority of citizens think such decisions are right.

Wealthy and poverty-stricken nations alike put up borders across which the poor are forbidden to pass. Rich nations will not let them in; poor governments will not let them out. We may have misgivings about this on both sides, but if we believe in immigration laws at all, we, on the richer side of the border, think that protecting our life style counts more than does their betterment, even if they just want to be better fed. If we let in anyone who wanted to enter the United States and gave them free passage, hundreds of millions would come. Already 30 percent of our population growth is by immigration, legal and illegal. Sooner or later we must fence them out or face the loss of prosperity that we value. We may not think this is always right, but when one faces the escalating numbers that would swamp the United States, it is hard not to conclude that it is sometimes right. Admitting refugees is humane, but it lets such persons flee their own national problems and does not contribute to any long-term solutions in the nations from which they emigrate. Meanwhile, people die as a result of decisions about who is admitted and who is not.

Some of these choices address the question of whether we ought to save nature if this causes people to die. Inside our U.S. boundaries, we have a welfare system that ostensibly refuses to let anyone starve. Fortunately, we are wealthy enough to afford this as well as nature conservation. But if push came to shove, we would think it wrong-headed to put animals (or art, or well-paid teachers) over starving people. Does that not show that, as domestic policy, we take care of our own? We feed people first—or at least second, after military defense. Yet we let foreigners die, when we are not willing to open up our five hundred wilderness areas, nearly 100 million acres, to Cubans and Ethiopians.

3. Hunger and Social Justice

The welfare concept introduces another possibility, that the wealthy should be taxed to feed the poor. We should do that first, rather than cut into much else that we treasure, possibly losing our wildlife, or wilderness areas, or giving up art, or underpaying the teachers. In fact, there is a way greatly to relieve this tragedy, could there be a just distribution of the goods of culture, now often so inequitably distributed. Few persons would need to go without enough if we could use the produce of the already domesticated landscape justly and charitably. It is better to try to fix this problem where it arises, within society, than to try to enlarge the sphere of society by the sacrifice of remnant natural values, by, say, opening up the wilderness areas to settlement. Indeed, the latter only postpones the problem.

Peoples in the South (a code word for the lesser developed countries, or the poor) complain about the overconsumption of peoples in the North (the industrial rich), often legitimately so. But Brazil has within its own boundaries the most skewed income distribution in the world. The U.S. ratio between personal income for the top 20 percent of people to the bottom 20 percent is 9 to 1; the ratio in Brazil is 26 to 1. Just 1 percent of Brazilians control 45 percent of the agricultural land. The biggest 20 landowners own more land between them than the 3.3 million smallest farmers. With the Amazon still largely undeveloped, there is already more arable land per person in Brazil than in the United States. Much land is held for speculation; 330 million hectares of farm land, an area larger than India, is lying idle. The top 10 percent of Brazilians spend 51 percent of the national income.[6] This anthropocentric inequity ought to be put "at the centre of concern" when we decide about saving nature versus feeding people.

Save the Amazon! No! The howler monkeys and toucans may delight tourists, but we ought not to save them if people need to eat. Such either-or choices mask how marginalized peoples are forced onto marginal lands; and those lands become easily stressed, both because the lands are by nature marginal for agriculture, range, and life support, and also because by human nature marginalized peoples find it difficult to plan for the long range. They are caught up in meeting their immediate needs; their stress forces them to stress a fragile landscape.

Prime agricultural or residential lands can also be stressed to produce more and more, because there is a growing population to feed; or to grow an export crop, because there is an international debt to pay. Prime agricultural lands in southern Brazil, formerly used for growing food and worked by tenants who lived on these lands and ate their produce as well as sent food into the cities, have been converted to lands that use mechanized farming to grow coffee as an export crop, to help pay Brazil's massive debt, contracted by a military government since overthrown. Peoples forced off these lands were resettled in the Amazon basin, aided by development schemes fostered by the military government, resettled on lands really not suitable for agriculture. The integrity of the Amazon, to say nothing of the integrity of these peoples, is being sacrificed to cover for misguided loans. Meanwhile, the wealthy in Brazil pay little or no income tax that might be used for such loan repayment.

The world is full enough of societies that have squandered their resources, inequitably distributed wealth, and degraded their landscapes; and who will be tempted to jeopardize what natural values remain as an alternative to solving hard social problems. The decision about social welfare, poor people over nature, usually lies in the context of another decision, often a tacit one, to protect vested interests, wealthy people over poor people, wealthy people who have exploited nature already, ready to exploit anything they can. At this point in our logic, en route to any conclusion such as let-people-starve, we regularly reach an if-then, go-to decision point, where before we face the people-over-nature choice we have to reaffirm or let stand the wealthy-over-poor choice.

South Africa is seeking an ethic of ecojustice enabling 5 million privileged whites and 29 million exploited blacks (as well as several million underprivileged "coloureds") to live in harmony on their marvelously rich but often fragile landscape.[7] Whites on average earn nearly ten times the per capita income of blacks. White farmers, 50,000 of them, own 70 percent percent of farmland; 700,000 black farmers own 13 percent of the land (17 percent is held by others). Black land ownership of land was long severely restricted by law. Forced relocations of blacks and black birth rates have combined to give the homelands, small areas carved out within the South African nation, an extremely high average population density. When ownership patterns in the homelands are combined with those in the rest of the nation, land ownership is as skewed as anywhere on Earth. Compounding the problem is that the black population is growing, and is already more than ten times what it was before the Europeans came.

The land health in South Africa is poor. South African farmers lose twenty tons of topsoil to produce one ton of crops. Water resources are running out; the limited wetlands in an essentially arid nation are exploited for development; water is polluted by unregulated industry. Natal, one of the nation's greenest and most glorious areas, is especially troubled with polluted winds. Everywhere, herbicides float downwind with adverse human, vegetative, and wildlife effects on nontarget organisms.

With an abundance of coal, South Africa generates 60 percent of the electricity on the African continent, sold at some of the cheapest rates in the world, although less than a third of South Africans have electricity. The Eskom coal-burning power plants in the Transvaal are the worst offenders in air pollution, leaving the highveld as polluted as was East Germany, and threatening an area producing 50 percent of South Africa's timber industry and 50 percent of the nation's high potential agricultural soils. As a result of all this, many blacks go poorly nourished; some, in weakened condition, catch diseases and die.

What is the solution? South Africa also has some of the finest wildlife conservation reserves in Africa. Some are public; some are private. They are visited mostly by white tourists, often from abroad. One hears the cry that conserving elitist reserves, where the wealthy enjoy watching lions and wildebeest, cannot be justified where poor blacks are starving. What South Africa needs is development, not conservation. In an industry-financed study, Brian Huntley, Roy Siegfried, and Clem Sunter conclude: "What is needed is a much larger cake, not a sudden change in the way it is cut."[8] One way to get a bigger cake would be to take over the lands presently held as wildlife reserves.

But more cake, just as unequally cut, is not the right solution in a nation that already stresses the carrying capacity of its landscape. Laissez-

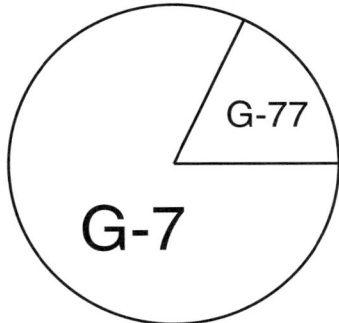

Figure 1. Proportionate Production and Consumption among Nations.

faire capitalists propose growth so that every one can become more prosperous, oblivious to the obvious fact that even the present South African relationship to the landscape is neither sustainable nor healthy. They seem humane; they do not want anyone to starve. The rhetoric, and even the intent, is laudable. At the same time, they want growth because this will avoid redistribution of wealth. The result, under the rubric of feeding people versus saving nature, is in fact favoring the wealthy over the poor.

What is happening is that an unjust lack of sharing between whites and blacks is destroying the green. It would be foolish for all, even for white South Africans acting in their own self-interest, further to jeopardize environmental health rather than to look first and resolutely to solving their social problems. It would not really be right for South Africa to open up their magnificent wildlife reserves, seemingly in the interests of the poor, while the cake remained as inequitably divided as ever. Fortunately, many South Africans have realized the deeper imperative, and the recent historic election there and efforts toward a new constitution promise deep social changes. This, in turn, will make possible a more intelligent conservation of natural values.[9]

In the more fortunate nations, we may distribute wealth more equitably, perhaps through taxes or minimum wage laws, or by labor unions, or educational opportunities, and we do have in place the welfare systems referred to earlier, refusing to let anyone starve. But lest we seem too righteous, we also recall that we have such policies only domestically. The international picture puts this in a different light. There are two major blocs, the G-7 nations (the Group of 7, the big nations of North America, Europe, and Japan, "the North"), and the G-77 nations, once 77 but now including some 128 lesser developed nations, often South of the industrial North. The G-7 nations hold about one-fifth of the world's five billion persons, and they produce and consume about four-fifths of all goods and services. The

G-77 nations, with four-fifths of the world's people, produce and consume one-fifth. (See Figure 1.) For every person added to the population of the North, twenty are added in the South. For every dollar of economic growth per person in the South, twenty dollars accrue in the North.[10]

The distribution problem is complex. Earth's natural resources are unevenly distributed by nature. Diverse societies have often taken different directions of development; they have different governments, ideologies, and religions; they have made different social choices, valued material prosperity differently. Typically, where there is agricultural and industrial development, people think of this as an impressive achievement. Pies have to be produced before they can be divided, and who has produced this pie? Who deserves the pie? People ought to get what they earn. Fairness nowhere commands rewarding all parties equally; justice is giving each his or her due. We treat equals equally; we treat unequals equitably, and that typically means unequal treatment proportionate to merit. There is nothing evidently unfair in the pie diagram, not at least until we have inquired about earnings. Some distribution patterns reflect achievement. Not all of the asymmetrical distribution is a result of social injustice.

Meanwhile, it is difficult to look at the distribution chart and not think that something is unfair. Is some of the richness on one side related to the poverty on the other? Regularly, the poor come off poorly when they bargain with the rich; and wealth that originates as impressive achievement can further accumulate through exploitation. Certainly many of the hungry people have worked just as hard as many of the rich.

Some will say that what the poorer nations need to do is to imitate the productive people. Unproductive people need to learn how to make more pie. Then they can feed themselves. Those in the G-7 nations who emphasize the earnings model tend to recommend to the G-77 nations that they produce more, often offering to help them produce by investments that can also be productive for the G-7 nations. Those in the G-77 nations do indeed wish to produce, but they also see the exploitation and realize that the problem is sharing as well as producing. Meanwhile the growth graphs caution us that producing can be as much part of the problem as part of the solution. One way to think of the circular pie chart is that this is planet Earth, and we do not have any way of producing a bigger planet. We could, though, feed more people by sacrificing more nature.

Meanwhile too, any such decisions take place inside this 1/5-gets-4/5ths, 4/5ths-gets-1/5 picture. So it is not just the Brazilians and the South Africans, but all of us in the United States, Europe, and Japan as well that have to face an if-then, go-to decision point, reaffirming and or letting stand the wealthy-over-poor division of the Earth's pie that we enjoy. This is what stings when we see the bumper sticker ethical injunction: "Live simply that others may simply live."

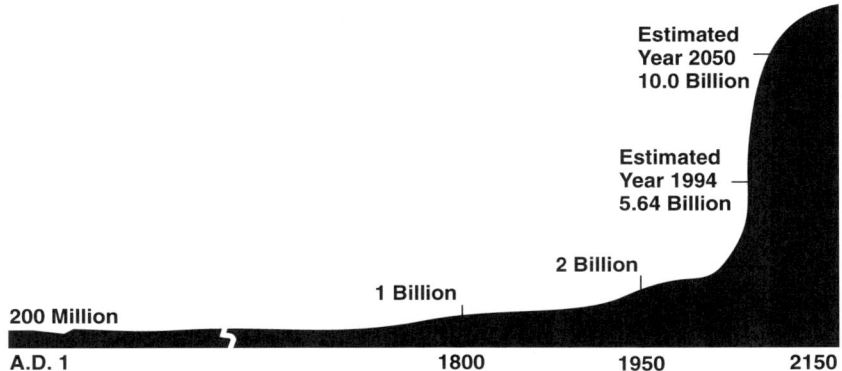

Figure 2. World Population Growth. Adapted from data in U.S. Bureau of the Census, Statistical Abstract of the United States: 1994, 114th edition (Washington, DC: 1994), p. 850.

4. Escalating Human Populations

Consider human population growth. (See Figure 2.) Not only have the numbers of persons grown, but also their expectations have grown so that we must superimpose one exploding curve on top of another. A superficial reading of such a graph is that humans really start winning big in the twentieth century. There are lots of them, and they want, and many get, lots of things. If one is a moral humanist, this can seem a good thing. Wouldn't it be marvelous if all could get what they want, and none hunger and thirst anymore?

But when we come to our senses, we realize that this kind of winning, if it keeps on escalating, is really losing. Humans will lose, and nature will be destroyed as well. Cultures have become consumptive, with ever-escalating insatiable desires, overlaid on ever-escalating population growth. Culture does not know how to say "Enough!" and that is not satisfactory. Starkly put, the growth of culture has become cancerous. That is hardly a metaphor, for a cancer is essentially an explosion of unregulated growth. The act of feeding people always seems humane, but, when we face up to what is really going on, feeding people without attention to the larger social results could mean we are feeding a kind of cancer.

One can say that where there is a hungry mouth, one should do what it takes to get food into it. But when there are two mouths the next day, and four the day after that, and sixteen the day after that, one needs a more complex answer. The population of Egypt was less than 3 million for over five millennia—fluctuating between 1.5 and 2.5 million. Today the population of Egypt is about 55 million. Egypt has to import more than half its food. The effects on nature, both on land health and on wildlife, have been adversely proportional.

If, in this picture, we look at individual persons caught up in this uncontrolled growth, and if we try to save nature, some persons will go hungry. Surely that is a bad thing. Would anyone want to say that people ought not to sacrifice nature, if need be, to alleviate such harm as best they can? From the people's perspective, they are only doing what humans have always done, making a resourceful use of nature to meet their own needs. Isn't that a good thing anymore? People are doomed, unless they can capture natural values.

But here we face a time-bound truth, where too much of a good thing becomes a bad thing. We have to figure out where people are located on the population curve, and realize that a good thing when human numbers are manageable is no longer a good thing when such people are really more cells of cancerous growth. That sounds cruel, and it is tragic, but it does not cease to be true for these reasons. For a couple to have two children may be a blessing; but the tenth child is a tragedy. When the child comes, one has to be as humane as possible, but one will only be making the best of a tragic situation; and if the tenth child is reared, and has ten children in turn, that will only multiply the tragedy. The quality of human lives deteriorates; the poor get poorer. Natural resources are further stressed; ecosystem health and integrity degenerate; and this compounds the losses again—a lose-lose situation. In a social system misfitted to its landscape, one's wins can only be temporary in a losing human ecology.

Even if there were to be an equitable distribution of wealth, the human population cannot go on escalating without people becoming all equally poor. Of the 90 million new people who will come on board planet Earth this year, 85 million will appear in the Third World, the countries least able to support such population growth. At the same time, each North American will consume 200 times as much energy, and many other resources. The 5 million new people in the industrial countries will put as much strain on the environment as the 85 million new poor. There are three problems here: overpopulation, overconsumption, and underdistribution. Sacrificing nature for development does not solve any of these problems, none at all. It only brings further loss. The poor, after a meal for a day or two, perhaps a decade or two, are soon hungry all over again, only now poorer still because their natural wealth is also gone.

To say that we ought always to feed the poor first commits a good-better-best fallacy. If a little is good, more must be better, most is best. If feeding some humans is good, feeding more is better. And more. And more! Feeding all of them is best? That sounds right. We can hardly bring ourselves to say that anyone ought to starve. But we reach a point of diminishing returns, when the goods put at threat lead us to wonder.

5. Endangered Natural Values

Natural values are endangered at every scale: global, regional, and local; at levels of ecosystems, species, organisms, populations, fauna and flora, terrestrial and marine, charismatic megafauna down to mollusks and beetles. This is true in both developed and developing nations, though we are discussing here places where poverty threatens biodiversity.

Humans now control 40 percent of the planet's land-based primary net productivity, that is, the basic plant growth that captures the energy on which everything else depends.[11] If the human population doubles again, the capture will rise to 60–80 percent, and little habitat will remain for natural forms of life that cannot be accommodated after we have put people first. Humans do not make effective use of the lands they have domesticated. A World Bank study found that 35 percent of the Earth's land now has now become degraded.[12] Daniel Hillel, in a soils study, concludes, "Present yields are extremely low in many of the developing countries, and as they can be boosted substantially and rapidly, there should be no need to reclaim new land and to encroach further upon natural habitats."[14]

Africa is a case in point, and Madagascar epitomizes Africa's future. Its fauna and flora evolved independently of the mainland continent; there are thirty primates, all lemurs; the reptiles and amphibians are 90 percent endemic, including two-thirds of all the chameleons of the world; and there are 10,000 plant species, of which 80 percent are endemic, including a thousand kinds of orchids. Humans came there about 1,500 years ago and lived with the fauna and flora more or less intact until this century. Now an escalating population of impoverished Malagasy people rely heavily on slash-and-burn agriculture, and the forest cover is one-third of the original (27.6 million acres down to 9.4 million acres), most of the loss occurring since 1950.[14] Madagascar is the most eroded nation on Earth, and little or none of the fauna and flora is safely conserved. Population is expanding at 3.2 percent a year; remaining forest is shrinking at 3 percent, almost all to provide for the expanding population. Are we to say that none ought to be conserved until after no person is hungry?

Tigers are sliding toward extinction worldwide. Populations have declined 95 percent in this century; the two main factors are loss of habitat and a ferocious black market in bones and other body parts that are used in traditional medicine and folklore in China, Taiwan, and Korea—uses that are given no medical credence. Ranthambhore National Park in Rajasthan, India, is a tiger sanctuary; there were forty tigers there during the late 1980s, but they have been reduced to 20–25 tigers today by human pressures—illicit cattle grazing and poaching. There are 200,000 people within three miles of the core of the park—more than double the population when the park was launched twenty-one years ago. Most of these people depend on wood from the 150 square miles of park to cook their food. They graze in and around the park some 150,000 head of scrawny cattle, buffalo, goats, and camels. The cattle impoverish habitat and carry diseases to the ungulates that are the tigers' prey base. In May 1993, a young tigress gave birth to four cubs; that month 316 human babies were born in the villages surrounding the park.[15]

The tigers may be doomed, but ought they to be? Consider, for instance, that there are minimal reforestation efforts, or that cattle dung can be used for fuel with much greater efficiency than is being done, or that, in an experimental herd of Jersey and Holstein cattle there, the yield of milk increased ten times that of the gaunt, free-ranging local cattle, and that a small group of dairy producers has increased milk production 1,000 percent in just three years. In some moods we may insist that people are more important than tigers. But in other moods these majestic animals seem the casualties of human inabilities to manage themselves and their resources intelligently, a tragic story that leaves us wondering whether the tigers should always lose and the people win.

6. When Nature Comes First

Ought we to save nature if this results in people going hungry? In people dying? Regrettably, sometimes, the answer is yes. In twenty years Africa's black rhinoceros population declined from 65,000 to 2,500, a loss of 97 percent; the species faces extinction. Again, as with the tigers, there has been loss of habitat due to human population growth, an important and indirect cause; but the primary direct cause is poaching, this time for horns. People cannot eat horns; but they can buy food with the money from selling them. Zimbabwe has a hard-line shoot-to-kill policy for poachers, and over 150 poachers have been killed.[15]

So Zimbabweans do not always put people first; they are willing to kill some, and to let others go hungry rather than sacrifice the rhino. If we always put people first, there will be no rhinos at all. Always too, we must guard against inhumanity and take care, so far as we can, that poachers have other alternatives for overcoming their poverty. Still, if it comes to this, the Zimbabwean policy is effective. Given the fact that rhinos have been so precipitously reduced, given that the Zimbabwean population is escalating (the average married woman there *desires* to have six children),[17] one ought to put the black rhino as a species first, even if this costs human lives.

But the poachers are doing something illegal. What about ordinary people, who are not breaking any laws? The sensitive moralist may object that, even when the multiple causal factors are known and lamented, when it comes to dealing with individual persons caught up in these social forces, we should factor out overpopulation, overconsumption, and maldistribution, none of which are the fault of the particular persons who may wish to develop their lands. "I did not ask to be born; I am poor, not overconsuming; I am not the cause but rather the victim of the inequitable distribution of wealth." Surely there still remains for such an innocent person a right to use whatever natural resources one has available, as best one can, under the exigencies of one's particular life, set though this is in these unfortunate circumstances. "I only want enough to eat, is that not my right?"

Human rights must include, if anything at all, the right to subsistence. So even if particular persons are located at the wrong point on the global growth graph, even if they are willy-nilly part of a cancerous and consumptive society, even if there is some better social solution than the wrong one that is in fact happening, have they not a right that will override the conservation of natural value? Will it not just be a further wrong to them to deprive them of their right to what little they have? Can basic human rights ever be overridden by a society that wants to do better by conserving natural value?

This requires some weighing of the endangered natural values. Consider the tropical forests. There is more richness there than in other regions of the planet; they contain half of all known species. South America, for example, contains one-fifth of the planet's species of terrestrial mammals (800 species) and there are one-third of the planet's flowering plants.[18] The peak of global plant diversity is in the three Andean countries of Colombia, Ecuador, and Peru, where over 40,000 species occur on just 2 percent of the world's land surface.[19] But population growth in South America has been as high as anywhere in the world,[20] and people are flowing into the forests, often crowded off other lands.

What about these hungry people? Consider first people who are not now there but might move there. This is not good agricultural soil, and such would-be settlers are likely to find only a short-term bargain, a long-term loss. Consider the people who already live there. If they are indigenous peoples, and wish to continue to live as they have already for hundreds and even thousands of years, there will be no threat to the forest. If they are *cabaclos* (of mixed European and native races), they can also continue the life styles known for hundreds of years, without serious destruction of the forests. Such peoples may continue the opportunities that they have long had. Nothing is taken away from them. They have been reasonably well fed, though often poor.

Can these peoples modernize? Can they multiply? Ought there to be a policy of feeding first all the children they bear, sacrificing nature as we must to accomplish this goal? Modern medicine and technology have enabled them to multiply, curing childhood diseases and providing better nutrition, even if these peoples often remain at thresholds of poverty. Do not such people have the right to develop? A first answer is that they do, but with the qualification that all rights are not absolute, some are weaker, some stronger, and the exercise of any right has to be balanced against values destroyed in the exercise of that right.

The qualification brings a second answer. If one concludes that the natural values at stake are quite high, and that the opportunities for development are low, because the envisioned development is inadvisable, then a possible answer is: No, there will be no development of these reserved areas, even if people there remain in the relative poverty of many centuries, or even if, with escalating populations, they become more poor. We are not always obligated to cover human mistakes with the sacrifice of natural values.

Again, one ought to be as humane as possible. Perhaps there can be development elsewhere, to which persons in the escalating population can be facilitated to move, if they wish. Indeed, this often happens, as such persons flee to the cities, though they often only encounter further poverty there, owing to the inequitable distribution of resources that we have lamented. If they remain in these areas of high biological diversity, they must stay under the traditional life styles of their present and past circumstances.

Does this violate human rights? Anywhere that there is legal zoning, persons are told what they may and may not do in order to protect various social and natural values. Land ownership is limited ("imperfect," as lawyers term it) when the rights of use conflict with the rights of other persons. One's rights are constrained by the harm one does to others, and we legislate to enforce this (under what lawyers call "police power"). Environmental policy may and ought to regulate the harms that people do on the lands on which they live ("policing"), and it is perfectly appropriate to set aside conservation reserves to protect the cultural, ecological, scientific, economic, historical, aesthetic, religious, and other values people have at stake here, as well as for values that the fauna and flora have intrinsically in themselves. Indeed, unless there is such reserving of natural areas, counterbalancing the high pressures for development, there will be almost no conservation at all. Every person on Earth is told that he or she cannot develop some areas.

Persons are not told that they must starve, but they are told that they cannot save themselves from starving by sacrificing the nature set aside in reserves—not at least beyond the traditional kinds of uses that leave the biodiversity on the landscape. If one is already residing in a location where development is constrained, this may seem unfair, and the invitation to move elsewhere a forced relocation. Relocation may be difficult proportionately to how vigorously the prevailing inequitable distribution of wealth is enforced elsewhere.

Human rights to development, even for those who are poor, though they are to be taken quite seriously, are not everywhere absolute but have to be weighed against the other values at stake. An individual sees at a local scale; the farmer wants only to plant crops on the now-forested land. But environmental ethics sees that the actions of individuals cumulate and produce larger-scale changes that go on over the heads of these individuals. This ethic will regularly be constraining individuals in the interest of some larger ecological and social goods. That will regularly seem cruel, unfair to the individual caught in such constraints. This is the tragedy of the commons; individuals cannot see far enough ahead, under the pressures of the moment, to operate at intelligent ecological scales. Social policy must be set synoptically. This invokes both ecology and ethics and blends them, if we are to respect life at all relevant scales.

These poor may not have so much a right to develop in any way they please, as a right to a more equitable distribution of the goods of the Earth that we, the wealthy, think we absolutely own.

Our traditional focus on individuals and their rights can blind us to how the mistakes (as well as the wisdom) of parents can curse (and bless) the children—as the Ten Commandments put it, how "the iniquity of the fathers is visited upon the children to the third and fourth generation" (cf. Exod. 20:5). All this has a deeply tragic dimension, made worse by the coupling of human foibles with ecological realities. We have little reason to think that misguided compassion that puts food into every hungry mouth, bearing whatever consequences they may, will relieve the tragedy. We also have no reason to think that the problem will be solved without wise compassion, balancing a love for persons and a love for nature.

Ought we to feed people first and save nature last? We never face so simple a question. The practical question is more complex:

- If persons widely demonstrate that they value many other worthwhile things over feeding the hungry (Christmas gifts, college educations, symphony concerts);
- and if developed countries, to protect what they value, post national boundaries across which the poor may not pass (immigration laws);
- and if there is unequal and unjust distribution of wealth, and if just redistribution to alleviate poverty is refused;
- and if charitable redistribution of justified unequal distribution of wealth is refused;
- and if one-fifth of the world continues to consume four-fifths of the production of goods and four-fifths consumes one-fifth;
- and if escalating birth rates continue so that there are no real gains in alleviating poverty, only larger numbers of poor in the next generation;

- and if low productivity on domesticated lands continues, and if the natural lands to be sacrificed are likely to be low in productivity;

- and if significant natural values are at stake, including extinctions of species;

then one ought not always to feed people first, but rather one ought sometimes to save nature.

Many of the "ands" in this conjunction can be replaced with "ors" and the statement will remain true, though we cannot say outside of particular contexts to what extent. The logic is not so much that of implication as of the weighing of values and disvalues, natural and human, and of human rights and wrongs, past, present, and future.

Some will protest that this attitude risks becoming misanthropic and morally callous. The Ten Commandments order us not to kill, and saving nature can never justify what amounts to killing people. Yes, but there is another kind of killing here, one not envisioned at Sinai, where humans are a superkilling species. Extinction kills forms (*species*)—not just individuals; it kills collectively, not just distributively. Killing a natural kind is the death of birth, not just of an individual life. The historical lineage is stopped forever. Preceding the Ten Commandments is the Noah myth, when nature was primordially put at peril as much as it is today. There, God seemed more concerned about species than about the humans who had then gone so far astray. In the covenant reestablished with humans on the promised Earth, the beasts were specifically included. "Keep them alive with you . . . according to their kinds" (Gen. 6:19–20). There is something ungodly about an ethic by which the late-coming homo sapiens arrogantly regards the welfare of his own species as absolute, with the welfare of all the other five million species sacrificed to that. The commandment not to kill is as old as Cain and Abel, but the most archaic commandment of all is the divine, "Let the earth bring forth" (Gen. 1:24). Stopping that genesis is the most destructive event possible, and we humans have no right to do that. Saving nature is not always morally naive; it can deepen our understanding of the human place in the scheme of things entire, and of our duties on this majestic home planet.

Notes

1. *Rio Declaration on Environment and Development*. 1992. Principle 1, UNCED document A/CONF.151/26, vol. 1, pages 15–25.
2. *Rio Declaration*, Principle 4.
3. *Rio Declaration*, Principle 5.
4. Insurance Institute for Highway Safety, Arlington, Virginia, *Status Report*, 29 (10) (September 10, 1994):3.
5. Ruth Leger Sivard, *World Military and Social Expenditures*, 15th ed. (Washington, DC: World Priorities, Inc., 1993).
6. Jonathan Power, "Despite Its Gifts, Brazil Is a Basket Case," *The Miami Herald*, June 22, 1992: 10A.
7. The empirical data below are in: Brian Huntley, Roy Siegfried, and Clem Sunter, *South African Environments into the 21st Century* (Cape Town: Human and Rousseau, Ltd., and Tafelberg Publishers Ltd., 1989); Rob Preston-Whyte and Graham House, eds., *Rotating the Cube: Environmental Strategies for the 1990s* (Durban: Department of Geographical and Environmental Sciences and Indicator Project South Africa, University of Natal, 1990); and Alan B. Durning, *Apartheid's Environmental Toll* (Washington, DC: Worldwatch Institute, 1990).
8. Huntley, Siegfried, and Sunter, *South African Environments*, p. 85.
9. Mamphela Ramphele, ed., *Restoring the Land: Environment and Change in Post-Apartheid South Africa* (London: Panos Publications, 1991).
10. The pie chart summarizes data in the *World Development Report 1991* (New York: Oxford University Press, 1991).
11. Peter M. Vitousek, Paul R. Ehrlich, Anne H. Ehrlich, and Pamela A. Matson, "Human Appropriation of the Products of Biosynthesis," *BioScience* 36 (1986): 368–373.
12. Robert Goodland, "The Case That the World Has Reached Limits," in Robert Goodland, Herman E. Daly, and Salah El Serafy, eds., *Population, Technology, and Lifestyle* (Washington, DC: Island Press, 1992), pp. 3–22.
13. Daniel Hillel, *Out of the Earth* (New York: Free Press, Macmillan, 1991), p. 279.
14. E. O. Wilson, *The Diversity of Life* (Cambridge, MA: Harvard University Press, 1992), p. 267; Alison Jolly, *A World Like Our Own: Man and Nature in Madagascar* (New Haven: Yale University Press, 1980).
15. Geoffrey C. Ward, "The People and the Tiger," *Audubon* 96 (4) (July–August 1994): 62–69.
16. Joel Berger and Carol Cunningham, "Active Intervention and Conservation: Africa's Pachyderm Problem," *Science* 263 (1994): 1241–42.
17. John Bongaarts, "Population Policy Options in the Developing World," *Science* 263(1994):771–776.
18. Michael A. Mares, "Conservation in South America: Problems, Consequences, and Solutions," *Science* 233 (1986): 734–39.
19. Wilson, *The Diversity of Life*, p. 197.
20. Ansley J. Coale, "Recent Trends in Fertility in the Less Developed Countries," *Science* 221 (1983): 828–32.

Part 3

Struggle Up Close:
Current Conflicts in Environmental Theory and Practice

12

Ecofascism:
A Threat to American Environmentalism?

Michael E. Zimmerman

In 1994, the U.S. Congress saw the introduction of "takings" bills that, under certain conditions, would compel the government to provide monetary compensation to wetlands owners if recently adopted federal regulations prevent those owners from "developing" their wetlands property for economic gain.[1] These bills maintain, in part, that such regulations violate the "takings" clause of the Fifth Amendment to the U.S. Constitution, according to which private land shall not be "taken for public use without just compensation."[2] Though the regulations do not involve outright appropriation of private property, they do restrict the *use* of property in a way that may deprive owners of potential profit. In addition to complaining about regulatory complexity, landowners maintain that federal agencies have recently used questionable criteria to define as "wetlands" large tracts of land that many people would not count as wet.

Louisiana Congressman Billy Tauzin, whose state is losing twenty-five square miles of wetlands every year, is one of the most outspoken proponents of the "takings" bills. Tauzin compares public anger over intrusive wetlands regulation with the resentment against "taxation without representation" that led to the Boston Tea Party in 1773.[3] As one journalist notes, Congressman Tauzin's crusade for property rights has "an indis-

Copyright 1995 by *Social Theory and Practice*, Vol. 21, No. 2 (Summer 1995).

putable appeal in America, where a distrust of big government, the mythos of rugged individualism, and the sanctity of private property are powerful political forces."[4] The American environmental movement now seems threatened by the same anti–Big Government forces that generated the Republican victory in November 1994.

Environmentalists insist that people have already destroyed ("developed") vast amounts of land that everyone would agree is wet, including ecologically vital tidal marshes, swamps, and bogs. Moreover, despite the complaints voiced by property rights advocates, evidence suggests that wetlands regulations have failed appreciably to slow the development of wetlands, even in such ecologically important and closely scrutinized areas as the Chesapeake Bay. Hence, environmentalists regard the "takings" bills as dangerous instances of the environmental backlash being fostered by demagogues like Rush Limbaugh and by antiecological groups such as the "Wise Use Movement."[5] Some supporters of "takings" bills, including Ron Arnold of the Wise Use Movement, have gone so far as to describe environmental activists as romantic, irrational, nature-worshipping "ecofascists," who want the government to seize private property in order to protect plants and animals, thereby preventing honest human beings from making a living.[6] Even some environmental philosophers have accused their more radical colleagues of promoting ecofascism, by elevating the needs/interests of the organic "whole" (the "land") above the needs/interests of individual organisms, including animals and humans. In this respect, such critical environmental philosophers find themselves siding with anthropocentric market liberals and Marxists who have traditionally disparaged "nature-loving" romantics for having reactionary tendencies that may be consistent with fascism.

Of course, depicting *all* those who seek to protect nature as ecofascists is ludicrous, since contemporary politicians of virtually every stripe purport to be in favor of a healthy environment. Presumably, even Ron Arnold (a former Sierra Club member) retains some concerns about environmental degradation, though he now regards much of environmentalism as misguided and dangerous. Nevertheless, increasingly widespread public resentment against the tangled thicket of federal environmental regulations allows him and his allies to put the environmental lobby on the defensive by polarizing public opinion: either one is a citizen favoring people, private property, and the U.S. Constitution; or one is a radical environmentalist favoring nature, communal ownership, and ecofascism.

The "radical" ecology that Arnold has in mind promotes biocentrism or ecocentrism, according to which humankind is not a privileged species but rather one member of the complex biotic community. One type of radical ecology, "Deep Ecology," claims that the consequences of human arrogance—manifest in rampant industrial technology and mindless consumerism—are threatening the integrity of the biosphere. According to deep ecologists, the longer that democratic societies postpone making the difficult political decisions needed to solve environmental problems, the more drastic will be the political measures that may have to be taken later on to save remnants of humankind and the biosphere. One can imagine what such measures might entail: not only the seizure of private property, but perhaps also harassment, internment, torture, deportation, and worse, those designed to force people to comply with centrally imposed regulations (in areas ranging from consumption to reproduction) purporting to deal with an "ecological emergency."

One could certainly describe such practices as draconian or tyrannical, but not necessarily as fascist. Fascism gains its power by claiming to restore dignity, nobility, purpose, and privilege to some unique people or race whose members feel that their original mystical-organic social unity and their ties with the homeland are degenerating because of the insidious influence of alien races and foreign ideas. Moreover, fascism involves "an understanding of society in essentially military terms that stress struggle, heroism, leadership, masculinity, and youth."[7] To merit the name "ecofascism," then, a radical ecology movement would have to do more than demand that ecologically vital private property be protected from those who would despoil it. In addition to portraying ecological despoliation as a threat to the racial integrity of the people, an ecofascist movement would have to urge that society be reorganized in terms of an authoritarian, collectivist leadership principle based on masculinist-martial values. In this essay, I argue that even though the great majority of American environmentalists cannot be described as ecofascist, there are legitimate reasons to be concerned about the possible emergence of something like ecofascism in years to come.

Concern about the potential link between real (and perceived) ecological problems, on the one hand, and authoritarian politics, on the other, is not limited to critics who may use the term "ecofascism" hyperbolically. This concern has also been voiced about the Third World, where ecological problems, overpopulation, and socioeconomic woes are leading to nightmarish consequences. Recently, Robert D. Kaplan asserted:

> It is time to understand "the environment" for what it is: the national security issue of the early twenty-first century. The political and strategic impact of surging populations, spreading disease, deforestation and soil erosion, water depletion, air pollution, and, possibly, rising sea levels in critical, overcrowded regions like the Nile Delta and Bangladesh—developments that will prompt mass migrations and, in turn, incite group conflicts—will be the core foreign-policy challenge from which most others will ultimately emanate,

arousing the public and uniting assorted interests left over from the Cold War.[8]

Kaplan bases some of his remarks on the work of Thomas Fraser Homer-Dixon, who argues that rapidly mounting populations and resulting material scarcity are the breeding ground for Third World "hard regimes," whose leaders will have little patience with democratic principles and practices.[9] Some critics dismiss such ideas as the latest version of Malthusianism, but Kaplan and Homer-Dixon raise legitimate concerns about the social, political, and ecological consequences of unparalleled human population growth, which is occurring in the context of rising material expectations.[10] Much of the projected doubling of human population in the next forty years will occur in Third-World countries whose sociopolitical institutions, economic structures, and environmental bases are in some cases already fragile. Since many of the "countries" in question are the residual constructs of western colonialism, they may seem to lack a nationalistic sense needed to give rise to "fascist" regimes properly so-called. Yet if resurgent tribalism were harnessed by a charismatic leader who urged his people to recover their "homeland" and to protect it from environmental ruin, and if that leader had access to modern military equipment and communication facilities, he (or she) might generate a tribal ecofascism. Another possibility (one that already seems to be occurring) is that those tribes themselves will disintegrate into roving bands of youths, craving macho power and intoxication, and equipped with weapons seized from the defunct national armies that were formerly supplied by the superpowers.

If ecofascism does emerge, however, it would probably occur in countries already possessing a long-term sense of national identity that could be construed as "racial." In European countries, right-wing politicians have already been calling for a halt to immigration, and even for the expulsion of those aliens (principally, but not only nonwhite) who threaten the nation's cultural identity, along with its social, political, and economic well-being. As hordes of desperate people from ecologically devastated and politically disintegrating Third-World countries continue to pour into First-World countries, some politicians will call for harsh measures to exclude immigrants so as to prevent them from reproducing the same ecological and political crises that they (allegedly) created in their abandoned homelands. Ignoring the role played by western colonialism in creating those crises, European neofascist leaders may argue that dark-skinned immigrants are not only destroying native soil but are also polluting native blood. The question is whether—under mounting political and social stress—radical (and even mainstream) ecology movements in advanced societies, including the United States, will avoid the risk of aligning themselves with these dark forces.

1. A Nonfascist Ecosophy

In answering this question, let us begin by examining the "ecosophy" (Arne Naess's term for a "philosophical ecology") developed by a noted European environmentalist, Dr. Walther Schoenichen. Schoenichen's diagnosis of and prescription for the ecological problems of industrial countries are interesting in part because he can be read as distinguishing between environmental "reformism" and Deep Ecology. The former holds that environmental problems can be solved by making some changes within the existing system, for example, by controlling industrial pollution. Many Deep Ecologists maintain that such reforms cannot solve the ecological problems generated by the interlocking economic, social, political, and cultural paradigms of liberal and Marxist modernity. Those paradigms need revolutionary changes, if ecological disaster is to be averted.

Even before the Greens emerged in West Germany, Schoenichen was instrumental in organizing the German nature protection (*Naturschutz*) movement. Sounding like a Deep Ecologist, he writes that ecological destruction ensues when industrial societies ignore nature's inherent lawfulness and worth, and treat it instead as merely a "meaningless play of atoms."[11] Manifestly, since we are a part of the natural world, we cannot survive if we destroy the biological conditions that sustain us. Capitalism and communism alike pollute air and water, annihilate farmland and forests, destroy the habitats necessary for the preservation of species diversity, and exterminate native peoples.[12] Though recognizing that humans must intervene in the natural environment to survive and prosper, Schoenichen insists that high employment and a sound economy are consistent with environmentally sound practices, whereas unsound environmental practices undermine social and economic well-being. Industrial planning should be informed by biological and ecological sciences, which could warn people whether a proposed production method is ecologically destructive.[13] Crucial for long term human and environmental health is to replace exploitative economics (*Raubwirtschaft*) with sustainable economic development.[14]

To protect nature from the pressures of industrial economies, Schoenichen urges that national legislation be based on principles that acknowledge the importance of ecological integrity and natural beauty. He adds that government and industry should prepare what today would be called an "environmental impact statement" before undertaking projects that would significantly affect the land.[15] Anticipating contemporary American views about "environmental compensation," Schoenichen maintains that when someone profits at the expense of degrading the land, some profits should go for preserving natural monuments, for establishing protected nature regions, or for supporting the nature protection movement in general.[16] Moreover, landscape specialists should have major input into the planning of large construction sites.[17]

Like deep ecologist Arne Naess, Schoenichen recommends dividing the land into three regions: urban/industrial, mixed use, and wilderness. Urban/industrial regions would be given over to heavy human use, though even here pollution would be strictly limited to protect surrounding farmland, forests, and wild areas. In mixed-use areas, where some environmentally sustainable human intervention is permissible (for example, forestry, farming, and light industry), environmental and aesthetic planning would be involved.[18] Finally, nature protection regions would be located behind a kind of "protective hedge" (*Hege*) that would prevent human intervention, apart from nonintrusive activities such as hiking and scientific investigation. Nature protection areas should be of two types: the first would be guided by the principle of "preservation," the other by the principle of "conservation" (English in original). Preservation, "nature protection in the strict sense," involves establishing regions that are "hermetically sealed," so that humans do not interfere with the "free play of natural forces."[19] Conservation goes beyond mere protection by intervening to assist endangered natural phenomena. Schoenichen warns, however, that conservation practices run the risk of becoming so intrusive that the "natural" phenomenon being preserved virtually becomes a human artifact.[20]

Though eschewing a purely instrumentalist attitude toward nature, Schoenichen recognizes that "the development of culture is bound up with the constantly increasing mastery of nature," which has reached such a point that today only climate, earthquakes, volcanoes, and cosmic events seem to elude human control.[21] If self-interest has led to human domination of plants and animals, however, enlightened self-interest has also helped to justify the U.S. national parks that Schoenichen so admires. These parks not only protect wild nature from human abuse but also offer a place of recreation, refreshment, and aesthetic appreciation for urban families and workers who are usually deprived of direct contact with nature. In addition to providing a site for the scientific study of relatively undisturbed ecosystems, national parks also give voice to "the boundless pride in the land, which is shared by every real American."[22]

Encounters with the beauty of nature help to counter the prevailing view that nature is nothing more than an object for commercial exploitation. Like Paul Shepard in *Nature and Madness*, Schoenichen suggests that Christianity's antinaturalism, along with western rationalism and materialism, combined to disclose nature as something "other" than and inferior to humankind. To overcome this dissociative attitude, Schoenichen recommends that all German children be allowed to tend plants, so that "by witnessing all the wonder of growth and becoming—there can arise the first, still not comprehended surmise of the mysteries of divine Creation."[23]

In referring to Creation, Schoenichen emphasizes that the nature protection movement is motivated by more than utilitarian and prudential concerns. Many reform environmentalists assume that "The moral is what benefits mankind," but Schoenichen doubts that moral law can be derived from purely biological and utilitarian concepts.[24] Self-interest leads humans to protect species that happen to be useful or pleasing, but what about species that are useless or ugly? Don't they, too, deserve protection?[25] Though a scientist, not a philosopher, Schoenichen tries to solve the thorny problem of human duties to nonhuman beings by affirming that there is a principle higher than biological survival. Whatever this principle may be called—Providence, the Almighty, God—it has ultimate responsibility for maintaining all of Creation, especially species, that have worth in themselves.[26] Though our capacity for science and technology shows that divine nature wills "that man be master of all its realms," nature also "sinks into his soul the feeling of responsibility" for that over which man has gained dominion.[27] The moral law demands that the victor display mercy toward the vanquished, including the natural world.[28]

For Schoenichen, the most senseless destruction involves annihilating an entire species: "The downfall of a noteworthy type of animal or plant awakens in man's soul the feeling of a deeply deplorable loss," which is so painful because we sense that "all the different kinds in the realm of living things are linked to one another through secret bonds, all of which have the same origin and which beget all new life."[29] Like Aldo Leopold, Schoenichen writes that the land is an organism whose parts are constituted by species.[30] Before the only species left on Earth are cultivated plants and domestic animals, nations must join together to make sure that "in the future never again will an animal species be made extinct or be brought near to dying out because of the influences of civilization."[31] Schoenichen also condemns Europe's shameful history of subjugating primitive peoples, from whom we have much to learn.[32]

Schoenichen's ecosophy strikes a familiar chord with contemporary environmentalists who call for basic changes in the attitudes and practices of industrial societies. At times appealing to utilitarian grounds to justify better treatment of nature, Schoenichen elsewhere affirms that humans ought to respect nonhuman life, especially wild nature, for its own sake. Though sounding like a Deep Ecologist in calling for a more profound identification with all life, Schoenichen's position would best be described as the "weak anthropocentrism" of mainstream environmentalism, which ascribes a leading role to human life but also affirms the inherent worth of all life. Further, despite his pantheistic tendencies, Schoenichen's call to respect Creation is consistent with much of Christian ecotheology. Though Schoenichen's tendencies toward nature mysticism may alarm some critics, there is no necessary relation between such mysticism and fascism.

2. Schoenichen's Ecosophy as Ecofascism

The similarities between Schoenichen's ecosophy and contemporary environmentalism may prove disturbing in view of the following fact: he explicitly portrayed his ecosophy as consistent with the *Blut und Boden* (blood and soil) racist ideology of National Socialism. Indeed, one of his two books from which I have been citing is called *Nature Protection in the Third Reich* (1934). In having deliberately omitted reference to passages in which Schoenichen shows how his ecosophy is consistent with Nazism, I have tried to indicate that his ecosophy can stand on its own, without reference to Nazi ideology. A third postwar book, *Nature Protection, Homeland Protection* (1954), makes no mention of such ideology, a fact that suggests that he himself did not see any necessary connection between Nazism and the nature protection movement.[33] Indeed, before 1933, many branches of that movement were neither militaristic, territorially expansionistic, nor overtly racist in the way that Nazism was.[34] As a member of the *völkisch* branch of the nature protection movement, however, Schoenichen helped to move it toward a racist ecofascism consistent with Nazism. In 1928, as Raymond H. Dominick has noted, Schoenichen took control of a leading conservation journal, *Naturschutz*, and changed the subtitle to the politically provocative "Monthly for all Friends of the German *Heimat*." Soon after the Nazi takeover, he published an issue whose cover shows a German youth carrying a Nazi flag on a country trail. In an article in the same issue, he stated that "between Aryans and non-Aryans there exist fundamental, unbridgeable differences, as for example especially in the areas of worldviews, sexuality, the relationship to nature, etc." He also praises Hitler for rejecting foreign elements that were corrupting the German *Volk*.[35]

Although the following account of Schoenichen's ecofascism may be painful to read, environmentalists must become aware of the potentially dark side of ecological politics. Many American environmentalists are not aware that the close bonds forged between Nazism and the nature protection movement prevented the "Green" movement from surfacing in Germany until the 1970s. Indeed, contemporary Green movements throughout Europe must constantly struggle against those who wish to redirect those movements in a protofascist, ultranationalist direction.

Nature protection and National Socialism stand "in a tight connection," according to Schoenichen, because the *Führer* wills a new German folk-community (*Volksgemeinschaft*), whose foundation is drawn from "blood and soil, i.e., from the primordial forces of life and soul that are proper to our race, and from the nature-willed bond that subsists between us and the sod of the homeland."[36] The German race draws its strength from being rooted in its own native land. Because modern forces—including "the malicious poison" of liberalism, capitalism, communism ("materialism"), democracy, rationalism—uproot the German people and lead to racial degeneration (*Entartung*), such sinister foreign forces must be expelled from the homeland. "Adolf Hitler," we read, "demands that man must understand the basic necessity of the rule of nature and must also grasp how much his existence is subjugated from above to these laws of eternal struggle and contest."[37] Like Hitler, the nature protection movement wants

> to remind and warn our *Volk*, indeed all of mankind, that the primordial forces of life, the instincts willed by Nature, decide about the being and non-being of peoples and races, and that no technology, no rationalism can protect us, together with our economy and our civilization, from going under [*Untergang*] if we detach ourselves from the natural foundation of life.[38]

Although celebrating mystical union with Nature, Schoenichen pointedly remarks that nature protection is not "otherworldly rapture," but instead "shows full understanding of the practical demands of life."[39] Recognizing that Nazism had no intention of abandoning industrialism, Schoenichen notes that the Nazi "Beauty of Labor" and "Strength Through Joy" movements agree that aesthetic considerations are crucial for the well-being of workers, especially for heavy laborers in urban areas.[40] Laboring all day at the industrial plant, and being passively entertained at night by radio and mass sporting events, the Nazi worker needs to encounter the natural beauty of self-creative nature. Without sinking his roots into native soil, the German worker may be ruined like workers in countries run by the cosmopolitan, antinaturalistic worldviews of egoistical liberalism and materialistic Marxism. Having conquered the socially disintegrating forces of party politics and class warfare, the Nazi movement will restore the German *Volk* and land to health. In this process, the nature protection movement can play a crucial role, by helping to give the *Volk* a deeper "racial imprinting" by offering people the necessary opportunities—such as hiking through forests and fields—to have a mystical, felt experience (*Erlebnis*) of their native land. The German *Volk*'s sensibility was long shaped by its experience within the ancient forests, now mostly extirpated by greedy capitalists.[41] Crucial for the rerooting process is the Nazi version of what today is called "bioregionalism." Hitler Youth, we are told, should be reminded of the inner connection between the landscape and the manners, customs, songs, and practices of the various German tribes. By partaking in nature festivals, which "awaken the feeling of a close, common bond with the mother ground of home [*mit dem heimatlichen Mutterboden*]," young people will "envision the highest goal and highest task of all being solely in this: to be allowed to live and die for Germany."[42]

Schoenichen praises National Socialism for having passed such innovative nature protection laws as the world's first comprehensive nature protection act (June 26, 1935), the law protecting plants and animals (March 18, 1936), and the law protecting mother-soil (November 16, 1939). In addition to enhancing the *Volk*'s social and economic welfare, these nature protection laws also help protect *Volksgeist* from corrosive foreign influences and modernist attitudes. The romantic dimension of the nature protection movement is not sentimental nature worship, but "meditation on our *völkisch* essence and on the springheads, from which it always draws new force and inner quality."[43] In view of the fact that "respect for the creations of Nature lies in the blood of the northern peoples [*Völkern*]," Schoenichen expresses surprise that there is no native German word for "Nature."[44] Hence, he recommends that the nature protection movement should speak of the "primal landscape [*Urlandschaft*]," which should be protected against further human intrusion, since it generated the German soul. Though World War I and the corrupt Weimar Republic set back early nature protection, the way was finally cleared "Only through the [Nazi] revolution of the year 1933, which once again reached back to the essential foundations of the German *Volk*. . ."[45]

According to the Nazi worldview, "the landscape is in the first place the living space [*Lebensraum*] of the *Volk*."[46] Hence, even urban, industrial, and agricultural lands must be protected, so far as is possible, from destructive human activities. Moreover, Germany's beautiful nature areas must be protected not only from highway billboards and radios blaring "jazz music and Negro noise," but also from the intrusion of hotels, gas stations, and other commercial ventures (though, of course, small merchants can be accommodated). Just as there must be healthful, beautiful, inspiring, and bountiful *Lebensraum* for the *Volk*, so *Lebensraum* is needed for indigenous plants and animals.[47] Citing Hermann Göring, Schoenichen asserts that the animal world is "the living soul of the landscape."[48]

The Nazis justified their aggressive foreign policy by saying that the German *Volk* had been unjustly deprived of its native land either by other white peoples or else by allegedly inferior races. In Eastern Europe, the Nazis planned to displace the Slavs so that German settlers would gain needed *Lebensraum*. Postwar relations with the advanced white races of northern Europe (for example, the French, English, Danish, Belgians, and Scandinavians) would be a different story, however. Once historical territorial disputes had been resolved in Germany's favor, those peoples would be encouraged to re-establish their roots in their own native soils, thereby reinvigorating all the worthy tribes of Europe's great white race. In this respect, Nazi Germany portrayed itself as the savior not just of the German *Volk*, but of all the European *Völker*. Pointedly remarking that Germany had been excluded from international meetings concerned about protecting endangered species, Schoenichen envisages a time when cooperation with people from northern European countries would again be possible.

If endangered species need protecting, so do primitive peoples who represent humankind's distant past. According to the French naturalist, Paul Sarasin, "The white man is the great destroyer of creation, the meddler in earthly paradise."[49] Money-driven liberal societies have been particularly responsible for exploiting these peoples. To be sure, Schoenichen reminds his readers, "The subjugation of the indigenous population—though certainly not the inhumanity with which this was all too often accomplished—was in many cases surely a necessary consequence of the struggle for existence [*Kampfes um das Dasein*] and resulted from the expansionist strivings of the European races."[50]

European colonial practices differ from one another, however. For example, the French act according to their Jacobin revolution's foolish universalist principles, including the "theory of human rights, which bases itself on the supposition of the equality of all men. Although this [supposition] has been long contradicted by biological research, the misguided French spheres hold fast to their assimilation theory."[51] Supposedly seeking to assimilate native peoples to French language, law, and custom, the French in fact turn them into cannon fodder. The English are at least wise enough to reject this policy of assimilation, choosing instead to make use of native traditions and customs whenever possible in the process of establishing British rule. Though recognizing "the fact of the eternal differences of blood- and species-determined races, which can never be made equivalent by education," the English unfortunately used this knowledge to justify their dominance over all other peoples and races, and only secondarily to justify a sense of responsibility for primitive peoples.[52] Nazi Germany offers better treatment of primitive peoples. Renouncing "the politics of subjugation and extermination" [!], as well as "foolish ideals" of assimilation, Germany encourages the development of indigenous people according to their own racial heritage.[53] For the Nazis, of course, some races were superior to others. Though the colored races might develop skills needed for greater interchange with Europeans, interracial marriages and extramarital intercourse must be prohibited: "From the standpoint of the National Socialist worldview, both prohibitions are self-evident."[54]

While earlier suggesting that our submission to the Divine constrains us to respect all of creation, Schoenichen offers a different justification when it comes to the politically more dangerous topic of protecting other races. Such protection is to be afforded not because of "sentimentality," "compassion," or "in the name of a deformed idea of humanity, which would like to throw Hottentots, Botekudes, Jews, and Aryans into the same pot."[55] Instead, nonwhite races are to be protected because the Germans must exhibit a loyal responsibility to science, which is among "the highest and noblest spiritual goods which our culture has created."[56] Surely, he insists, these primitive races ought not be treated with any less scientific responsibility than apes and chimpanzees [!].

Having begun the present essay by considering the charge that current wetlands regulations constitute ecofascism, I now turn to Schoenichen's justification of analogous measures, including those specified in the Nazi Nature Protection Law (1935). On several occasions, he asserts that the Nazis would rid Germany of the destructive influence of Roman law that, imposed in part by Napoleon's invasion, subjected Germany to Enlightenment concepts about the absolute right of individuals to use private property however they see fit. The contrary Nazi worldview is expressed in the slogan "The Common Good Takes Precedence over the Private Good." A folk comrade owns something—especially the land—not as an object over which he has unlimited authority, but rather as a common good that he is duty-bound to manage according to the needs of the *Volk* and *Reich*.[57] The claims of economy, trade, and settlement, then, must achieve a harmonious relationship with the equally important claims of culture, particularly the ideal of homeland: "Hence the individual is not the major issue, but rather the great whole."[58] Comparable to a living organism, this glorious whole—race and land, blood and soil—"assigns to every part of the landscape its function and gives it form in such a way that for men the highest spiritual and economic values will be achieved."[59] According to the Nazi worldview, then, biological considerations justify the conviction that the *Volk*'s collective needs and "original right"[60] always trump individual rights.

Though compromises will sometimes be necessary in the complex process of nature protection, they must be made in accordance not with egoistical private interests or with the conflicting interests of political parties, but rather in accordance with the overall good of German *Blut und Boden*. Of course, "only in states governed by authoritarian rule can there be fulfilled all the presuppositions necessary for a really far-reaching organization of space [or land]."[61] Hence, only *Volk* leaders—attuned to the organic needs of the *Volksgemeinschaft* and advised by scientific, economic, military, and cultural experts—can best dispose of all German land. If preserving some forest area were deemed crucial for the well-being of the *Volk*, for example, that area would be preserved, even if this involved expropriating it from a private owner who hoped to develop it for private economic gain. (Here, it is worth noting that even though Nazi leaders allowed big industrialists and the armed forces to exploit the land, ideological considerations led those leaders to take steps—including land expropriation—to guarantee a sound, healthy, organically pure, racially homogeneous, and beautiful homeland.)

3. Ecofascism in U.S. Environmentalism?

Turning now to the issue of whether ecofascism can be found in American environmentalism, I begin by noting that there are a number of parallels between the American environmental movement and the German nature protection movement, from which the Nazis derived some of their "green" vocabulary. Both American environmentalism and German nature protection were fueled first by public health threats posed by urban wastes and industrial pollution, and second by scientific claims that the practices of industrial civilization were destabilizing the "balance of nature." In addition, a certain nature-romanticism bordering on pantheism provides other parallels between the two movements. Mid-nineteenth-century American and German Romantics suggested that industrialism was destroying not only the natural world but also the human body and spirit. In addition to camping out with Muir at Yosemite, President Theodore Roosevelt proclaimed that contact with wild nature was needed for a vital citizenry. *Völkisch* leaders in Germany insisted that destruction of German soil was bound up with degeneration of German blood.

This was the dark side to the widespread fin-de-siècle concern about healthy bodies and healthy nature. Some social Darwinists feared "colored" races would triumph if industrial practices and consumer goods "softened" the white race, made it less "manly," and deprived it of contact with vital natural instincts. Such fear explains, in part, the transatlantic success of Oswald Spengler's book, *The Decline of the West*, shortly after World War I. Concern about racial degeneration was so strong in the United States during the first part of this century that American scientists became the world leaders in eugenics research. Nazi officials relied heavily on that research in devising their own eugenics measures (including euthanasia, sterilization, and murder) designed to "purify" and "regenerate" the German population. Though Nazi social horrors caused eugenics to lose credibility in postwar America, most Americans do not realize that the Nazis combined eugenics with mystical ecology into the perverted "green" ideology of *Blut und Boden*. Though the postwar wave of American environmentalism spoke not of race and tribe, but instead of Muir and Leopold's celebration of nature's beauty, market liberals and Marxist critics understandably feared the potential connection between nature worship and fascism.

Despite such legitimate fears, is there any evidence that American environmentalism has been influenced by racist, ultranationalist, or potentially fascist ideals? Or is the term "ecofascism" simply political rhetoric used by certain unscrupulous members of property-rights groups? Turning to the second question first, I would note that government appropriation of land and restriction of its use are scarcely new phenomena. For example, privately held land has been appropriated for freeways or urban renewal, often with nightmarish environmental consequences. Even though the Constitution entitles a landowner to compensation for such appropriation, it does not always cover the owner's real costs. There is only limited public sentiment for absolute right to control private land. People usually tolerate appropriation of property if there is compensation and if public "benefits" of the proposed projects outweigh their private "costs." If government appropriation of land is too heavy-handed, short-sighted, or corrupt, however, one may hear cries of "fascism." I wish to emphasize, however, that in a truly fascist state, one could not publicly condemn the action, seek redress through the courts, organize political campaigns to oust those responsible for the appropriation, or otherwise seek to overturn the action democratically.

When the beneficiary of government appropriation and/or regulatory restriction is not another human being or a human institution, but instead is a forest, watershed, swamp, marsh, or bog, property owners and other private citizens may protest more vociferously, especially if compensation is not forthcoming. Those deprived of their property rights complain that their human needs are being trumped by the needs of nonhuman beings that lack human interests, rights, and legal standing. Property-rights defenders assert that it is bad enough to have one's land sharply regulated because of the "needs" of shrimp and water grasses, birds and mucky ground, but even worse to have such regulations imposed on land that is manifestly not "wet" or environmentally crucial. Wallace Kaufman argues that if some public good is to be achieved by protecting private land from development, it is unfair to make the economic burden of such a good fall on landowners.[62] Instead, state or federal governments should purchase such land, thereby spreading the financial burden across the general tax base in a way that invites public debate about the wisdom of such purchases.

Though some pro-property lobbying groups, such as the Wise Use Movement, represent large-scale extractive industries (forestry, mining, grazing) instead of private citizens, other such groups represent the legitimate concerns of landowners who are often not antienvironmentalists, but who do resent what they regard as unconstitutional government incursions into the use of private property. Unless environmental activists address more effectively the legitimate concerns of property owners, and find ways to work with market forces in protecting vital natural areas, yesterday's environmental gains will be in for trouble, especially in a political climate emphasizing the American tradition of individual rights. Though critical of the natural destruction that this tradition has at times justified, American environmentalists should ponder the fact that ecofascism demands that the individual good be sacrificed for the good of the whole.

A crucial difference between much of American environmentalism and German ecofascism is that, for the former, "wild" nature often symbolizes rugged individualism and personal liberty, while for the latter "wild" nature symbolized the instinctual bond between the vital blood of the *Volk* and its land, to both of which individuals were expected to submit themselves. By way of contrast, American environmentalism has often argued that individual liberty is undermined by unhealthy water and air, constrained by the loss of the wild nature that symbolizes personal freedom, and diminished by the despoliation of natural beauty reminiscent of individual self-expression.

After making great strides in the 1960s and 1970s, environmentalism's political fortunes changed dramatically in 1980 with the election of President Reagan, who helped to initiate the antiregulatory, anti–Big Government mood that culminated in the recent national elections. Progressive critics charge that such a mood also represents a dangerous rightward turn reflected in recent anti-immigration actions. Nevertheless, environmental sentiment remains strong in public opinion polls. Hence, the following scenario is possible: As America's growing population continues to cause environmental problems, and if economic scarcity becomes a reality for more and more people, environmental sentiments may become mixed with or displaced by fears that fast-breeding illegal immigrants will destroy American land just as they have destroyed their own.

Such considerations lead us to address the first question posed earlier: Does racism play a role in American environmentalism? Unfortunately, anti-immigrationist attitudes, occasionally mixed with racist and xenophobic sentiments, have converged at times in utterances made by some radical ecologists, including certain members of Earth First! Deep Ecologists in particular have been criticized by socialists and market liberals alike as being potentially ecofascist. In *Le nouvel ordre écologique*, for instance, French liberal Luc Ferry argues that by sharing romanticism's sentimental aestheticism, according to which "true nature" must be wild, pure, virgin, savage, and irrational, Deep Ecology and Nazism regard modernity as a profound mistake that produced human alienation and ecological destruction.[63] In fact, most deep ecologists believe that a spiritual revolution is needed to save humankind and the planet from the evils of capitalism and communism, which are allegedly different expressions of the same acquisitive anthropocentric drive for total control of planet Earth. Though calling for global political intervention to save endangered ecosystems and native peoples from destruction, and though perhaps willing to cancel some individual property rights in an "ecological emergency," Deep Ecologists prefer that urgent ecological matters be dealt with now, in the context of democratic practices.

Despite tendencies that some may read as ecofascist, Deep Ecologists and many Earth First!ers have more in common with anarchists than with fascists. (Indeed, the young Dave Foreman, cofounder of Earth First!, was an avid supporter of Barry Goldwater and a member of Young Americans for Freedom.) Emphasizing democratic procedures, asserting that ecosystems and individual beings alike have inherent worth, and celebrating cultural diversity, Deep Ecologists like Arne Naess reject the Nazis' totalizing concept of *Gleichschaltung*, according to which social, political, economic, and cultural structures were to be aligned with the aims of the movement.[64] Despite criticizing modernity's ecologically destructive and socially alienating dark side, then, Deep Ecology may be read as adhering to a broadly "progressive" interpretation of human history, according to which humankind can develop a deeper sense of identification with all life. Such identification would make it possible to expand modernity's emancipatory ideal beyond the human to all forms of life, thereby realizing Deep Ecology's vision of "letting things be."[65]

Rejecting this progressive reading of Deep Ecology, libertarians, leftists, and some ecoanarchists accuse it of indulging in the same kind of antihumanistic and anti-individualist nature mysticism that paved the way for Nazi victory in a period of social, political, and "ecological emergency." But Deep Ecologists are scarcely alone in fearing that human population growth, resource decline, and environmental degradation are undermining modernity's utopian hopes. In view of such problems, William Ophuls asserts that "the golden age of individualism, liberty, and democracy is all but over."[66] Moreover, economist Robert Heilbroner confesses that only one social system can effect the "profound and painful adaptations" that future generations must make in the face of unparalleled ecological, population, and economic problems: "This is a social order that will blend a 'religious' orientation and a 'military' discipline."[67] Arguably, National Socialism was in part just such a militaristic society inspired by a "religion of nature."[68]

An argument in favor of something like an ecofascist view may be found in early essays by a leading American radical environmentalist, J. Baird Callicott, who once believed he could derive this view from Aldo Leopold's "Land Ethic." Having great respect for Callicott's philosophical acumen and his contributions to environmental philosophy, I wish to emphasize that he has not only purged his writings of any "ecofascist" vocabulary, but also maintains that he was wrong in claiming to be able to derive apparently ecofascist ideas from Leopold's writings.[69] Today, Callicott insists that his more mature and responsible respect for individual human rights is not, in relation to the Land Ethic, an ad hoc addition. Instead, such respect flows from the philosophical foundations common both to a communitarian version of human ethics and to the land ethic itself.[70] Even though now repudiated at least in part by their author, Callicott's early essays on Leopold remain interesting because they suggest how the claims of scientific ecology may be used to justify policies that demand that individual rights conform to the needs of the organic community.

The early Callicott sought to draw from Leopold's "Land Ethic" an environmentally sound, though "misanthropic" social system. Leopold helped to popularize the (now contested) ecological idea that plants and animals are not independent entities (somehow akin to the atomistic individuals of liberal political theory), but instead are internally related phenomena whose existence is tightly interwoven with the larger, interdependent constituents of the biotic or ecological community.[71] This community, "the land," refers to the enormously complex, ontologically constitutive relations among the organic and inorganic elements—plants, animals, and insects, as well as soil and water—of a particular biome or region. Leopold sometimes described these elements as being analogous to the organs of an organism. To survive, an organism's organs must cooperatively limit their behavior in ways that serve the higher good of the whole organism. Hence, Leopold's Land Ethic states that "a thing is right when it tends to preserve the integrity, stability and beauty of the biotic community. It is wrong when it tends otherwise."[72] Individual organisms lack ethical importance, for they are merely temporary instantiations of enduring species whose interlocking relationships constitute "the land." Hence, to the chagrin of many environmentalists, Leopold was an avid hunter who took on the role once played by now-scarce predators, in order to sustain the population balance needed to preserve "the land."

Until now, Leopold notes, ethical consideration had been limited to human affairs, but now consideration should be expanded to include the elements of the biotic community in which humankind is imbedded, for our species will perish unless it establishes ethical constraints regarding treatment of "the land": "All ethics so far evolved rest upon a single premise: that the individual is a member of a community of interdependent parts."[73] Leopold here assumes the validity of the Darwinist assumption that all human behavior (including ethical comportment) amounts to adaptive strategy:

> An ethic, ecologically, is a limitation on freedom of action in the struggle for existence. An ethic, philosophically, is a differentiation of social from antisocial conduct. These are two definitions of the same thing. The thing has its origins in the tendency of interdependent individuals or groups to evolve modes of cooperation.[74]

Reconstructing and defending Leopold's Land Ethic, Callicott explains that David Hume's communally oriented ethic of sympathy influenced Darwin's contention that human ethical attitudes were elaborations of the successful strategies of animals made "fit" by the adaptive strategy of social cooperation. Darwin's work, in turn, shaped Leopold's ideas that humans are members of a still larger community, the land, and that humankind will lose its "struggle for existence" if it fails to develop cooperative (ethical, limiting) relations with the land. Depicted in this blunt manner, the Land Ethic would seem to be merely a prudential imperative ("Do right by the land if you want to survive!") rather than an ethical imperative ("Do right by the land because it is morally right to do so!"). Though admitting that, at bottom, "ethical" behavior is utilitarian and adaptive, Callicott argues that because human animals live in a linguistic-cultural community, they develop mythical and religious explanations for moral conduct that minimize or altogether conceal its utilitarian-adaptive dimension, and emphasize instead its ethical-normative dimension. Just as moral persuasion was useful long ago in the (adaptive) process by which warring tribes came to recognize their shared humanity, so too moral persuasion will be useful today in persuading people to have sympathy for and to respect the larger biotic community. The Land Ethic "creates obligations over and above self-interest."[75] Just as conscience obliges us to value other people for their own sakes, so too can it be cultivated to oblige us to value the land for its own sake, thereby benefiting humankind and the whole biotic community.

Callicott admits that because human society differs significantly from the ecological community, "the term 'ecological community' has at best an analogical sense."[76] The ecological community is best understood "as an economic system par excellence," whose members play specific functional roles (producers, consumers, decomposers) in the flow of material and energy within an ecosystem ("the land"). Humans are "members" of this community in the sense that just as we depend economically on bankers, farmers, and auto mechanics, so too do we depend on "phytoplankton, forests, earthworms, honey bees, etc., etc., etc., in a way that is *formally the same*."[77] Callicott may push this "formal similarity" too far, however. Emphasizing the holistic character of ecological systems, he argues that just as we have the moral duties of self-preservation and social preservation, so too we have the moral duty of helping to preserve the biotic community.[78] Seen from the perspective of the biotic community, individual needs, interests, and rights are secondary. Consider the human organism: its "interests" are not to be confused with the interests of its cells, added up and averaged out: "For the sake of the person taken as a whole, some parts may be, as it were, unfairly sacrificed."[79] Now moving to human social organization, Callicott remarks that

> the interests of society may not always coincide with the interests of its parts. Discipline, sacrifice, and individual restraint are often necessary in the social sphere to maintain social integrity as within the

bodily organism. A society, indeed, is particularly vulnerable to disintegration when its members become preoccupied totally with their own particular interest, and ignore those distinct and interdependent interests of the community as a whole.[80]

From the position of moral individualism, Callicott notes, a truly biocentric land ethic may appear to be misanthropic, for it "does not exempt *Homo sapiens* from moral evaluation in relation to the well-being of the community of nature taken as a whole."[81] He maintains that the enormous human population "is at present a global catastrophe ... for the biotic community."[82] Refusing to read the land ethic as a means of "managing nature" for human ends, and thus as ascribing to humans "an ultimate value essentially different from that of his 'resources',," Callicott approvingly cites Garrett Hardin's "lifeboat" and "survival" ethics as an indication of the harsh measures that may be needed to bring human population into line with the obligations imposed by the Land Ethic. "The extent of misanthropy in modern environmentalism thus may be taken as a measure of the degree to which it is biocentric."[83] Further, unlike the humanistic ethical theories of modernity,

> the land ethic manifestly does not accord equal worth to each and every member of the biotic community; the moral worth of individuals (including, n.b., human individuals) is relative, to be assessed in accordance with the particular relation of each to the collective entity that Leopold called "land."[84]

In the social holism of Plato's *Republic*, Callicott finds analogies to the anti-individualistic, organic demands of the land ethic. Plato makes many "apparently inhuman recommendations"—ranging from infanticide and destruction of the nuclear family, to "a program of eugenics" and "the utter abolition of private property"—because he "seems to regard individual human life and certainly human pain and suffering with complete indifference."[85] Just as the separate interests of the parts of the body had to be subordinated to the health and well-being of the whole body, and just as some elements of the human psyche had to be subjugated by others for the good of the whole psyche, so the separate interests of individuals had to be subordinated to the "well-being of the community as a whole."[86] Given the formal parallels between Plato's ethical system ("foreign" though it may seem to modern ethicists) and the Land Ethic, Callicott concludes that it, too, is an ethical philosophy "in relation to environmental virtue and excellence."[87]

Implementing the Land Ethic, Callicott admits, would "require discipline, sacrifice, retrenchment, and massive economic reform, tantamount to a virtual revolution in prevailing attitudes and life styles."[88] Our currently selfish, wasteful, and unecological way of life would give way to "a renaissance of tribal cultural experience" that would prize manly hardness, courage, resiliency, simple diet, rigorous exercise, capacity for pain, and social responsibility.[89] "Personal, social, and environmental health would, accordingly, receive a premium rather than comfort, self-indulgent pleasure, and anesthetic insulation from pain. Sickness would be regarded as a worse evil than death."[90]

Callicott was surely unaware that his diagnosis of and prescriptions for curing an ecologically unsound society mirror, even more than do Schoenichen's writings, many aspects of Nazism's military ethos of masculinity, hardness, courage, and ruthlessness. Though avoiding racist remarks, Callicott's contrast between the healthy and the sickly is uncomfortably reminiscent of Nazi justifications for eugenic laws designed to strengthen Aryan blood by sterilizing the insane; by killing Jews, Gypsies, homosexuals, and communists; and by putting to death the "sickly"; since in the "struggle for existence" the "health" of the organic whole—"blood and soil"—could only be won by repudiating the bourgeois and Jewish-Christian values of mercy and compassion. In this context, Callicott's appeal to Plato's thought is made more problematic by Hans Sluga's observation that

> for the Nazi philosophers, Plato became the most authoritative political thinker and the *Republic* the most widely read work on political theory. His critique of democracy, his militarism, his belief in the need for a strict political order, and his commitment to the existence of different social ranks all fitted the prevailing political mood.[91]

I wish to emphasize that even when presenting the views discussed above, Callicott had no sympathy for National Socialism, and that his current views emphasize the moral standing of human individuals. My point in showing the proximity between his early views and those of National Socialism is to stress the need for environmentalists to realize that in recent history, during a moment of social, political, and ecological "crisis," Nazi ideologues drew both on ecological science and on thinkers like Plato to justify their racist policies and murderous practices. Despite its awful consequences, National Socialism remains an object of widespread fascination, a fact that should alert us to the possibility that in periods of ecological stress and social breakdown, preachers of a "green" fascism might once again find a sympathetic audience. There can be no doubt that the messy, pragmatic, time-consuming, and unsatisfying processes of democratic politics pale in comparison with the ecstatic promises of fascist leaders who appeal to those who are repelled by the social disintegration and ecological destruction associated with modernity, and who yearn for an ethnically unified, prosperous, and beautiful society living in harmony with the "laws" of nature.[92] Arguably, some elements of those conservative political movements that espouse anti-immigrationist attitudes are attracted to more of these themes than are environmentalists, many of whom celebrate cultural diversity. Conservative (and libertarian) support for individual rights, including property rights, however, does stand in the way of fascism's collectivist impulse.

Though not present in U.S. environmentalism, a type of ecofascism might gain a future foothold if social, economic, political, and ecological problems increase. To develop nonecofascist solutions to such problems, environmentalists must educate themselves about two different temptations. The first involves romantically reidentifying with nature in a way that calls for organic recollectivization and retribalization. For peoples armed with modern weapons, a move in this direction would probably lead to unimaginable ecological and social disasters. One motivation for becoming reabsorbed in "natural" rhythms is to free oneself from the anxiety and moral responsibility experienced by self-conscious, historical agents. As humanity's technological power becomes ever greater, however, people must be more willing than ever to accept responsibility for making the difficult choices that lie ahead, including deciding which species are worth fighting to preserve, and which species must be allowed to perish, no matter how regrettable this may be.[93] Environmentalists must be willing to re-examine their own understanding of the relation between "humanity" and "nature." Sometimes regarding "genuine" nature as untouched by human hands, many environmentalists fail to grasp that even the Amazonian rain forests have been significantly altered by native human populations. Since humans are natural beings interacting with their natural environment, the question is how to guide such interaction in a way that minimizes ecological destruction while providing for human well-being.

The second temptation is to conclude that centralized government is the best institution available to guide the ever-expanding human-nature interaction. The ecologically devastating consequences of state socialism in the former U.S.S.R., the widespread contamination left by U.S. government nuclear weapons facilities, and the unsustainable forestry and grazing taking place on U.S. public lands demonstrate that governments often have the worst environmental records of any institution. Though government regulations are necessary to limit environmental damage by public and private enterprise, environmentalists are often naive in the confidence they display in government. Moreover, by failing adequately to initiate viable partnerships with other concerned parties (including ranchers, farmers, hunters, and industries), and by supporting environmental regulations that do not adequately respect property rights, American environmentalists have unwittingly encouraged the misguided charge of "ecofascism."

In view of my earlier examination of Schoenichen's ecofascism, however, I hope that the reader sees why current federal wetlands regulations cannot fairly be conceived of as an instance of it. Those regulations did not spring from a government controlled by militant racists who believe that preventing further "development" of wetlands will preserve the purity of the white (or some other) race and its native soil. Instead, those regulations arose from hotly contested deliberations about how to achieve a workable balance between those who would benefit (for example, vacation-home developers) and those who would suffer (for example, fishermen) in the long run if wetland "development" were allowed to proceed unabated. The debates often focused on anthropocentric cost-benefit issues. Of course, many environmentalists continue to argue that threatened ecosystems are worthy of protection for their own sakes, not merely for their value as economic resources. Moreover, wetland regulations are sometimes enforced overzealously by regulators with their own environmental ax to grind. Even though asserting that people are part of an interdependent ecological order, however, most environmentalists do not speak in terms of a new racism or tribalism, especially one justifying a dictatorship that would redesign or dismantle modern industrial society in order to preserve precious "blood and soil."

Possibly, the new Congress not only will pass "takings" bills to compensate owners who have been deprived of the use of their wetlands, but also will scale back or simplify other kinds of environmental regulations. In contesting such legislation, environmentalists must consider replacing "gloom and doom" vocabulary with more constructive discussions of how to reconcile individual freedom with social and ecological well-being. If there is "no turning back" from modern technology and market economies, environmentalists must become more imaginative as the boundaries between the "natural" and the "social" are increasingly erased.[94] The laudable goals of human emancipation and ecological well-being cannot be achieved, however, simply by relying on biotechnological innovations that claim to merge the human and the natural, any more than human freedom and ecological well-being were furthered by fascist fantasies about plunging back into "nature" in a way that surrendered moral responsibility and historical self-consciousness.

Full human emancipation will be achieved only in connection with further stages in the consciousness "evolution" that generated modernity's noble ideal of universal human rights.[95] Such evolution will not only require the reintegration of what western "man" has dissociated (including the biosphere, the human body, the emotions, and the repressed "female") and will not only need to prepare for the eventuality that human artifacts will acquire "human" characteristics, but will also need to deepen humanity's sense both of its moral responsibility in the historical realm and of its nonhistorical relation to the transcendent dimension. It remains to be seen whether humankind, in the face of increasingly difficult social and ecological circumstances, will be able to provide for itself the time required for the evolution of the sensibility needed to move beyond the limited self-understanding that has helped to produce today's global problems.

Notes

1. *Sierra* magazine 79 (4) (July/August 1994), reviews current antienvironmental legislation, including "takings" amendments. Thanks to Cris Feldman for helpful discussions about this issue.
2. Judicial decisions are tending to support the claims of private property holders. On the "takings" issue, see Gary Varner, "Environmental Law and the Eclipse of Land as Private Property," in Frederick Ferré and Peter Hartel, eds., *Ethics and Environmental Policy* (Athens: University of Georgia Press, 1994); and Richard A. Epstein, *Takings* (Cambridge: Harvard University Press, 1985).
3. "Environmental Backlash: The Land War," *New Orleans Times-Picayune*, July 3, 1994, p. 4.
4. Ibid.
5. On the "Wise Use Movement," see "Any Color but Green," *U.S. News and World Report* Vol. 111 (October 21, 1991): 74–76; "Meet the Anti-Greens," *The Progressive* Vol. 55 (October 1991): 21–23, "Gunning for the Greens," *Time* 139 (February 3, 1992): 50–52; "Whose Agenda for America?" *Audubon* 94 (September/October, 1992): 80–91.
6. Ron Arnold, cited by Jessica Matthews of World Resources Institute, *International Herald Tribune*, February 25, 1992, p. 6.
7. Hans Sluga, *Heidegger's Crisis: Philosophy and Politics in Nazi Germany* (Cambridge: Harvard University Press, 1993), p. 129.
8. Robert D. Kaplan, "The Coming Anarchy," *The Atlantic Monthly* (February 1994): 44–76; citation is from p. 58.
9. See Thomas Fraser Homer-Dixon, "On the Threshold: Environmental Changes as Causes of Acute Conflict," *International Security* 16 (1991): 76–116, and "Environmental Scarcities and Violent Conflict," *International Security* 19 (1994): 5–40.
10. In "Must It Be the Rest Against the West?" *The Atlantic Monthly* 274 (6) (December 1994): 61–84, Matthew Connelly and Paul Kennedy reinforce the idea that Third-World population growth is threatening overall human well-being. See Virginia Abernethy's reply in the same issue, "Optimism and Overpopulation," pp. 84–91.
11. Walther Schoenichen, *Naturschutz im Dritten Reich: Einführung in Wesen und Grundlagen zeitgemässiger Naturschutz-Arbeit* (Berlin: Hugo Vermühler Verlag, 1934), pp. 5–6 (hereafter "NDR").
12. Walther Schoenichen, *Naturschutz als völkische und internationale Kulturaufgabe* (Jena: Gustav Fischer Verlag, 1942), p.19 (hereafter "NK").
13. NK, pp. 19–20.
14. NDR, pp. 13–14.
15. Ibid., p. 80.
16. NK, p. 80.

17. Ibid., p. 87.
18. Ibid., pp. 16ff.
19. Ibid., pp. 3, 14.
20. Ibid., p. 15. American environmentalists have used the terms "preservation" and "conservation" somewhat differently.
21. Ibid., p. 75.
22. Ibid., p. 46. But Schoenichen condemns the huge hotels, advertising signs. and other such intrusive phenomena that despoil American national parks and German natural monuments (NK, pp. 48ff).
23. NDR, p. 88.
24. NK, p. 76.
25. Ibid.
26. Ibid., p. 77.
27. Ibid., p. 404.
28. Ibid., p. 77.
29. Ibid., p. 90.
30. Ibid., p. 16.
31. Ibid., p. 405. In NDR, p. 19, Schoenichen notes that already in 1853, Wilhelm Heinrich Riehl was speaking of "the right of wilderness [*das Recht des Wildes*]."
32. NK, pp. 405ff, 416.
33. Schoenichen, *Naturschutz, Heimatschutz* (Stuttgart: Wissenschaftliche Verlagsgesellschaft, 1954).
34. On this topic, see Raymond H. Dominick, *The Environmental Movement in Germany: Prophets and Pioneers 1871–1971* (Bloomington: Indiana University Press, 1992.
35. Ibid., p. 99.
36. NDR, pp. i, 7.
37. Ibid., p. 7.
38. Ibid.
39. Ibid., p. 15.
40. Ibid., pp. 4, 74–75.
41. Ibid., pp. 26ff.
42. Ibid., p. 90.
43. NK, p. 35.
44. Ibid., p. 1.
45. Ibid., p. 45.
46. Ibid., p. 83.
47. NDR, p. 58.
48. Ibid., p. 46.
49. NK, p. 411.
50. Ibid.
51. Ibid., p. 412.
52. Ibid., p. 413.
53. Ibid.
54. Ibid.
55. Ibid., pp. 414, 408.
56. Ibid., p. 414.
57. Ibid., p. 84.
58. Ibid., p. 30.

59. Ibid.
60. NDR, p. 36; see also p. 80.
61. NK, p. 85.
62. Wallace Kaufman, *No Turning Back: Dismantling the Fantasies of Environmental Thinking* (New York: Basic Books, 1994), pp. 109–40. This is an insightful work.
63. Luc Ferry, *Le nouvel ordre écologique* (Paris: Bernard Grasset, 1992), pp. 180ff. In this interesting, though flawed, book I first encountered an analysis of Schoenichen's work. See also the discussion in Dominick, *The Environmental Movement in Germany*.
64. See Arne Naess, "The Encouraging Richness and Diversity of Ultimate Premisses in Environmental Philosophy," *The Trumpeter* 9 (Spring 1992): 53–60.
65. For this reading of deep ecology, see Zimmerman, "Rethinking the Heidegger–Deep Ecology Relationship," *Environmental Philosophy* 15 (1993): 195–224, and *Contesting Earth's Future: Radical Ecology and Postmodernity* (Berkeley and Los Angeles: The University of California Press, 1994).
66. Cited by Robert C. Paehlke, *Environmentalism and the Future of Progressive Politics* (New Haven: Yale University Press, 1989), p. 66.
67. Cited in ibid., p. 66.
68. See Robert A. Pois, *National Socialism and the Religion of Nature* (London: Croom Helm Publishers, 1986); and Zimmerman, "Rethinking the Heidegger–Deep Ecology Relationship."
69. Years later, he remarked ruefully that "given their magnitude and monstrosity, these derivations [from Leopold's work] would constitute a *reductio ad absurdum* of the whole land ethic enterprise." See J. Baird Callicott, "The Conceptual Foundations of the Land Ethic," in *In Defense of the Land Ethic* (Albany: SUNY Press, 1989), p. 92.
70. Personal communication.
71. Callicott's essay, "The Metaphysical Implications of Ecology," *Environmental Ethics* 8 (1986): 301–16, has been criticized by Karen J. Warren and Jim Cheney in "Ecosystem Ecology and Metaphysical Ecology: A Case Study," *Environmental Ethics* 15 (1993): 99–116. See Callicott, "On Warren and Cheney's Critique of Callicott's Ecological Metaphysics," *Environmental Ethics* 15 (1993): 373–74.
72. Aldo Leopold, *The Land Ethic* (New York: Oxford University Press, 1949), pp. 224–25.
73. Ibid., p. 203.
74. Ibid., p. 202.
75. Ibid., p. 223, cited by Callicott, "Animal Liberation: A Triangular Affair, *Environmental Ethics* 2 (1980): 311–38.
76. Callicott, "Elements of an Environmental Ethic: Moral Considerability and the Biotic Community," *Environmental Ethics* 1 (1979): 71–81. Citation from p. 80; emphasis in original.
77. Ibid, p. 81, emphasis in original.
78. Callicott, "Animal Liberation," p. 322.
79. Ibid., p. 323.
80. Ibid., pp. 323–24.
81. Ibid., p. 326.
82. Ibid.
83. Ibid.

84. Ibid., p. 327.
85. Ibid., p. 328.
86. Ibid., p. 329.
87. Ibid.
88. Ibid., p. 338.
89. Ibid., p. 334.
90. Ibid.
91. Sluga, *Heidegger's Crisis*, p. 175. Partly reacting to Nazi appropriation of Plato's *Republic*, Karl Popper wrote his controversial book, *The Open Society and its Enemies* (London: G. Routledge and Sons, Ltd., 1947).
92. See Pois, *National Socialism and the Religion of Nature*.
93. See Gregory Stock, *Metaman* (Toronto: Doubleday Canada, 1993).
94. See Kaufman, *No Turning Back*; Kevin Kelly, *Out of Control* (Reading, MA: Addison-Wesley, 1994; Stock, *Metaman*; and the final chapter of Zimmerman, *Contesting Earth's Future*.
95. See Ken Wilber, *Sex, Ecology, and Spirituality* (Boston: Shambhala, 1995), and *Up From Eden* (Boston: Shambhala, 1981).

13

Materialists, Ontologists, and Environmental Pragmatists

Andrew Light

This paper will briefly review an earlier distinction I made between materialists[1] and ontologists[2] in environmental political theory, and then proceed to formulate a strategy for negotiating their competing claims. I will call this new formulation "environmental pragmatism." The main thrust of my argument is that the urgency of the environmental crisis necessitates some form of metatheoretical compatibilism on the part of environmental materialists, ontologists, and other political ecologists. One would hope that this would be an uncontroversial position for such theorists to take, as a premise of their approaches is that environmental concerns are the prepolitical conditions around which a political theory must be formed. Metatheoretical pragmatism can give us the compatibilism necessitated by dire environmental conditions, without requiring the surrender of the philosophical commitments of different forms of political ecology. However, the form of pragmatism defended is one importantly predicated as *environmental*. It is a reminder that for us there is a foundational role that environmental concerns play in directing how radical ecologists decide to form bonds of agreement toward ecological renewal in practice.

Copyright 1995 by *Social Theory and Practice,* Vol. 21, No. 2 (Summer 1995).

I

For both environmental materialists and ontologists, environmental problems require much deeper analysis than classical liberal policy-making approaches would suggest. Materialists, like Murray Bookchin's school of social ecologists and Herbert Marcuse's version of environmental critical social theory, see the crisis of environmental degradation, and human suffering as a result of that degradation, as presupposed by the material conditions of a capitalist (or state capitalist) economy. Included in their primary concerns are the technological processes that are part of these economies as well as the political systems that sustain them globally.[3] From such an analysis, Bookchin, Marcuse, and other environmental materialists are led to the conclusion that the solutions to these problems should begin with an analysis of the possible range of alternatives that a political or economic system can sustain with an eye toward pushing the boundaries of these systems and changing the material conditions of society as a whole. Changes in individual consciousness, while sometimes necessary for sustaining new social conditions, will, on this general theory, follow material changes. By demonstrating that environmental problems emanate from material conditions of society, environmental materialists identify systemic facets of environmental problems that go deeper than the evaluations found in most liberal policy analyses.[4]

Contrast this overall framework with environmental ontologists, who attempt to place humans and nature in some sort of philosophically informed symbiotic context. This context is not grounded in some argument for a uniquely human material social history (as we find in social ecology) but is instead based on the inseparable ontological roots of humans and nonhuman nature. Such theorists argue that humans should be identified with nature not as a separable organism or set of organisms, but as an integrated part of a larger life/world system.

Environmental ontologists focus their critique of mainstream environmentalism on the need for more analysis of and changes in individual human consciousness with respect to the relationship between humans and the nonhuman natural world. For ontologists, significant solutions to environmental problems are also to be found primarily in a redescription of human ontology that then serves as the basis for environmental practice and policy. The focus for political reform then is in a reform of the self as expressed in individual identity, rather than in the materialist focus on the social group or the institution as the primary mechanism of environmental renewal. Perhaps the clearest example of an ontological theory can be seen in the development of Deep Ecology, originally by Arne Naess and later by his American and Australian disciples.

Deep Ecology, if we formulate it through Naess's original statements on the theory, certainly primarily posits an ontological and not a materialist criticism of human interaction with the nonhuman natural world. Naess takes great pains to distinguish his theory even from a theory of environmental ethics. He argues that the experience of the world provides the belief and identity basis for Deep Ecology, an identity that makes it more than simply an ethical system. According to Naess: "If deep ecology is deep it must relate to our foundational beliefs, not just to ethics. Ethics follow from how we experience the world. If you articulate experience then it can be a philosophy *or* a religion."[5]

I emphasize the "or" in this last passage to highlight a source of tension between materialists and ontologists, particularly two of its respective representatives. The failure to distinguish between the sorts of understanding and reasoning at work in religious belief, and that at work in philosophical explanation, is at the root of Murray Bookchin's worries about Deep Ecology as a theory and as a practice. In fact, Bookchin is very quick to lump all forms of ecospiritualism into one foul mass, usually characterized by its most outrageous representatives and their most counterintuitive claims.[6]

Bookchin's worries are focused on the spiritual dimensions of the theory, and, I would suggest, their environmental ontological commitments. As a consequence of such an emphasis, Deep Ecology is notable, on Bookchin's account, for its "absence of reference to social theory." A clear struggle therefore emerges for social ecologists:

> In America, the rapidly forming Green movement is beset by a macho cowboy tendency [the reference is to deep ecologists of the Dave Foreman/ Earth First! variety] that has adopted Malthusianism with its racist implications as a dogma, an anti-humanism that among some of the wilderness oriented "campfire" boys has become a brutalized form of misanthropy, and a "spiritualist" tendency that tends to extol irrationalism and view ecology more as a religion than a form of health naturalism. It has become primarily the task of American eco-anarchists to develop a sustained resistance to these primitivistic, misanthropic, and quasi-religious tendencies.[7]

Bookchin's overall attitude toward Deep Ecology, characterized by this lumping and splitting of the opposition, is exemplary of the anticompatibilism that makes environmental political theory difficult to translate into practice. This is particularly problematic when practice may require large-scale coalitions across a broad array of philosophical commitments. But the blame cannot be laid at one doorstep: Bookchin and Naess are both guilty of a kind of essentializing of the theoretical ground of their positions (the former in Kropotkinian evolutionary theory and the latter in religious ecospiritualism) that would make any kind of communication between the two difficult. But it is not only at the theoretical level that divisions are drawn. It is clear that Bookchin really does see the practices of ecoanarchists to be directed as much against Deep Ecologists as they are against liberal environmentalists and growth-oriented polluters of the earth.

But once we move beyond the particular example of Bookchin versus Naess, the materialist/ontological distinction gets more complicated and difficult to sustain as a basis for hard and fast theoretical divisions. In Marcuse's work there is at least an acknowledgment of the role of individual ontological transformations in the struggle of humans and nature against debilitating forms of technological rationalism and instrumentalism.[8] This should not be so surprising. On the other side of the environmental spectrum, as I am describing it, even Naess's version of Deep Ecology includes a dimension that is clearly materialist. In his opus *Ecology, Community and Lifestyle*, he has several chapters discussing the social and political order, the types of economic organization that are best suited to a "no growth" society and references to the types of policy changes that are needed to enact a deep ecology program.[9] While I still want to categorize Naess as an "environmental ontologist" because of his strong commitment to the priority of ontological changes in the transformation to a lasting shift in the human relationship with the nonhuman natural world, we should acknowledge that he also makes a strong contribution to the materialist dimension of political ecology both in theory and practice.

The distinction between materialists and ontologists in environmental political theory is still useful despite this overlap, but more as an heuristic tool that aids in analysis rather than as a hard and fast division that points to necessary contentions of deep disagreement. The distinction will help us to predict the propensity for certain serious points of contention, but it need not be the case that such disagreements are necessary for the distinction to work. The most important part of the distinction becomes where one's priorities fall when it comes to general questions of reform given bad environmental conditions.

II

Why not simply combine the positions?[10] Unfortunately, the respective contemporary theorists just discussed do not usually see the merits of such a move. While I think that both of these approaches to environmental problems are useful and have recognizable advantages over mainstream policy approaches to environmental issues, their utility is limited by the importance their proponents give to the construction of their theoretical props. Both materialists and ontologists get caught up in arguing with each other over the differences between the theoretical implications of each others' philosophies that in turn may weary the public to their mutual concerns.[11] Some may claim that the cause of the social ecology–deep ecology debate is the strong personalities involved. Other explanations however are needed. For Bookchin and Naess there is a strong foundationalism at work that, though commendable as an attempt at theoretical rigor, may in practice hold up attempts at broad organizational unity.[12]

But even if we could come up with some definitive explanation of the differences between the intuitions and priorities of materialists and ontologists, at some point as political activists, in addition to being theorists, we must negotiate a settlement between these two firmly established camps in order to get things done. Sigmund Kvaløy, the most materialist of the theoretical forerunners of Deep Ecology, has said something similar in his formulation of "ecophilosophy." He argues that we theorists "should strive to be as wide in scope as the attack on the life struggle of the ecosystem and human society is today."[13]

Even though I count myself as a straightforward environmental materialist and am committed to its theoretical development, I believe that some theoretical questions (and sometimes some strategic materialist principles), while valuable, often get in the way of an attempt to formulate a broad-based radical plan to solve environmental problems. In order to achieve that goal, some sort of mutual toleration of competing theories is demanded by the overwhelming need for action on the environmental front. But such compatibilism, even though it may appear to violate the foundational claims of some materialists and ontologists, need not lead inevitably to a facile philosophical relativism. I propose then a principle of tolerance in the form of a pragmatic position that would require radical environmentalists to leave some questions that divide them to private dispute. At the same time this pragmatism would require theorists and practitioners to publicly communicate a straightforward position that endorses the trumping ethical and political environmental considerations on which they agree and the practices that expedite their mutually desired goals.

There have been several attempts to express the connections between classical American philosophy and environmental ethics,[14] and a general pragmatist account of some debates in environmental ethics,[15] but no comprehensive attempt as of yet to specifically outline the structure of a politically pragmatist and compatibilist account of ecological theory and practice. Part of the reason for this gap is explained by the failure of those interested in pragmatism and environmental thought to distinguish between two formulations of the relationship between environmentalism and pragmatism. The first form is found in the strategy of using philosophical pragmatism directly to set up substantive positions against other theories in environmental philosophy. Much of the pragmatism in contemporary environmental ethics (of both sorts just described) gets directed into the project of creating an ethical pluralist stance as a response to what is now the mainstream moral monism of the field.[16] Such moves do not necessarily contribute to the politically pragmatist principles that I am looking for here and may wind up adding to the number of intractable sides in environmental ethics debates.[17] Unfortunately, many theorists seem to take this to be the sole contribution that pragmatism can make to environmental thought.

In contrast, a second application emerges out of other contemporary pragmatist-inspired sources that have evolved as answers to theoretical log jams in other areas. Applied to the concerns here, this position takes the theoretical predisposition of a generalized philosophical pragmatism to weed through debates in environmental ethics and environmental political and social theory, but would not attempt to express a full and complete ethical or political theory on its own.

I take the first project to be a direct philosophical use of pragmatism, and the second project to be a more metaphilosophical use of pragmatism. I will focus here on the latter form of pragmatism.

Metaphilosophically inclined environmental pragmatists would argue that we need to give up on some of the debates in political ecology for no other reason than the fact that there is much that we do agree on (whether we are materialists or ontologists, and for that matter, holists or individualists, anthropocentrists or nonanthropocentrists) that has not yet been effectively put into policy or communicated to the public. From this metatheoretical perspective, environmental pragmatists are not wedded to any particular theoretical framework from which to evaluate specific problems, but can choose the avenue that best protects the long-term health and stability of the environment, regardless of its theoretical origin.

But this position needs to be fleshed out. How does one "give up" on certain debates and continue others while still retaining the integrity of one's political-philosophical position? In what remains of this paper I want to sketch out one way of constructing a version of metaphilosophical environmental pragmatism, specifically as a guidepost for political ecology, which tries to achieve this theoretical-practical balance. Along the way I will give a concrete example of the sorts of choices that result from embracing such a pragmatism.

III

For an environmental pragmatist, as is true with all political ecologists, concern for the environment becomes a prepolitical condition that any future politics must have the ability to address. That is, these theories suggest that political thought in general, and specific political decisions in particular, can only be determined within the context of their effects on the state of the environment. The subject of ecology is inherently part of the public sphere for such theories, and the health of the public sphere is the object of any political theory. Such a connection between environmental issues and political organization has in some ways characterized some Green politics already, particularly in Europe. Christopher Manes has pointed out, "With the looming threats of radiation, acid rain, and toxic waste in their own backyard, European environmentalists did not have the *luxury of separating ecology from politics.*"[18] Of course there are many ways to embrace a specifically ecological political theory. One could construct such a theory from within the bounds of an authoritarian system out of the same motivation to resolve the environmental crisis, as we have seen in the egalitarian visions of Bookchin and Naess.[19] I will not seriously consider the authoritarian option for political ecology, so all of the theories considered here can more accurately be described as examples of democratic political ecology (DPE), where "democratic" does not denote a specific form of government but does indicate a rejection of antiegalitarian structures. A specifically democratic political ecology, then, is one that also takes as a presupposition the necessity of maintaining some form of participatory political order within the public sphere. The rules of the formation of political ecology, as it were, must emerge within some kind of participatory context, and the implementation of the policies derived from political ecology must accord with reasonable egalitarian principles. So what marks the DPE theorist who embraces my form of environmental pragmatism?

The environmental pragmatist takes the commitment to stopping environmental destruction and increasing the likelihood of sustaining and expanding existing organic environmental systems a bit further than the nonmetaphilosophically pragmatic political ecologist or DPE theorist. Here, the ecological-political claim is extended from its role as a pre-political condition to serve the function of a regulative ideal ranging over the use and function of theory in political ecology. Difficulties in political ecological practice lead us therefore to new considerations of the status of our theories. Again, I find an ally to this approach in Sigmund Kvaløy, who, not surprisingly, Peter Reed and David Rothenberg describe as closest to Marx among the Norwegian environmental philosophers in linking practice to theory thorough action (thus appropriately challenging their separation from the beginning).[20] Kvaløy interestingly enough finds a plethora of good action and a dearth of good theories coming out of the first decade of environmental activism (in which he played a major role) in Europe. He maintains that out of this process, "We need ... to sit down and do ecophilosophy again, to philosophize under direct influence of the reaped experience."[21]

In accordance with this goal, metaphilosophical environmental pragmatism is a good principle to embrace. One strategy within this framework is for the pragmatist not to be concerned in public with the outcome of certain debates among environmentalists and ecologists within a certain range of issues. Some questions would be shelved for now, such as: What is the intrinsic value of species that fuels a duty to insure their diversity? or, What is the ontological relationship between humans and nonhuman animals? In these cases the environmental pragmatist may search for answers to these questions in private while publicly pursuing the best possible solutions to practical environmental questions. Such solutions may skirt the resolution of these questions altogether. The pragmatist also recognizes that the current state of the world, either economically, politically, or ontologically, is the result of a contingent history: existing individual or social relationships with the nonhuman natural world are not assumed to be the only ones that must be worked within in looking for solutions to environmental problems. There are no universal statements on the state of nature available for the environmental pragmatist (or rather, the pragmatist does not permit such statements in public); descriptions of the relationship between humans and nature are descriptions of states of affairs that could have been otherwise. And because they could have been otherwise, pragmatists do not essentialize their conceptions of the identity of nature or our duty to it that would limit their ability to reconceptualize nature in such a way that it can be most expeditiously protected.

What sort of choices are the result of this view? One example. A metaphilosophical environmental pragmatist would bear no public theoretical grudge against ecological restorations—public works projects that attempt to reconstruct previously damaged lands. Such projects, like the prairie restorations at the University of Wisconsin's Arboretum, which may require some hard choices about what state of the prairie to restore to (pre-Columbian or not, for example), would be fully supported. Restoration makes sense because on the whole it results in many advantages over mere preservation of ecosystems that have been substantially damaged by humans.[22] Therefore, an environmental pragmatist at best would find it only mildly intellectually interesting, and at worst morally irresponsible, to contribute to the raging debate in environmental ethics over the question of whether ecological restoration results in a full restoration of the intrinsic value of nature.[23]

What is the point of such arguments now, asks the pragmatist? And what answer do we offer the hard-working volunteer restorationists (of which there are thousands) who ask whether environmental philosophers support what they are sacrificing their time in doing?

"Well, yes of course we support what you do," says the nonpragmatist engaged in this debate, "but the problem is that no matter what you think you're doing, you can never really *restore* nature (in its full normative sense), because nature has certain definitive value, and the best arguments concerning that value suggest that a priori restoration can never really restore nature." Cold comfort it seems for the environmental do-gooder who is engaged in a truly radical practice. This point is particularly cogent when such a restorationist is compared with the reform environmentalist, who sometimes seems content to either ignore restoration all together or leave it in the hands of those who will do it for profit and never really care about the result for the land itself.[24] For purposes of interacting with the world of environmental practice, the pragmatists let the demands of the environmental crisis regulate their interaction with the practice of restoration: assume for the moment that there is no essential normative issue about the moral status of the restored land as a part of the larger assumption; for now, that there is no necessary value of nature that need concern us here.

Privately though, the pragmatist may hold definite views concerning the normative status of restored nature and may even wish to publish these views in obscure academic journals. But the pragmatist will understand that to publicly voice such worries leads to too much confusion in the practice of a project that almost everyone can agree is a good idea: restoring damaged lands to something approximating their former native species diversity to the point where the land can once again regenerate itself. I qualify this description of the restored landscape to indicate what I hope is obvious—being a pragmatist on questions like this does not mean that one lends full support to every act of restoration. Sometimes, environmental pragmatists will want to publicly be very vocal about what should count as a good restoration (they may, as I do, want to exclude corporate-sponsored restorations as good restorations). In that case, it is the duty of the ecophilosopher, following Kvaløy, to become deeply involved in the debate on this issue with practitioners rather than only privately discussing the problem. The philosopher or political theorist certainly may have much to add to a full analysis of the relevant questions concerning the practice.

The version of the public/private distinction employed in this discription of environmental pragmatism is drawn from Richard Rorty's *Contingency, Irony, and Solidarity*,[25] but in a very qualified sense that I will explain soon. In general, Rorty argues that modern philosophy's focus on metaphysical and transcendental truth has not been useful in the pursuit of freedom and instead has distracted from attempts to increase the autonomy of individuals or communities. Rorty has said in other discussions that we should "look for regulative ideals, . . . stick to freedom and forget about truth and rationality."[26] Instead of pursuing foundations for knowledge—or, in the case I am taking up here, foundations for conceptions of what nature really is and what we are obligated to do based on that positive definition of nature—Rorty argues that liberal culture needs "an improved self-description." This description amounts to an openness to redescribe our beliefs in terms that do not provide us with a reason as to why we might be wrong about things through any sort of *normative* force, but rather as suggestions as to how things might have been otherwise.

But such redescriptions are limited for Rorty. In public, the agents of redescription are concerned with problems of humiliation and suffering, and how bonds of solidarity can be forged between persons to limit such suffering (by redescribing others in terms that place them as agents with whom we can easily feel solidarity). In private we can redescribe others in whatever terms we please—terms according to Rorty, "which have nothing to do with my attitude toward your actual or possible suffering. . . . My private purposes, and the part of my final vocabulary which is not relevant to my public actions, are none of your business."[27]

Environmental pragmatists employ a similar strategy, with the exception that their public redescriptions are concerned with the suffering of all of nature rather than only with the suffering of humans. (Some may find it contentious to speak of nature as "suffering," but whether nature really does suffer is, again, not important for now for the metaphilosophical pragmatist.) Since they are compelled to speak in public of redescriptions, environmental pragmatists will not fall into the trap of articulating essentialist positions about what nature really is, or our duties to it, for the purposes of engaging in theoretical debates only.

But environmental pragmatists should hold this view only as a metaphilosophical position and not as a concrete philosophical stance. Though there is not time to explore the question adequately here, I do not think environmental pragmatists ought to embrace moral relativism. The crisis of the environment drives this metaphilosophical view, not necessarily an antecedent commitment to philosophical relativism or postmodernism.

Pragmatists, on my account, may disagree about the framework in which to explore environmental questions. Some may start with inclinations toward communitarian political ideas and socialist economic frameworks, while others may have leanings toward market solutions and various shades of liberalism. Some will be materialists, others ontologists, and others something else. But what my pragmatists agree on is that the truth of these approaches is not always fundamental for purposes of environmental practice, and the appropriateness of any one theory in a particular case is contingent on historical, cultural, social, and resource conditions. What is constant is the environmental pragmatist's acknowledgment that environmental needs are the foreground and the background of such contingencies and the controlling interest in deciding which framework is most appropriate in deciding how to solve environmental problems.[28] Importantly, the environmental pragmatist, because she is not committed to a dogmatic defense of her starting frameworks (which may direct her research or activist interests at first), and because she is not looking for a totalizing definition of the true state of the natural world when so many practical considerations are at hand, is open to acknowledging that at some time her framework may be inappropriate for obtaining the goals of preservation and protection of the environment. On some issues the ontologist must defer to the materialist, for example, when questions of social structures are on the table. Materialists generally have stronger traditions of considering such questions from which they can draw.

Environmental pragmatists should be concerned with generating multiple overlapping descriptions of the value of nature, as well as appeals for overlapping methods of valuation—a pluralism with respect to value theories. This strategy will serve the interests of building a critical mass of people on specific goals for improving environmental quality. And this makes sense not only given the strategic considerations raised here, but also given the character of the thing being valued. Gary Varner argues than any holistic conception of the value of nature must of necessity reject moral monism in favor of a theoretical pluralism:

> It is because an ecosystem has no welfare of its own, that a holistic environmental ethic must be pluralistic. If it is plausible to say that ecosystems (or biotic communities as such) are directly morally considerable—and that is a very big *if*—it must be for a very different reason than is usually given for saying that individual human beings are directly morally considerable (and, perhaps, higher animals or all individual living organisms).[29]

For strategic purposes, though, pragmatists in general and environmental pragmatists in particular also understand that inquiry into social problems need not come to an end in order for action to take place. As Rorty suggested some time ago, "Reference to action . . . can take place at any step in the eternally incomplete series of interpretations."[30] When we feel the urgency to act, in order to provide aid to nature, we will find temporary stopping places in our ongoing conversations on how best to act for nature and how best to interpret the needs of nature.[31] For pragmatists as I conceive them, there may be no public solutions to some problems other than "pragmatic" ones, as we have given up on the idea of a totalizing discourse for now about conceptions of or duties to nature. But pragmatic solutions are not something we settle on, they are things we strive for while privately pursuing, if we choose, our individual redescriptions of nature in positive, totalizing, or hegemonic terms.

IV

This picture, however, is not complete. While it is true that Rorty shares with Marx a desire to connect philosophy with practical critique,[32] his politics still is decidedly not radical (and probably reformist), and his philosophical neopragmatism is too weak as a whole to sustain the sort of environmental political theory I have in mind. So, if in the first part of this paper I am celebrating the environmentalism and the radicalism of materialist and ontological radical political ecologists, how can I then embrace a version of Rorty's notorious liberalism at the end? The answer is that it is not Rorty's political theory that I am endorsing but only his theories' organizational principles.

Rorty's liberal neopragmatism, by itself, is too restrictive for the purposes of political ecologists. At the end of the day, his position is clear: Liberal democracies seem to be the best form of government created in history so far (even with their more obvious faults), so there is no compelling reason to break from their way of doing things.[33] But following Bookchin, Naess, Marcuse, the early Bahro, Gorz, Vandana Shiva, Jim O'Connor, Enrique Leff, and others, I think there are lots of reasons why liberal democracies, embedded in capitalist growth-oriented structures, are insufficient by themselves for achieving the goals of long-term environmental sustainability. Classical liberalism, per se, is insufficient as a principle within which to compose a political theory that acknowledges the priority of the environment in political structures, planning, and decision making that emerges from a DPE.

I hope I have demonstrated, however, that the compatibilism that is generated as a part of Rorty's neopragmatist philosophy for constructing bonds of solidarity between different people (and specifically the public-private split as part of that strategy) can be very useful for structuring a metaphilosophical position within which radical environmental materialists and ontologists can work together. At least Naess realizes the importance of some sort of political tolerance within the Deep Ecology movement: "Those working on any goal of global dimensions should have certain principles in common. But these principles should not imperil deep differences in ultimate metaphysical or religious views."[34] Why not use part of Rorty's neopragmatism then to extend Naess's call for principles beyond the borders of Deep Ecology?

Finally, I want to suggest that this pragmatist strategy may not only work within the bounds of DPE, but also seems to be a good way to get at a workable DPE that all political ecologists can live with. Why is this the case? Metaphilosophical pragmatists, of whatever origin, when looking at a political-ecological situation, take politics in general to be a constantly shifting contested terrain. This view follows from their nonessentialist approach to practical political questions. But again, the environmental part of this pragmatism forces such assessments within the context of constructing the best approach that accords with established ecological principles. Looking out at the wealth of political theories, the environmental pragmatist (here concerned with constructing a DPE) is going to use the theory that best embraces the fact of the contested terrain of politics toward the goal of laying the foundation of egalitarian practices that are in the best interests of the long-term health of the environment. Nothing in principle restricts the pragmatist from such broad considerations and choices, while at the same time the pragmatist is obligated to undergo such a search.

This idea reminds us that neither materialists nor ontologists are completely constitutive of the terrain of democratic political ecology. DPE exists on a different theoretical plane and in a different relationship to materialists and ontologists than they have to each other. The relationship is similar to that of Radical Political Economics as a whole to both Althusserian-influenced Social Structures of Accumulation theorists, and classical labor theory of value Marxists: the last two are both forms of radical political economics, and neither exclusively amounts to what counts as a radical economic theory. Environmental pragmatism, as I imagine it, stands in a relationship to DPE and to materialists and ontologists, as providing something like rules for the game of how the last two should work in practice as forms of DPE.[35]

Notes

1. Andrew Light, "Rereading Bookchin and Marcuse as Environmental Materialists," *Capitalism, Nature, Socialism* 4 (1993).
2. Andrew Light, "The Role of Technology in Environmental Questions: Martin Buber and Deep Ecology as Answers to Technological Consciousness," *Research in Philosophy and Technology* 14 (1992).
3. I mean here by materialism a thin interpretation of the term. Rather than the thick version of materialism advocated by Marx and most Marxists, which entails a strong form of ontology, I mean to suggest simply an analysis of social problems that takes into account and looks primarily at the physical structures and institutions that make up a society.
4. See, for example, Bookchin's critique of environmental liberals in his *Rethinking Society: Pathways to a Green Future* (Boston: South End Press, 1990); and Marcuse's general arguments concerning the chilling effect of liberal society on activism in his *One-Dimensional Man* (Boston: Beacon Press, 1972). For an exchange on the utility of the materialist/ontologist distinction and a defense of that utility, see Bookchin's "Response to Andrew Light's 'Bookchin and Marcuse as Environmental Materialists'," and my "Which Side Are You On? A Rejoinder to Murray Bookchin," *Capitalism, Nature, Socialism* 4 (1993).
5. Cited in Rothenberg's introduction to Naess's *Ecology, Community and Lifestyle*, trans. David Rothenberg (Cambridge: Cambridge University Press, 1989), p. 20; emphasis added.
6. See the first chapter of *Rethinking Society*, which lumps indiscriminately goddess worship with Deep Ecology. While some Deep Ecologists may invite this comparison, certainly not all warrant it—see, for example, David Rothenberg's remarks

on the differences between various branches of theoretical Deep Ecology in his introduction to *Ecology, Community and Lifestyle*.

7. Bookchin, "New Social Movements: The Anarchic Dimension," in David Goodway, ed., *For Anarchism: History, Theory, and Practice* (London: Routledge Press, 1989), p. 273.

8. See chapter 2 of Marcuse's *Counterrevolution and Revolt* (Boston: Beacon Press, 1972), and his "Ecology and the Critique of Modern Society," *Capitalism, Nature, Socialism* 3 (1992).

9. Also see Naess's "The Politics of the Deep Ecology Movement," reprinted in Peter Reed and David Rothenberg, eds., *Wisdom in the Open Air: The Norwegian Roots of Deep Ecology* (Minneapolis: University of Minnesota Press, 1993).

10. George Bradford tries to do just that in his "Toward a Deep Social Ecology," reprinted in Michael Zimmerman et al., eds. *Environmental Philosophy* (Englewood Cliffs, NJ: Prentice Hall, 1993).

11. National Green gatherings have been notoriously known to end in debilitating arguments between Deep and Social Ecologists. Disputes have also often spilled over into the popular press in Europe and America in magazines like *The Nation* and *Z*, as well as some national newspapers.

12. One cannot stress enough the need for pragmatic conciliation at times (within the bounds of maintaining moral commitment) in order to get things done, as has been evidenced in the alliances formed around the Redwood Summer activities in Northern California. For a fascinating account of the environmental politics there, see the interview with Judi Barry in issue 17 of *Capitalism, Nature, Socialism* 5 (1994).

13. Sigmund Kvaløy, "Complexity and Time: Breaking the Pyramid's Reign," in Reed and Rothenberg, *Wisdom in the Open Air*, p. 119.

14. See, for example, William Chaloupka's "John Dewey's Social Aesthetics as a Precedent for Environmental Thought," *Environmental Ethics* 9 (1987); Bob Taylor's "John Dewey and Environmental Thought," *Environmental Ethics* 12 (1990); and Robert Fuller's "American Pragmatism Reconsidered: William James's Ecological Ethic," *Environmental Ethics* 14 (1992).

15. See, for example, Anthony Weston's "Beyond Intrinsic Value: Pragmatism in Environmental Ethics," *Environmental Ethics* 7 (1985), as well as his "Before Environmental Ethics," *Environmental Ethics* 14 (1992); and Kelly Parker's "The Value of a Habitat," *Environmental Ethics* 12 (1990).

16. The prejudice toward monism can be seen, for example, in J. Baird Callicott's remarks refuting a suggestion that Holmes Rolston is not a monist: "Given that even Rolston is not really a pluralist after all, one begins to wonder why *our best, most systematic, and thoroughgoing* environmental philosophers cling to moral monism." In "The Case against Moral Pluralism," *Environmental Ethics* 12 (1990); p. 109, my emphasis. Obviously, many very well-respected theorists in this field may have some disagreement with such a claim.

17. See, for example, the Katz-Weston exchange in *Environmental Ethics* 10 (1988).

18. My emphasis, from Manes's *Green Rage* (Boston: Little, Brown, 1990), p. 124.

19. See for example William Ophuls's political ecology in his *Ecology and the Politics of Scarcity* (San Francisco: W. H. Freeman and Co., 1977).

20. Peter Reed and David Rothenberg, "Sigund Kvaløy," in *Wisdom in the Open Air*, p. 114.

21. Kvaløy, "Complexity and Time," p. 116.
22. For an overview of this example of ecological restoration and others, see Anthony Baldwin, Judith DeLuce and Carl Pletsch, eds., *Beyond Preservation* (Minneapolis: University of Minnesota Press, 1994).
23. See Robert Elliot's "Extinction, Restoration, Naturalness," *Environmental Ethics* 16 (1994); Alastair Gunn's "The Restoration of Species and Natural Environments," *Environmental Ethics* 13 (1991); as well as Eric Katz's "The Big Lie: Human Restoration of Nature," *Research in Philosophy and Technology* 14 (1992), reprinted in part as "Restoration and Redesign: The Ethical Significance of Human Intervention in Nature," *Restoration and Management Notes* 9 (1991). Elliot and obviously Katz are the perpetrators here.
24. For a good critique of some of Robert Elliot's arguments on restoration and an argument for the literal terrain up for grabs in restoration as a practice, see Eric Higgs's "A Quantity of Engaging Work to be Done: Ecological Restoration and Morality in a Technological Culture," *Restoration and Management Notes* 9 (1991). A critique of the liberal tolerance of corporate-sponsored restoration can be found in my "Hegemony and Democracy: How the Politics in Restoration Informs the Politics of Restoration," *Restoration and Management Notes* 12 (1994).
25. Richard Rorty, *Contingency, Irony and Solidarity* (Cambridge: Cambridge University Press, 1989).
26. Richard Rorty, "Truth and Freedom: A Reply to Thomas McCarthy," *Critical Inquiry* 16 (1990): 634.
27. Rorty, *Contingency, Irony and Solidarity*, p. 91.
28. Such a nontheoretical approach to encountering nature has been an integral part of some radical political ecology. Christopher Manes describes John Seed's (one of the early leaders of the Australian environmental movement) conception of nature as "an intense, tangible reality, not a theoretical issue involving resource depletion and land management" (*Green Rage*, p. 120).
29. Gary Varner, "No Holism without Pluralism," *Environmental Ethics* 13 (1991): 179.
30. Richard Rorty, "Pragmatism, Categories and Language," *Philosophical Review* 70 (1961): 219.
31. I am indebted to Meredith Garmon for pointing out this last passage in Rorty's work. For a much more complete discussion of Rorty's position with respect to action, see Garmon's *Pragmatist Critiques of Jurisprudence*, unpublished dissertation, University of Virginia, May 1992.
32. The strongest version of this argument is that Rorty's is a call not so much for a new philosophy (à la Marx) but a claim that any theory that is disconnected from practical critique does not count as philosophy. See Garmon, *Pragmatist Critiques*.
33. See Richard Rorty, "The Priority of Democracy to Philosophy," reprinted in Alan Malachowski, ed., *Reading Rorty* (Oxford: Blackwell, 1990).
34. Arne Naess, *Is it Painful to Think* (Minneapolis: University of Minnesota Press, 1993), p. 136.
35. I am indebted to Avner de-Shalit, Eric Higgs, Eric Katz, and an anonymous referee at *Social Theory and Practice* for helpful comments.

14

Challenging Pluralism:
Environmental Justice and the Evolution of Pluralist Practice

David Schlosberg

Both pluralism and environmentalism in the United States have evolved significantly in this century. The development of the environmental movement into a major interest group in the liberal pluralist model was crucial to the establishment of much of the United States' environmental policy. Recently, however, many grassroots actors have come to criticize the organizational model and focus of mainstream environmentalism. A central issue in the growing environmental justice movement is the numerous definitions and experiences of "environment"—a diversity, they argue, that is ignored and devalued by the major environmental organizations. In the development of the environmental justice movement, the acknowledgement of this multiplicity has led to a form of organization, methods of communication, and institutional designs that are quite distinct from the conventional model of liberal pluralism.

A new generation of pluralism is pushing to recognize the validity of diverse understandings of environmental problems and to open the political process to positions previously excluded. The argument here is that this new form of organization is still pluralism—ideas and practices are mirrored in both earlier and more recent versions of pluralist theory. In addition, this form of pluralism not only proposes a remedy to the limitations of the conventional model, but also is more able to confront changes in the nature of power, capital, and the political oversight of environmental problems.

Defining the Conventional Model: Liberal Pluralism in Theory and Practice

The model of liberal pluralism (Banfield 1961; Dahl 1961, 1967; Polsby 1963; Truman 1960) is defined by its emphasis on groups as the key unit of political action; these groups act out of the desire for self-interested economic rewards, competing against one another before the authority of the state. The model also stresses open access—the penetrability and heterogeneity of the political system (Dahl 1961). If interests are organized well and are supported by ample resources, participants with those interests will find success.

It is not my task here to offer a full critique of this generation of pluralism, as this has been done both from outside and, later, from within its ranks (see, for example, the collections edited by Wolff, Moore, and Marcuse 1965; Connolly 1969; as well as Bachrach and Baratz 1962; Lindblom 1977, 1982; Lowi 1969; Manley 1983). These authors have exposed the limitations of the vision of this stage of pluralism: its narrow notion of interest, the limited understanding of political action, the exclusion of some groups, the dismissal of the political power and economic resources of elites, and its flawed concept of tolerance. The point here is to focus on the elements of those criticisms that are relevant to a discussion of the environmental movement.

Lowi's (1969) critique of interest group liberalism, which he calls a "vulgarized" version of pluralism, laments the fact that the model negates the traditional mode of political life, a mode that focuses on philosophy and dialogue. Instead, this existing form of pluralism is based in the rituals of the competitive process. The management of interests has replaced the more noble notions and practices of politics. As the pluralist critic Wolff eloquently notes, "The genius of American politics is its ability to treat even matters of principle as though they were conflicts of interest" (Wolff 1965, 21).

This generation of pluralists also limited their attention to a particular type of experience—that in the economic realm. Truman (1960), for example, noted many times the diversity and overlapping nature of groups, yet focused almost exclusively on labor and economic identity. The varieties of political and social experience were forced through a filter that strained out all but particular economically based selfish interests.[1] Issues that could not be articulated in this manner were simply left out of the national conversation—remaining in the margins until they erupted into the social movements of the 1960s and after.

This is one reason why criticisms of the pluralist process focused on the denial of diversity and the exclusion of certain groups. Connolly noted the bias of a pluralism in which "some concerns, aspirations, and interests are privileged while others are placed at a serious disadvantage" (1969, 16). Lowi (1969) noted the conservatism of the model, as it effectively locked diversity out. And Wolfe (1965, 42) argued that liberal pluralism does not justify this, it simply ignores it. There was no protection, in the model or practice, for unrepresented minorities, whether racial, cultural, ideological, or class based.

In addition, critics focused on the nature of the interactions and participation within interest groups. As Kariel argues at the beginning of his study of pluralism, "The organizations which the early theorists of pluralism relied upon to sustain the individual against a unified government have themselves become oligarchically governed hierarchies" (Kariel 1961, 3–4). One of the original functions of interest groups—that of mediating between individuals and governmental decisions—became impossible as groups outgrew their originally envisioned size and mission.

Even as the theory of liberal pluralism was soundly attacked, and some of its original formulators began to have doubts and changes of heart (see especially Lindblom 1977, 1982), the political practice continued. As environmentalism exploded in its second wave of popularity in the 1960s and 1970s, groups in the movement explored political models. By the 1980s, numerous major groups had settled comfortably into both liberal pluralism and representative offices in Washington DC, styling themselves as an interest group among many.[2] As such, the mainstream has become quite successful—in terms of the model. Access to the decision-making process has increased, memberships and fundraising have grown, an increasing number of mainstream environmentalists have moved into various nodes in the process, and policies have been successfully negotiated.[3]

The zenith of the liberal model came when Clinton and Gore were elected in 1992. After twelve years of playing defense, the mainstream was ecstatic. The vice-president was the author of a best-selling book on the environment, and more than twenty staff members of the major environmental groups entered the new administration, including Bruce Babbitt, former head of the League of Conservation Voters, and George Frampton, former president of the Wilderness Society. Those inside were finally friendly, and the doors were left open for their old acquaintances.[4] As an interest group in the model of conventional pluralism, the environmental mainstream had reached its highest pinnacle of success. Or had it?

Criticism of Conventional Environmental Pluralism

While the mainstream environmental movement has focused on its improved position as an interest group in the liberal pluralist model, criticisms of the major groups and their organizing model have steadily increased. These criticisms had been growing long before the problems with the Clinton administration became obvious. Not surprisingly, they exemplify the problems earlier critics saw in the model of liberal pluralism. Most critical attention has been directd at the recent lack of success of the mainstream's campaigns, the organizational form of the major groups, and the lack of attention to the diversity of environmentalism as a whole.

As for the instrumental success of the mainstream model, one would have expected a new round of legislative victories with allies in the White House and Congress. Instead, the first two years of the new, environmentally friendly administration saw an Everglades plan favoring developers and sugar growers, support of bovine growth hormones, an attempt to repeal the Delaney Clause's zero tolerance for carcinogens, the ouster of Bureau of Land Management director Jim Baca, a political plan for Northwest forests that failed to take science seriously, the start-up of the WTI incinerator in Ohio, a retreat on grazing price reform, a retreat on elevating the Environmental Protection Agency to cabinet level, a defeat in attempts to reform the 1872 Mining law, the divisive battle on the North American Free Trade Agreement—and much more.[5] The 103rd congressional session was generally acknowledged to be the worst on environmental issues in the last twenty years. All of this, of course, was before the Democrats lost majority control of Congress.

In addition, the form that mainstream environmental organizing has taken has angered grassroots activists attentive to both resource and environmental justice issues. These critics make direct connections between the structure of the mainstream and the lack of democratic and community participation.[6] Specifically, there has been distrust of the professional atmosphere of the organizations; frustration with control by the major funding organizations rather than by memberships; and criticism of the centralized, hierarchical, professionalized organizations that are not accountable to memberships or local communities. The most common complaints include the limitation of participation of members to writing checks, the physical distance between mainstream groups and their members, the professionalization and careerism that came with the move to Washington, and attempts to silence dissent in the membership ranks. Along these lines, Jeff St. Clair, the maverick editor of *Wild Forest Review*, has offered a comprehensive lambasting of the movement:

> Somewhere along the line the environmental movement disconnected with the people. Rejected its political roots, pulled the plug on its vibrant tradition. It packed its bags, it starched its shirts and jetted to DC where it became what it once despised: a risk-aversive, depersonalized, overly analytical, humorless, access-driven, intolerant, statisti-

cal, centralized, technocratic, deal-making, passionless, sterilized, direct-mailing, jock-strapped, lawyer-laden monolith to mediocrity.
(St Clair 1995)

The result, finally, is that the larger mainstream organizations are seen as "alienated and estranged from their membership, and irresponsible to the needs of local communities" (Taylor 1992, 44).

Part of this disconnection is due to, and much of the critical reproach focuses on, the lack of diversity in the mainstream movement in terms of both ideas and participants. The insider, conciliatory nature of the beltway process left frustrated many individuals and groups that concentrate on resource and wilderness issues. Their exits from the process, beginning in the early 1980s, became the impetus for some of the more radical environmental organizations.[7] While many of the complaints from this radical school on the lack of diversity in the movement were limited—focusing solely on philosophical and strategic differences—the growing grassroots movements enthusiastically eschewed the previous pluralist model.

The emergent environmental justice community raised the issue of the lack of diversity to a new level, focusing on both the white, upper-middle-class nature of the mainstream organizations and their avoidance of issues of concern to people of color and urban dwellers (Shabecoff 1990; Taylor 1992). While the lack of minority representation in the offices and on the boards of environmental groups was a focus, the more telling complaint centered on the movement's focus on resources, wilderness, endangered species, and the like, rather than on toxics, public health, and the unjust distribution of environmental risks. Issues of interest to low-income communities and communities of color had been left off of the environmental agenda; by bringing them to the fore, the new movement helped to expose the bias of the mainstream's concerns (see Taylor 1992; Bullard 1994).

The central point of these criticisms of the environmental mainstream and its organizational model is that they have limited what counts as a valid environmental perspective. The result is that some notions of environmentalism and environmental action have been ignored or dismissed as invalid. In the development of environmental policy, this has kept a variety of "stakeholders" out of the conversation. For example, one of the major mainstream groups, the Environmental Defense Fund (EDF), stepped into the dispute between McDonald's and grassroots groups on the issue of styrofoam food containers. The EDF formed a task force that excluded grassroots leaders and reached an agreement with McDonald's that fell far short of the demands of those that had raised the issue in the first place (CCHW 1993, 12; Dowie 1995, 139–140). As one activist left out of those negotiations argued, the "established movement makes decisions and negotiates compromises for others while remaining isolated from where people actually live. . . . These communities strongly feel they can speak for themselves" (Newman 1993, 92).

All of these concerns—around the lack of success, the limitations of the organizational form, and the lack of diversity—have made the mainstream at best alienating, and at worst irrelevant, in the eyes of many grassroots activists and critics.

Critical Pluralism: An Evolving Theory

Thankfully, pluralism in both theory and practice has evolved beyond these limitations. A new generation of pluralist theory in the academy and pluralist practice in the grassroots has paid particular attention to the weaknesses in and criticisms of conventional liberal pluralism. A return to pluralist theory and practice might help to move beyond the limitations and exclusions of the liberal model.

"Return" is the operative term here, because much of what is being argued in the name of diversity and situated knowledges is quite similar to the ideas of pluralist theorists early in the century. William James's notion of pluralism began with a "radical empiricism" that opposed the more singular, monist position of rational absolutism. James argued that we must accept that "the constitution of reality is what we ourselves find empirically realized" in our everyday lives (James 1977, 145). "Knowledge of sensible realities . . . comes to life inside the tissue of experience. It is *made*; and made by relations that unroll themselves in time" (James 1976, 29). The point of James's pluralist universe is not just the recognition of difference, but the validation of varied experiences.

More recently, Donna Haraway's (1988) description of "situated knowledge" resurrects many of the arguments of James. Haraway uses the metaphor of vision to examine the multiple ways things can be seen, depending on one's experience, context, or, more generally, the view from one's body. She argues for "politics of epistemologies of location, positioning, and situating, where partiality and not universality is the condition of being heard to make rational knowledge claims" (589). Objectivity is based on the localized view; as such, it is multiple.[8]

For example, polluted water is not as singular an objective reality as it sounds. The Willamette River, flowing near my home, begins at one of the purest sources of water in the world; it ends, after converging with the Columbia, as one of the most polluted and carcinogenic rivers in the country. Yet the experience with pollution differs according to one's situation. Recent Asian immigrants in Portland, who fish for food, have a particular experience of the pollution as a threat to a way of life. Parents whose children play and swim downstream from a paper mill and its effluent worry about illnesses and the effects of dioxin on hormonal development. These experiences differ from the biologist who studies skeletal deformities in fish in the river, and from the lobbyist for a major environmental group in Washington, DC, who is reviewing proposed changes in the Clean Water Act. The knowledge of the same event differs, but, importantly, not all of the understandings are acknowledged equally. The radical empiricism of James and the situated knowledge of Haraway validate that diversity.[9]

The environmental justice movement takes this type of acknowledgement of diversity at its base; bringing out alternative environmental discourses that have *not* been equally heard is a basic project of environmental justice. As Barbara Deutsch Lynch argues:

> If environmental discourses are culturally grounded, they will differ in content along class and ethnic lines. Where power in society is unequally distributed, not all environmental discourses will be heard equally. Thus, questions of environmental justice must address not only the effects of particular land uses or environmental policies on diverse groups in society, but the likelihood that alternative environmental discourses will be heard and valued (1993, 110).

Environmental justice requires an understanding of the existence and importance of multiple perspectives and the need to validate that variety. The movement is constructed from difference and negates the importance of a singular perspective or ideology (ECO 1992).

Redefining Unity: The Development of Networks

A return to original notions of pluralism validates the diversity of experiences and knowledges that grow out of the variety of ways we are all situated in environmental degradation. For the first generation of political pluralists, this acknowledgement was just the first step. Mary Parker Follett, writing in 1918, argued that "life is a recognition of multitudinous multiplicity. Politics must be shaped for that" (291). The shift in concern was one thing; but the real question, as expressed by Follett, was "what is to be done with this diversity" (10). Follett's lament revolves around how to incorporate the reality of diversity into the necessity of coordinated political action.

Follett responded to her own question, advocating for a very particular type of "unity"—one differentiated from uniformity. Her concerns echo loudly in an era in which "unity in diversity" has become a slogan of a new generation of pluralism. "Unity, not uniformity, must be our aim. We attain unity only through variety. Differences must be integrated, not annihilated, nor absorbed" (Follett 1918, 35). In the end, it is heterogeneity, and not homogeneity, which makes a unity without uniformity.[10]

This acceptance of multiplicity as the precondition of political action is central to a new generation of critical pluralists. Theorists such as Rorty

(1989), Deleuze and Guattari (1987), and Haraway (1991) address the issue of how to bring together diverse experiences into a form of political organization without sacrificing the parts.[11]

Haraway's notion of a "cyborg community" is based on an elective construction of affinities. Affinities can be built across numerous issues, but they have the radical empiricism of James at their center. These constructions or solidarities are, importantly, not total but based on areas of partial overlap. Partial, locatable knowledges sustain "the possibility of webs of connections called solidarity in politics and shared conversations in epistemology" (Haraway 1988, 584). For Haraway, there is nothing in assertions against objectivity—and for partiality, local knowledge, and differentiated experience—that rules out the construction of a solid collective in the realm between individuals and the state.

In practice, organizing based on the validity of multiplicity at its base has increasingly taken on the form of alliances and networks. The environmental justice movement again serves as a model here, as it has consciously and deliberately organized in a manner distinct from the liberal pluralist model. The movement really has no center; rather, there are a number of identity- and issue-based organizations and networks, as well as regional networks working on a variety of issues.[12] All of these are constructed by local grassroots groups. Networks begin at the level of the community, with bases in everyday relationships at home, church, work, and play. The organization of networks takes these local realities seriously and continues the recognition and validation of diverse experiences, even as it links the multiplicity of peoples and issues into alliances.

The development of alliances based on a recognition of difference opens opportunities for collaboration and action across some fairly wide divides. The Community Alliance for the Environment (CAFE), organized by Latinos and Hasidim in the Williamsburg section of Brooklyn, New York, serves as an example here (Greider 1993). The Latino organization El Puente worked together with the Hasidic social service agency United Jewish Organizations to oppose construction of a massive garbage incinerator the city had planned for the neighborhood. In 1992, they organized the first environmental town meeting in Williamsburg; it included not only Latinos and Hasidim, but also the African Americans, Polish Americans, and Italian Americans from the neighborhood. By January 1993, CAFE was able to organize a multiethnic march of 1,500 local residents across the Williamsburg Bridge to protest the incinerator. While organizing around the unifying symbol of the local environment, CAFE leaders have taken pains not to offend, ignore, or marginalize anyone involved in CAFE activities. Respect for differences goes hand-in-hand with the building of an alliance.

The Southwest Network for Environmental and Economic Justice (SNEEJ) has organized a larger set of alliances, one example of which is the Electronics Industry Good Neighbor Campaign (EIGNC). Initially it tied together communities in Albuquerque, Austin, Phoenix, and San Jose; it has expanded to include groups in Portland and Eugene, Oregon, as well as groups across the border in Mexico. All of these communities have been sites of growth of the semiconductor industry and have shared issues of environmental hazards, worker protections, taxpayer subsidies, and growth. This collaboration has brought together a variety of environmental, labor, civil rights, land use, and taxpayer groups. While there are vast differences among the communities involved, the network provides strength in numbers, allows for the sharing of resources, and offers mutual support.

The Evolution of Pluralist Communication

Networks and alliances, however, deal only with form. Implicit in the discussion of the building of relations across differences is a practice of recognition and respect for those differences. I want to turn now more directly to this issue. How is it, exactly, that such solidarity can be constructed and maintained? How does a new generation of pluralism help with the development of a united, but not uniform, movement?

In the theoretical realm, agonistic respect (Connolly 1991, 1995), attempts at intersubjective understanding (Benhabib 1992; Habermas 1970a, 1970b; Honneth 1992), and inclusive, open discourse free from domination and the possibility of reprisals (Dryzek 1990; Forester 1989; Habermas 1984, 1987) are crucial to the theory of a new generation of pluralism.

An ethic of agonistic respect exceeds the bounds of past pluralist notions of tolerance as it moves to include an aspect of recognition. Some theorists and activists have argued that liberal toleration is too limited—it is simply inadequate to deal with contemporary diversity.[13] The equal "blindness" that characterizes toleration is not equalizing recognition (Brown 1993, 392). As Lorde argues, "We have no patterns for relating across our human differences as equals" (1984, 115).

Connolly has replaced discussions of tolerance with his discussions of agonistic respect (1991) and critical responsiveness (1995). Respect, and a cultivation of care for the different positions of others, becomes the necessary relationship between self and other, away from the detachment, isolation, and demonization of conventional notions of competitive pluralism.

An ethic of agonistic respect necessitates a more open and ongoing form of political discourse. A focus on the communicative process brings with it a suspicion of any completion of a discussion. The focus on a final consensus, or "generalized norms," that have been emphasized in both past notions of pluralism (e.g., Dahl 1961) and in Habermas's critical theory, are regarded as inhibitions to the further and continuing flowering of difference. Dryzek emphasizes that "a succession of discursive exercises held up to critical scrutiny could create and reinforce norms of free discourse" (1990, 87). Likewise, Benhabib argues that the "emphasis now is less on *rational agreement*, but more on sustaining those normative practices and moral relationships within which reasoned agreement *as a way of life* can flourish and continue" (1992, 38).

Theorists who examine the relationship between critical communicative theory and actual political practice have focused on one of two areas. Some turn their attention to the internal culture of new social movements (Epstein 1991; Evans and Boyte 1986; Melucci 1985) in order to examine the "prefigurative politics," "free spaces," and "symbolic challenges" of these movements. Other theorists, particularly within the fields of policy analysis and planning (Dryzek 1990; Fischer 1993; Fischer and Forester 1993; Forester 1989; Paehlke and Torgerson 1990), have focused on the possibilities for communicative transformations of political institutions.

Again, the environmental justice movement demonstrates the implementation of such practices. The movement has been critical of many of the communicative aspects of mainstream organizing: its top-down structure, the one-way nature of communication within the major groups, and the lack of attention to issues of public participation in policy making (see, for example, Bullard 1993; DeChiro 1992; Miller 1993; Moore and Head 1993; Taylor 1992). Environmental justice activists have also been subject to discrimination and disrespect in public hearings (see, for example, Hamilton 1994; Krauss 1994). Issues of communication, then, have been a central focus in the development and demands of environmental justice. Internally, the movement has attempted to employ more open discursive processes, paying particular attention to communication within and across diverse groups. Externally, the movement has made demands about issues of communication and more discursive and participatory policy making on local government, industry, and government agencies.

One example of the internal implementation of new communicative practices can be seen in the development of the Principles of Environmental Justice at the First National People of Color Environmental Leadership Summit in 1991. The process of the conference was designed explicitly to give voice, and a respectful ear, to all those who wished to contribute. Those involved reported that an atmosphere of trust, honesty, openness, and respect permeated the work of the appointed Drafting Committee. This climate "served to break down immense barriers and facilitate communication and cooperation which enabled it to arrive at a balanced articulation of the legitimate interests of the widely disparate conference delegates" (Madison, Miller, and Lee 1992, 51). The process exemplified an ongoing agonistic respect; it also showed the possibility of the development of a unity based in diversity rather than in a push toward uniformity. The entire experience was a lesson in the building of communicative tools necessary to construct the solidarity necessary for cooperation and collaboration across differences.

These forms of communication remain a central concern of the movement. SNEEJ notes that local groups that become part of the network are assured the right to be heard, the right to know, the right to be respected, and the right to be involved in the decision-making process of the network (SNEEJ 1993). In response, member individuals and groups are not only expected to share information and keep communication open, but are also reminded of the responsibility to recognize, respect, and be understanding of cultural diversity. It is only through these internal communicative and participative practices that the network has been able to expand its size and mission while remaining centered on the community concerns that led to its construction.

All of this concern with process and communication by environmental justice activists is focused on more than just inward concerns. The movement organizes as it does to strengthen its voice and demand its place in the development of governmental and industry policy. The principles adopted at the Leadership Summit call for "public policy [to] be based on mutual respect and justice for all peoples," "demands the right to participate as equal partners at every level of decision-making including needs assessment, planning, implementation, enforcement and evaluation," and "affirms the fundamental right to political, economic, cultural and environmental self-determination for all peoples" (Lee 1992, xiii–xiv). A shared and respected role in the decision-making process is a key demand of the environmental justice movement.

This demand surfaces in a number of ways. The movement demands information from industry and government in order to increase community discourse. Environmental Justice Centers, or "Communiversities," would emphasize "bilateral understanding and mutual respect between community residents and academicians" (Wright 1995, 64). Many argue the value of "participatory research" in examining environmental illness (Brown 1992; Brown and Tandon 1983; Bryant 1995; Gaventa 1991). A number of groups have also proposed "good neighbor agreements," which lay out community concerns and what companies and/or government agencies will do about them, and often call for the establishment of Community Advisory Committees to oversee ongoing monitoring, inspections, and environmental and safety audits of facilities (Lewis 1995).[14]

Much of the effort around expanding communication and public participation in environmental policy has also been focused on the U.S. Environmental Protection Agency (EPA). In 1992 the EPA made what it called an "initial step" in contributing to the national dialogue on environmental justice with a two-volume document on "environmental equity" (U.S. EPA 1992).[15] The environmental justice community, however, was quite critical of the report that, while mentioning the importance of communication, was prepared with no input from those who had been working on the issues for years. This, said critics, reflected the "disrespect and arrogance of the Agency" (2: 88). To many, the report appeared to be more a public relations strategy than a substantive effort to address the concerns of affected communities (2: 97, 101, 117).[16]

In response to many of the criticisms of the *Environmental Equity* report, the EPA established an Office of Environmental Justice. As a sign it was taking seriously some of the participatory concerns of critics in the movement after the change in administrations, the agency in 1993 established the National Environmental Justice Advisory Council (NEJAC). The NEJAC, the majority of whose members are leaders of environmental justice groups and resource centers, has been charged with advising, consulting with, and making recommendations to the administrator of the EPA on matters relating to environmental justice. NEJAC established a subcommittee on Public Participation and Accountability, and its chair, Peggy Saika of the Asian Pacific Environmental Network, has described issues of participation as the "linchpin" of all the NEJAC's recommendations to the EPA (Saika 1995).

Issues of communication, then, have been central to the environmental justice community. The movement has both exposed the weakness of and moved beyond the communicative limitations of mainstream environmental organizations. There has been insistence on open discourse, respect, and intersubjective understanding of difference within the environmental justice movement. And there have been demands made on local government, industry, and the EPA for participation in decisions affecting communities. These internal processes and external demands demonstrate that the movement sees improved communication—manifest in specific discursive practices—as both a key strategy of the movement and part of the definition of environmental justice.

Evaluating the Model: Openness and Effectiveness

The environmental justice movement represents an evolution in pluralism. It has pointed out the weaknesses of the liberal pluralist model—embodied by the mainstream environmental movement—and has developed forms of organization and communication more suited to its dedication to diversity. But this shift in pluralist emphasis begs two crucial questions. First, does a new critical pluralism invite everyone to the table? Second, and to the point, does this method of organizing actually *work*?

Just how far does a critical pluralism go in recognizing the value of respect for diverse participation? I see two ways of responding to this issue. First, the practices of a critical pluralism are both voluntary and reciprocal. We need not suffer gladly fools, fascists, fundamentalists, or anyone unwilling to reciprocate an agonistic respect. And there are, obviously, some positions based in discrimination, supremacy, and/or hate that cannot be reciprocally appreciated. But it is important that dismissals not be made out of hand. As Connolly sarcastically hints to those who would draw lines in the pluralist sands, "Is not Nazism exactly the doctrine that denies that almost everything counts for something?" (1995, 209).

One of the points of valuing a critical pluralism is that as difference confronts itself, change may occur. Approaching and questioning young Nazis or racists—engaging them directly in discourse—seems a better tactic for dealing with their closed fundamentalism than does meeting them with a similarly closed state of mind. Does a critical pluralism demand a respect for Nazis? As Nazis, no. As participants in an agonistic discourse that remains reciprocally respectful, maybe.

Maybe, because National Socialism is not very relevant to most policy discussions. While it is crucial to a critical pluralism that diverse positions get a seat at the table, there is no need in the real world to invite any Nazis to a discussion of salmon recovery on the Columbia River or toxic incineration in low-income neighborhoods. It is crucial, however, to open the discourse up to relevant stakeholders. One cannot, for example, have discussions about the allocation and management of land and resources in the western United States without addressing some of the issues brought up by the Wise Use and county rights movements. And bringing them to the table *will* entail confronting the race-based ideology that informs some of the organizations of the movement, as well as their connections to and support by the extractive industries (Helvarg 1994). An open discursive process will make these ideas and connections obvious and provide a place for what might be a productive exchange.

While Nazis at the table would be a rather rare occurrence, grassroots and environmental justice activists face a much more prevalent and difficult adversary: industry. Many in the movement see industry just as others see fascists: it would be an insult to sit with them, and impossible to talk to them anyway. But trying to make environmental policy without the representation of toxic polluters not only goes against any notion of pluralism but also is thoroughly unrealistic. Industry is a stakeholder and needs to be involved in discourse. The problem with the liberal pluralist model is that industry and, at times of strength, the mainstream environmental groups, have really been the *only* stakeholders that government agencies from the local to the national levels have involved. Numerous other positions have simply been excluded from discussions, just as critics of the liberal pluralist model have argued since the 1960s (Connolly 1969; Wolff 1965).

Critics who throw out the Nazi card or indignation with industry in order to dismiss the pluralist process miss a key issue. The central question for an evolution of pluralism is not how to avoid obnoxious voices like skinheads or destructive stakeholders like industry; instead the point is finally to get voices and positions recognized and involved that have not been a part of conventional pluralist discourse. The focus is not on who should be excluded in open discourse; rather it is on who *has* been excluded in a pluralist system of limited participation and how to bring them into the process.

In the environmental justice movement, for example, the central effort has been to give voice to and involve the victims of environmental destruction and industrial pollution in environmental discourse. One of the accomplishments of the movement has been a move toward the expansion of public hearings and participatory mechanisms. Previously, the interests of industry—e.g., the right to pollute, create illness, and make private decisions about communities—were simply assumed. With the expansion of voice and participation in a critical pluralist discourse, those interests are questioned and challenged. The movement has accepted many of the needs of industry, that is, to make products and profit; but it also demands an equal representation of its own needs: community participation, a healthy environment, and decent jobs. The point has been to expand policy conversations to include the diverse views, perspectives, and understandings that are to be affected and that have previously been excluded.[17]

Finally, we must ask whether this form of pluralism actually works—whether this model of pluralism is a useful one, especially in contrast to the conventional pluralist model. In addition to expanding the political discourse beyond the limited stakeholders now involved in environmental policy making, networks—such as those that make up the environmental justice movement—also provide a countermeasure against changes in power, capital, and politics. There are three interrelated issues here: changes in the understanding of power, changes in the nature of production, and changes in political oversight.

First, many theorists have discussed the relationship among various forms of power and the value of a diverse and linked response. Foucault (1978, 1980) has argued that power itself is a network that needs to be examined in its extremities—the everyday experiences of those subject to its reach. Laclau and Mouffe (1985) have also asserted that there is a litany of forms of power and antagonisms in the social realm, and networks can develop in response. Haraway (1991, 170) argues that an understanding of the weblike structure of power may lead to new couplings, new coalitions. Networks develop, then, not just out of pre-existing social relations but also out of an understanding of, and alliance around, how power links issues. Shared experiences are once again at the base, but these experiences are often ones of subjection.

Second, and obviously related, capital itself has taken on a more networked form. Corporate strategy has constructed production networks that link numerous companies and facilities over vast areas. Increasing regimes of "free trade" will continue this transition. Capital strategy now also includes the tactic of playing numerous communities off one another. Whether planning a waste incinerator or a semiconductor production plant, companies will select a number of sites and then wait for the community response. In the former, they are obviously looking for a site of least resistance; in the latter, for the most generous incentive package. To the extent that network organizing makes it possible to respond in numerous areas simultaneously, it is a more formidable opponent to such structures and strategies.

Third, and, again quite related, is the nature of the political sphere. Obviously, important environmental decisions are made at the state, county, and local levels. The current Congress would like to continue to de-emphasize the role of the national government in environmental decisions. However, the globalization of capital also minimizes the decision-making relevance of the nation-state as the market seeks to take its place. If traditional organizations, such as those that make up the mainstream environmental movement, continue a strategy that focuses on the national government, then they can address only a small portion of relevant political decisions. Citizen action is necessary on the regional and local levels, because that is where much of the control remains lodged; it is necessary on the global level because the institutions of governance there are so limited (and undemocratic). And it is necessary to network across each of these levels, as political power flows through them simultaneously.

Brecher and Costello (1994a, 1994b), have used the metaphor of Jonathan Swift's Lilliputians to describe this networked strategy. The little people used a web of hundreds of threads to capture Gulliver. "Similarly, facing powerful global forces and institutions, people need to combine their relatively modest sources of power with often very different sources of power available to participants in other movements and locations" (1994b, 758). A variety of local actions, woven together, creates a network strong enough to harness problems larger than any locality might be able to do on its own. In his analysis of what he calls "ecopopulism," Szasz (1994, 164) notes that "political activities that are confined to a single zone tend to yield disappointing results; in contrast, when political events occur simultaneously in several different zones, the interactions that ensue among them tend to generate real forward motion."

Conclusion

Pluralism has evolved in both theory and practice. A new generation of critical pluralism is a pluralism that acknowledges the validity of diverse knowledges and positions, that then opens the political process to positions previously excluded, and that is more able to confront changes in the nature of power, capital, and political oversight of environmental problems.

Surely research is called for to examine further the strategic value of the move to a critical pluralism. But just as importantly, we need to evaluate the "success" of this form of pluralism in more than just instrumental terms. The environmental justice movement has been able to construct a more inclusive and open movement in the past decade than the mainstream environmental movement has built in a century. It has responded to its own criticisms of the mainstream model—on issues of organizational structure, diversity, and voice—with the construction of a movement designed to take these issues seriously. No matter what the outcome of their tactical battles, the development of a movement and new prefigurative practices—a new form of pluralism—is surely a worthy feat.

Notes

1. And even economic articulation was not a guarantee of entry into the political sphere. This generation of pluralists explicitly avoided the elevation of individual and group economic interests to *class* interests in their dismissals of elite power theory.
2. For some histories and varying interpretations of this period of the movement, see Dowie (1995), Gottlieb (1993), Sale (1993), and Shabecoff (1993).
3. Very early in the process, before the mainstream was fully formed as a powerful interest group and lobby, landmark legislation began with the first Clean Air Act in 1955, the first Clean Water Act in 1960, and the Wilderness Act in 1964. With the growth of the movement came further legislation with the National Wild and Scenic Rivers Act in 1968, the National Environmental Policy Act in 1969, the creation of the Environmental Protection Agency in 1970, the Endangered Species Act in 1973, the Toxic Substances Control Act and the Resource Conservation and Recovery Act (RCRA) in 1976, and the creation of Superfund in 1980. While legislative victories were short during the Reagan and Bush years, the mainstream used its lobbying muscle to head off attempts to dismantle past accomplishments. While certainly victories for the movement, a closer examination of the history of many of these policy battles shows the immense influence of industry to curtail much of their potential. For a clearly constructed example, see Szasz (1994) for a description of industry influence on the final version of the RCRA in 1976.
4. Brock Evans, lobbyist for the Audubon Society, exemplified the euphoria of the time: "I can't tell you how wonderful it is to walk down the hall in the White House or government agency and be greeted by your first name" (quoted in Dowie 1995, 179).
5. See the discussion of the evolving relationship between the environmental mainstream and the Clinton administration in Dowie 1995, chapter 7.
6. Examples here are numerous. See Dowie (1991, 1995) and Gottlieb (1990, 1993) for discussions of these complaints. From within the movement, see Bullard (1994), Cockburn and St. Clair (1994), and Montague (1995).
7. Some of these are quite well-known, such as Earth First! and the Sea Shepherd Society; others are less so, but just as dedicated, such as the Native Forest Council.
8. This is not a validation of a position of complete relativism. Recognizing that varying positions are grounded in experience—often community experience—

offers a way out of the all-too-simple dichotomy of objectivism versus relativism. That is actually one of the more interesting connections between the first generation of pluralists and more recent theorists. Bernstein, for example, who makes the argument for varied, but locatable, knowledge claims in his *Beyond Objectivism and Relativism* (1983), edited and wrote the introduction to the reissue of James's *Pluralist Universe*.

9. This is not the place to get into a discussion of the complex philosophical issues surrounding radical empiricism or situated knowledges. The point in regard to pluralism is simple: that from varied perspectives, people may have different understandings of the same environmental event, and not all of those understandings have been granted stakeholder status in either the mainstream environmental movement or the environmental policy process. I am not asserting that any one of these perspectives should be privileged, nor am I arguing that any one perspective keeps people from understanding the position of others. As I argue in a later section, communication and respect across these varied perspectives is an important componant of the evolution of pluralism.

10. Follett was quite explicit about the difference in meanings here: "Good words: integrate, interpenetrate, . . . compound, harmonize, . . . coordinate, interweave, reciprocally relate. . . . Bad words: fuse, melt, amalgamate, assimilate, weld, dissolve, absorb, reconcile" (1918, 35 and Note 1).

11. Obviously, there are differences among these theorists. Rorty, for example, sees the possibility of connection only around similar experiences of pain and humiliation. See the commentaries by Bernstein (1987, 1990) and McCarthy (1990). Deleuze and Guattari's discussion of the rhizome as a model of political organization is more broad. The first three characteristics of the rhizome are the principles of connection, heterogeneity, and multiplicity (1987, 7–8). Rhizomatic organizing is based on making connections—recognizing patterns across both distance and difference. As Deleuze and Guattari argue, "the rhizome is alliance, uniquely alliance" (25).

12. Examples here include the Indigenous Environmental Network and the Asian Pacific Environmental Network; the Oilwatch Network and the Pesticide Action Network; and the Northwest Network for Environmental and Economic Justice and the Network for a Sustainable New York City.

13. For a thorough discussion of the way the grounds of toleration have changed, and an argument for toleration's inadequacy in the face of contemporary diversity, see McClure (1990).

14. Examples here include SWOP (Southwest Organizing Project) and the Mountainview Advisory Council's involvement in pushing for increased citizen involvement in the environmental policies at Kirtland Air Force Base (Montague 1989), and agreements signed between Rhone-Poulenc Basic Chemicals Company and Unocal Richmond Refinery and local community organizations—both of which establish ongoing monitoring and oversight by a community-based committee (Lewis 1995). It should be noted that industry and local governments just as often will reject such community proposals, as have the chip manufacturer Intel in Albuquerque, NM (SWOP 1995), and Hyundai in Eugene, OR.

15. Even with the choice of the term "environmental equity," the EPA showed its initial unwillingness to listen to, take seriously, and speak in the language of the grassroots community, which insisted on the use of the broader term "justice" rather than equity.

16. There has been debate about whether the EPA's response was genuine or not. A confidential memo released by the office of Congressman Henry Waxman reveals that some in the EPA were less interested in specific policy changes than in using the media and contacts with mainstream environmental groups to "win recognition" for the efforts of the EPA before "long-simmering resentment in the minority and native American communities about environmental fairness [becomes] one of the most politically explosive environmental issues yet to emerge." See the press release from Congressman Waxman, February 24, 1992.

17. It is crucial here, of course, to be aware of the power dynamics involved in any communicative structure that involves industry. Critics of environmental mediation, for example, have noted how power may currupt a discourse that is supposed to be undistorted. Access, authenticity, and discrepencies in negotiating experience may all be manipulated behind a conversational facade (see especially Amy 1987, 1990). In addition, power relations outside the conversation may remain untouched, though many in the environmental justice movement attempt to address this issue by calling for more public participation and third-party oversight.

References

Amy, Douglas. 1987. *The Politics of Environmental Mediation*. New York: Columbia University Press.

Amy, Douglas. 1990. "Environmental Dispute Resolution: The Promise and the Pitfalls." In Norman J. Vig and Michael E. Kraft, *Environmental Policy in the 1990s*. Washington, D.C.: Congressional Quarterly Press.

Bachrach, Peter and Morton Baratz. 1962. The Two Faces of Power. *American Political Science Review*, 56: 947–52.

Banfield, Edward. 1961. *Political Influence*. New York: Free Press.

Benhabib, Seyla. 1992. *Situating the Self: Gender, Community and Postmodernism in Contemporary Ethics*. New York: Routledge.

Bernstein, Richard. 1983. *Beyond Objectivism and Relativism*. Philadelphia: University of Pennsylvania Press.

Bernstein, Richard. 1987. "One Step Forward, Two Steps Backward: Rorty on Liberal Democracy and Philosophy." *Political Theory* 15: 538–563.

Bernstein, Richard. 1990. "Rorty's Liberal Utopia." *Social Research* 57: 31–72.

Brecher, Jeremy and Tim Costello. 1994a. *Global Village or Global Pillage: Economic Reconstruction From the Bottom Up*. Boston: South End Press.

Brecher, Jeremy and Tim Costello. 1994b. The Lilliput Strategy: Taking on the Multinationals. *The Nation* 259, No. 21: 757–760.

Brown, L. D., and R. Tandon. 1983. Ideology and Political Economy in Inquiry: Action Research and Participatory Research. *The Journal of Applied Behavioral Science* 19: 277–294.

Brown, P. 1992. Popular Epidemiology and Toxic Waste Contamination: Lay and Professional Ways of Knowing. *Journal of Health and Social Behavior* 33: 267–281.

Brown, Wendy. 1993. "Wounded Attachments." *Political Theory* 21: 390–410.

Bryant, Bunyan, ed. 1995. *Environmental Justice: Issues, Policies, and Solutions*. Covelo, CA: Island Press.

Bullard, Robert, ed. 1993. *Confronting Environmental Racism: Voices from the Grassroots*. Boston: South End Press.
Bullard, Robert. 1994. Environmental Justice at Home and Abroad. In Robert Bullard, ed., *People of Color Environmental Groups 1994–95 Directory*. Atlanta: Environmental Justice Resource Center.
Citizens Clearinghouse for Hazardous Waste (CCHW). 1993. *Ten Years of Triumph*. Falls Church, VA: Citizens Clearinghouse for Hazardous Waste.
Cockburn, Alexander and Jeffrey St. Clair. 1994. After Armageddon: Death and Life for America's Greens. *The Nation* 259, No.21: 760–765.
Connolly, William, ed. 1969. *The Bias of Pluralism*. New York: Atherton.
Connolly, William. 1991. *Identity\Difference: Democratic Negotiations of Political Paradox*. Ithaca, NY: Cornell University Press.
Connolly, William. 1995. *The Ethos of Pluralization*. Minneapolis: University of Minnesota Press.
Dahl, Robert. 1961. *Who Governs?* New Haven: Yale University Press.
Dahl, Robert. 1967. *Pluralist Democracy in the United States*. Chicago: Rand McNally.
Deleuze, Gilles, and Felix Guattari. 1987. *A Thousand Plateaus: Capitalism and Schizophrenia*. Minneapolis: University of Minnesota Press.
Di Chiro, Giovanna. 1992. Defining Environmental Justice: Women's Voices and Grassroots Politics. *Socialist Review* 22, No. 4: 93–130.
Dowie, Mark. 1991. American Environmentalism: A Movement Courting Irrelevance. *World Policy Journal* 9: 67–92.
Dowie, Mark. 1995. *Losing Ground: American Environmentalism at the Close of the Twentieth Century*. Cambridge: MIT Press.
Dryzek, John. 1990. *Discursive Democracy: Politics, Policy, and Political Science*. Cambridge: Cambridge University Press.
Environmental Careers Organization (ECO). 1992. *Beyond the Green: Redefining and Diversifying the Environmental Movement*. Boston: Environmental Careers Organization.
Epstein, Barbara. 1991. *Political Protest and Cultural Revolution: Nonviolent Direct Action in the 1970s and 1980s*. Berkeley: University of California Press.
Evans, Sara, and Harry Boyte. 1986. *Free Spaces: The Sources of Democratic Change in America*. New York: Harper and Row.
Fischer, Frank. 1993. "Citizen Participation and the Democratization of Policy Expertise: From Theoretical Inquiry to Practical Cases." *Policy Sciences* 26: 165–187.
Fischer, Frank and John Forester, eds. 1993. *The Argumentative Turn in Policy Analysis and Planning*. Durham, NC: Duke University Press.
Follett, Mary Parker. 1918. *The New State: Group Organization and the Solution of Popular Government*. New York: Longmans, Green and Co.
Forester, John. 1989. *Planning in the Face of Power*. Berkeley: University of California Press.
Foucault, Michel. 1978. *The History of Sexuality. Vol. 1, An Introduction*. New York: Random House.
Foucault, Michel. 1980. *Power/Knowledge*. Ed. Colin Gordon. New York: Pantheon Books.
Gaventa, John. 1991. *Participatory Research in North America*. New Market, TN: Highlander Center.
Gottlieb, Robert. 1990. An Odd Assortment of Allies: American Environmentalism in the 1990s. *Gannett Center Journal* 4, No. 3: 37–47.

Gottlieb, Robert. 1993. *Forcing the Spring: The Transformation of the American Environmental Movement*. Washington, D.C.: Island Press.
Greider, Katherine. 1993. Against All Odds. *City Limits* 18, No. 7: 43–38.
Habermas, Jurgen. 1970a. On Systematically Distorted Communication. *Inquiry*, 13: 205–218.
Habermas, Jurgen. 1970b. Toward a Theory of Communicative Competence. *Inquiry*, 13: 360–75.
Habermas, Jurgen. 1984. *The Theory of Communicative Action. Vol. 1: Reason and the Rationaliztion of Society*. Trans. Thomas McCarthy. Boston: Beacon.
Habermas, Jurgen. 1987. *The Theory of Communicative Action. Vol. 2: Life-world and System: A Critique of Functionalist Reason*. Trans. Thomas McCarthy. Boston: Beacon.
Hamilton, Cynthia. 1994. Concerned Citizens of South Central Los Angeles. In Robert D. Bullard, ed., *Unequal Protection: Environmental Justice and Communities of Color*. San Francisco: Sierra Club.
Haraway, Donna. 1988. Situated Knowledges: The Science Question in Feminism as a Site of Discourse on the Privilege of Partial Perspective. *Feminist Studies*, 14(3): 575–99.
Haraway, Donna. 1991 [1985]. A Cyborg Manifesto: Science, Technology, and Socialist-Feminism in the Late Twentieth Century. In *Simians, Cyborgs, and Women: The Reinvention of Nature*. New York: Routledge.
Helvarg, David. 1994. *The War Against the Greens: The "Wise-Use" Movement, the New Right, and Anti-Environmental Violence*. San Francisco: Sierra Club.
Honneth, Axel. 1992. Integrity and Disrespect: Principles of Morality Based on the Theory of Recognition. *Political Theory* 20: 187–201.
James, William. 1976 [1912]. *Essays in Radical Empiricism*. Cambridge, MA: Harvard University Press.
James, William. 1977 [1909]. *A Pluralist Universe*. Cambridge, MA: Harvard University Press.
Kariel, Henry S. 1961. *The Decline of American Liberalism*. Stanford, CA: Stanford University Press.
Krauss, Celene. 1994. Women of Color on the Front Line. In Robert D. Bullard, ed., *Unequal Protection: Environmental Justice and Communities of Color*. San Francisco: Sierra Club.
Laclau, Ernesto, and Chantal Mouffe. 1985. *Hegemony and Socialist Strategy: Toward a Radical Democratic Politics*. London: Verso.
Lee, Charles, ed. 1992. *Proceedings: The First National People of Color Environmental Leadership Summit*. New York: United Church of Christ Commission for Racial Justice.
Lewis, Sinclair. 1995. *Precedents for Corporate-Community Compacts and Good Neighbor Agreements*. Waverly, MA: The Good Neighbor Project.
Lindblom, Charles. 1977. *Politics and Markets*. New York: Basic Books.
Lindblom, Charles. 1982. The Market as Prison. *Journal of Politics*, 44: 324–36.
Lorde, Audre. 1984. *Sister Outsider*. Trumansburg, NY: Crossing Press.
Lowi, Theodore. 1969. *The End of Liberalism*. New York: Norton.
Lynch, Barbara Deutsch. 1993. The Garden and the Sea: U.S. Latino Environmental Discourses and Mainstream Environmentalism. *Social Problems* 40, No. 1: 108–124.
Madison, Isaiah, Vernice Miller, and Charles Lee. 1992. The Principles of Environmental Justice: Formation and Meaning. In Charles Lee, ed., *Proceedings: The*

First National People of Color Environmental Leadership Summit. New York: United Church of Christ Commission for Racial Justice.

Manley, John. 1983. "Neo:Pluralism: A Class Analysis of Pluralism I and Pluralism II." *American Political Science Review* 77: 368–383.

McCarthy, Thomas. 1990. "Ironist Theory as a Vocation: A Response to Rorty." *Critical Inquiry* 16: 644–655.

McClure, Kirstie. 1990. "Difference, Diversity, and the Limits of Toleration." *Political Theory* 18, No. 3: 361–391.

Melucci, Alberto. 1985. The Symbolic Challenge of Contemporary Movements. *Social Research*, Vol. 52, No. 4: 789–816.

Miller, Vernice D. 1993. Building on Our Past, Planning for Our Future: Communities of Color and the Quest for Environmental Justice. In Richard Hofrichter, ed., *Toxic Struggles: The Theory and Practice of Environmental Justice*. Philadelphia: New Society.

Montague, Peter. 1989. What We Must Do. *The Workbook* 14, No. 3: 90–110.

Montague, Peter. 1995. Big Picture Organizing, Part 5: A Movement in Disarray. *Rachel's Envrionment and Health Weekly*, No. 425.

Moore, Richard and Louis Head. 1993. Acknowledging the Past, Confronting the Future: Environmental Justice in the 1990s. In Richard Hofrichter, ed., *Toxic Struggles: The Theory and Practice of Environmental Justice*. Philadelphia: New Society.

Newman, Peggy. 1993. The Grassroots Movement for Environmental Justice: Fighting For Our Lives. *New Solutions* 3, No. 4: 87–95.

Paehlke, Robert and Douglas Torgerson, eds. 1990. *Managing Leviathan: Environmental Politics and the Administrative State*. Peterborough, Ontario: Broadview Press.

Polsby, Nelson W. 1963. *Community Power and Political Theory*. New Haven: Yale University Press.

Rorty, Richard. 1989. *Contingency, Irony, and Solidarity*. Cambridge: Cambridge University Press.

Saika, Peggy. 1995. Personal Communication.

St. Clair, Jeffrey. 1995. Cashing Out: Corporate Environmentalism in the Age of Newt. Paper given on panel on Foundation/Corporate Control Over Environmental Organizations, Public Interest Law Conference, Eugene, OR. March 11, 1995.

Sale, Kirkpatrick. 1993. *The Green Revolution: The American Environmental Movement 1962–1992*. New York: Hill and Wang.

Shabecoff, Philip. 1990. Environmental Groups Told They Are Racists in Hiring. *New York Times*. February 1, 1990.

Shabecoff, Philip. 1993. *A Fierce Green Fire: The American Environmental Movement*. New York: Hill and Wang.

Southwest Network for Environmental and Economic Justice (SNEEJ). 1993. *Southwest Network for Environmental and Economic Justice*. Albuquerque, NM: SNEEJ.

Southwest Organizing Project (SWOP). 1995. *Intel Inside New Mexico: A Case Study of Environmental and Economic Injustice*. Albuquerque: SWOP.

Szasz, Andrew. 1994. *EcoPopulism: Toxic Waste and the Movement for Environmental Justice*. Minneapolis: University of Minnesota Press.

Taylor, Dorceta. 1992. Can the Environmental Movement Attract and Maintain the Support of Minorities? In Bunyan Bryant and Paul Mohai, eds., *Race and the Incidence of Environmental Hazards: A Time for Discourse*. Boulder, CO: Westview Press.

Truman, David. 1960. *The Governmental Process: Political Interests and Public Opinion*. New York: Knopf.

United States Environmental Protection Agency. 1992. *Environmental Equity: Reducing Risk for All Communities*. Two Volumes. Washington, D.C.: Government Printing Office.

Waxman, Congressman Henry A. 1992. Environmental Equity Report is Public Relations Ploy: Internal Memoranda Reveal Report to be Misleading. Press release February 24, 1992.

Wolff, Robert, Barrington Moore, and Herbert Marcuse, eds. 1965. *A Critique of Pure Tolerance*. Boston: Beacon.

Wolff, Robert. 1965. Beyond Tolerance. In Wolff, Moore, and Marcuse, eds., *A Critique of Pure Tolerance*. Boston: Beacon.

Wright, Beverly. 1995. "Environmental Justice Equity Centers: A Response to Inequity." In Bunyan Bryant, ed., *Environmental Justice: Issues, Policies, and Solutions*. Covelo, CA: Island Press.

15

Environmental Justice, Neopreservationism, and Sustainable Spirituality

Mark I. Wallace

Radical Green politics in America today is divided between two camps: antitoxics groups, organized against environmental hazards in economically distressed communities, and conservation activists and scientists, who work toward the restoration of biodiversity in wilderness areas. Both camps consist of grassroots organizations that emphasize all persons' collective responsibility for healthy environments. Both camps, while generally not self-consciously Marxist or even New Leftist, recognize that the consumerist logic of the market-state—"grow or die"—will continue to result in the degradation of clean water and air, animal well-being, and human flourishing. As such, both camps are frontal challenges to the American liberal ideal that the pursuit of enlightened self-interest somehow guarantees that all members of the body politic will achieve a reasonable standard of living in relatively healthy home and work environments.

But the affinities between antitoxics and biodiversity activists are initially difficult to discern in the face of the deep disagreements between the two camps. The antitoxics movement has its origins in the plight of human communities—urban, suburban, and rural—precariously situated close to health hazards such as waste dumps, polluted water supplies, contaminated soil sites, and toxic storage plants. Antitoxics argue that large industrial polluters in collusion with local public officials look for economically distressed areas in which to build hazardous facilities that promise immediate economic gains for the area's inhabitants. In urban areas, more often than not, poor people of color are most directly impacted by these new economic initiatives; in many suburban and rural areas, low-income whites are often disproportionately affected by the use and abuse of their environment and its resources. "Numerous studies have found that those who live in close proximity to noxious facilities are disproportionately people of color or of low income, and race has been found to be the stronger indicator of the two."[1] The antitoxics movement, therefore, is primarily concerned with environmental justice for *disenfranchised persons* who have suffered from historic class and racial discrimination and now have been deprived of their right to live and work in safe and healthy environments.

The new preservationist movement focuses primarily on the exigency to restore ecological richness and vitality in under- and nondeveloped areas that have not been irredeemably damaged by the influx of human populations. Here the emphasis falls on rehabilitating wildlife and wilderness areas for the sake of biodiversity rather than on the promotion of justice as such for disadvantaged human communities that have suffered environmental degradation. Otherwise disparate groups and movements such as Greenpeace, the Sea Shepherd Society, Earth First!, and Deep Ecology are united by their vigorous bioregional attempts to recover the integrity of nonhuman species by preserving their habitats. One such movement, the Wildlands Project, states that its mission is "to help protect and restore the ecological richness and native biodiversity of North America through the establishment of a connected system of reserves."[2] From this perspective, the best way to address the degraded environments of impoverished human cities and towns is to do so indirectly through the promotion of wild spaces that ensure the welfare of *all* life, not just human life.

At first glance, then, the differences between the antitoxics and the new conservationists appear stark and irreconcilable: either the focus falls on enabling disenfranchised human communities to overcome historic economic and environmental degradation, or it is on protecting the ecosystemic integrity of all beings without assigning any special concern to the needs of human beings. The understandable but unfortunate continuation of this disagreement further fragments an already divided environmental movement.

In light of this division within contemporary Green populism, what role if any can an environmentally nuanced spirituality play in healing this breach? Can champions of wilderness preservation and antitoxics activists find common ground in a "sustainable spirituality," to use

Charlene's Spretnak's felicitous phrase, that both seeks to protect nature for its own sake and fight social injustice?[3] I define *sustainable spirituality* as a nonsectarian spiritual vision concerning the deep interrelationships of all life-forms on the planet and the concomitant ethical ideal of preserving the integrity of these relationships through one's social and political praxis. While different historic religious traditions have articulated this vision in their own idiom—for example, the Jewish and Christian idea of the "Spirit" as binding all things to one another; or the Buddhist notion of "dependent origination," the belief that no entity, human or otherwise, is ontologically separate from any other entity—such a vision is not the province of any one tradition. On the contrary, sustainable spirituality is a generic sensibility available to all persons interested in crafting a holistic vision of life on the planet. This mode of spiritual awareness neither entails (nor precludes) belief in God (or the gods) nor subscription to any particular creed or ritual practice. Its roots are deep in the rich soil of various earth-friendly spiritualities. Sustainable spirituality offers its practitioners a powerfully useful root metaphor—the image of all life as organically interconnected—that can enable a fresh reappraisal of the debate between biocentric conservationists and advocates for environmental justice.

This essay is divided into three parts. Parts one and two use a case-study approach to explicate the agendas of antitoxics groups and contemporary conservation coalitions, respectively. Part three considers the role of sustainable spirituality in mediating the differences that now divide the two movements. In light of this mediation, I conclude with suggestions concerning the challenge of Green populism to the market mentality of the late capitalist West.

Toxic Sacrifice Zones and the Quest for Justice

Many local economies in urban and rural America today are dependent upon the production and management of toxic wastes. In economically distressed communities, the promise of a stabilized tax base, improved infrastructure, and jobs for underemployed residents is almost impossible to resist. The waste management industry offers an immediate quick fix to chronic poverty and instability in declining cities and neighborhoods that can no longer attract government and private investment. The price for allowing the storage and treatment of biohazardous materials in one's community may be long-term environmental problems. But people in the grip of poverty and joblessness have few options when their very survival, materially speaking, is contingent upon the construction of a trash incinerator or chemical dump in their neighborhood.

Corporate investors know a good thing when they see it. Waste management facilities cannot be sited where politically empowered middle-

and upper-class residents will fight the establishment of such facilities through the courts. Close proximity to hazardous industries immediately depresses property values in residential areas where virtually no one wants to risk endangering his or her physical and economic well-being by allowing such a liability to be built in their own backyard. And in those rare instances where such facilities have come on line in high-income areas, the residents have the means and mobility to "'vote with their feet' and move away from a high risk place of residence."[4]

Recent popular movements of resistance to the expansion of the toxics industry into various communities—poor and middle class alike—is surprisingly resilient. The conflict at Love Canal, New York, in the 1970s is the best known example of a successful grassroots response to callous irresponsibility in the powerful waste industry. A citizens' movement led by Love Canal homeowner-activist Lois Gibbs protested Hooker Chemical's disposal of toxic chemicals into the ground on which homes and schools were later built. The Love Canal homeowners convincingly documented the deleterious health affects that had resulted from living in the middle of a chemical dump and persuaded officials to buy out and permanently relocate town residents.[5] Other local antitoxics campaigns of the 1980s and 1990s are also notable, if not always as successful: the protest against siting a PCB landfill in Warren County, North Carolina; the movement against building a waste incinerator by the Mothers of East Los Angeles; the campaign by Native American activists against building a waste-to-fertilizer plant on native lands in Vian, Oklahoma.[6]

The problems and prospects of antitoxics campaigns in blighted urban areas is graphically evident in the resistance to a series of waste management plants in Chester, Pennsylvania, a postindustrial city just west of Philadelphia. Chester is an impoverished, predominantly African-American community in an almost all-white suburb, Delaware County. Its median family income is 45 percent lower than the rest of Delaware County; its poverty rate is 25 percent, more than three times the rate in the rest of Delaware County; and its unemployment rate is 30 percent. Chester has the highest infant mortality rate and the highest percentage of low-weight births in the state.[7] Chester would appear to be the last place to build a constellation of hazardous facilities. Nevertheless, three waste and treatment plants recently have been built on a square-mile site surrounded by homes and parks in a low-income, African-American neighborhood in Chester. The facilities include the Westinghouse trash-to-steam incinerator, the Delcora sewage-treatment plant, and the Thermal Pure Systems medical-waste autoclave. A fourth waste processing plant devoted to treating PCB-contaminated soil has recently received a construction permit. The clustering of waste industries only a few yards from a large residential area has made worse the high rate of asthma and other respiratory and health problems in Chester; it has brought into the neighborhood an infestation of rodents, the omnipresence of five hundred trucks a day at all hours, soot and dust covering even the insides of people's homes, and waves of noxious odors that have made life unbearable.[8] In a landmark health study of the environmental degradation of Chester, the EPA found that lead poisoning is a significant health problem for the majority of Chester children; that toxic air emissions have raised the specter of cancer to two-and-a-half times greater than the average risk for area residents; and the fish in Chester waters are hopelessly contaminated with PCBs from current and previous industrial abuses.[9]

The EPA study has made public what many Chester residents have long known: the unequal dumping of municipal wastes in Chester has permanently undermined the health and well-being of its population. Chester is a stunning example of environmental racism: 100 percent of all municipal solid waste in Delaware County is burned at the Westinghouse incinerator; 90 percent of all sewage is treated at the Delcora plant; and close to a hundred tons of hospital waste per day from a half-dozen nearby states is sterilized at the Thermal Pure plant.[10] As Jerome Balter, a Philadelphia environmental lawyer puts it, "When Delaware County passes an act that says all of the waste has to come to the city of Chester, that *is* environmental racism."[11] Or as Peter Kostmayer, former congressman and head of the EPA's midatlantic region says, high levels of pollution in Chester would "not have happened if this were Bryn Mawr, Haverford or Swarthmore [nearby well-to-do white suburbs]. I think we have to face the fact that the reason this happened is because this city is largely—though not all—African American, and a large number of its residents are people of low income."[12] *Chester has become a "local sacrifice zone," where the disproportionate pollution from its waste-industrial complex is tolerated because of the promise of economic revitalization.*[13] But the promise of dozens of jobs and major funds for the immediate areas around the existing toxics industries have never materialized. Indeed, of the $20 million the Westinghouse incinerator pays to local governments in taxes, only $2 million goes to Chester while $18 million goes to Delaware County.[14]

Chester is Delaware County's sacrifice zone. The surrounding middle-class, white neighborhoods would never allow the systematic overexposure of their citizens to such a toxics complex. The health and economic impact of siting even one of the facilities now housed in Chester would likely be regarded as too high of a risk. But to build a cluster of such complexes in nearby Chester is another matter. Nevertheless, many in Chester have tried to fight back against this exercise in environmental apartheid. The Chester Residents Concerned for Quality Living, led by community activist (or as she prefers, "reactivist") Zulene Mayfield, has used nonviolent resistance tactics—mass protests, monitoring of emissions levels, protracted court actions, and so forth—to block the expansion of the complex. In opposition to granting a permit for operation for the fourth waste facility to be built in the area, the soil remediation plant, former Chester democratic mayor Barbara Bohannan-Sheppard concluded her remarks at a public hearing with the following:

> Chester should not and will not serve as a dumping ground. A dumping ground for what no other borough, no other township, or no other city will accept. Yes, Chester needs the taxes, Chester needs the jobs. But, Chester also needs to improve its image and not be a killing field.[15]

Hope is not lost in Chester. There is a growing awareness of the injustice being done to low-income, often minority communities that have suffered from the unequal distribution of environmental hazards in their neighborhoods. Bill Clinton recently signed an executive order mandating all federal agencies to ensure the equitable location of polluting industries across race and economic lines.[16] But the signs are not good that the Chester Residents organization can successfully combat the expansion of the waste industry in their area. Ms. Bohannan-Sheppard recently lost her reelection bid and was replaced by a proindustry mayor and city council. No major environmental organization has taken up the Chester cry against environmental racism as its own. And time is running out as the investors in the fourth envisioned waste plant are preparing to overcome the last legal hurdles to bringing the soil remediation firm on line.

What role if any can Green spirituality play in the struggle against environmental racism in areas like Chester, Pennsylvania? In response, it should first be noted that few people see it as in their interests to express solidarity with disadvantaged communities that have suffered the brunt of unequal distribution of environmental risks. Many people have become inured to the gradual environmental degradation of their home and work environments and most likely consider the development of occasional toxic "sacrifice zones" and "killing fields" to be a tragic but necessary result of modern technological life and its attendant creature comforts. If everyone has the right to pursue his or her own material self-interests, and if some persons are better able to do this on the basis of their natural advantages because of family or national origin, socioeconomic class, and so forth, then it follows that some disadvantaged groups will be marginalized in the human struggle for increased wealth, security, and power. Green spirituality challenges this liberal assumption by affirming instead that all persons are fundamentally equal and that everyone has the right to family stability and meaningful work in a healthy environment regardless of one's racial, cultural, economic, or sexual identity. *Moreover, Green spirituality affirms the common interdependence of all persons with each other—indeed, of all species with each other—as we all struggle to protect the integrity of the life-web that holds together our planet home.* In religious terms, Green religion testifies to the bond of unity that unites all God's children together on a sacred earth. As the participants of the First National People of Color Environmental Leadership Summit put it: "Environmental justice affirms the sacredness of Mother Earth, ecological unity and the interdependence of all species, and the right to be free from ecological destruction."[17] Earth-centered religion values the interconnections between all members of the biosphere in contradistinction to the liberal ideal of maximizing self-interest.

I envision Green spirituality as a distillation of the earth-centered sensibilities within different world religions. It is not a reductionist syncretism of all global spiritualities into one totalizing perspective but rather a selective and self-conscious interpretation of many different religious traditions for the sake of renewing the earth and its inhabitants. The earth-centered mythologies of different world religions make up the content of sustainable spirituality. Depending upon one's religious and cultural background and interests, possible religious ideas, among many others, that could be candidates for inclusion in such a spiritual vision are the following: the Jewish narrative of a common creation story where all species possess inherent worth as the handiwork of the Creator;[18] the Christian idea of the Holy Spirit, the animating power of life in the universe who unifies and sustains all things;[19] the Chinese doctrine of *Ch'i*— the vital force within nature that dynamically integrates all forms of life into common flow patterns;[20] and the Amerindian and neopagan imagery of the earth as our Great Mother which entails the values of care and respect for the "body" of our common parent.[21] Alternately theistic and nontheistic, scriptural and preliterate, eastern and western, these earth-friendly religious traditions offer a body of rich stories and images for enabling the quest for environmental justice.[22]

As a Green hermeneutic of these traditions (and many others could be mentioned as well), sustainable spirituality is an exercise in rhetorical reason rather than a scientific enterprise in the narrow sense of that term. Its goal is to motivate all persons to live responsibly on the earth; its aim is not to prove through observation and experimentation that the doctrines and beliefs of green religious traditions are incorrigibly certain. The point of sustainable spirituality is not to demonstrate empirically that the world really *is* just as Green spirituality figures it to be (though there is compelling evidence to support the claim that the earth is an interconnected living organism, a claim consistent with the spiritual vision adumbrated here). Rather, the point is to imagine the world as a communitarian family of beings that mutually depend upon one another in order to liberate sisterly feelings for the many life-forms that populate the earth. Neither disinterested nor value free in orientation, Green spirituality does not claim to provide scientific or metaphysical descriptions of the physical world; instead, it offers spiritually nuanced refigurations of the world that can set free a primal sense of identification with all forms of life—to set free, as Jonathan Edwards wonderfully puts it, the union of heart with Being as such.[23]

In the struggle against environmental injustice, Green spirituality can serve an important role: the inculcation of a comprehensive world view concerning the underlying unity of all things that can sustain communities of resistance over the long haul. While this model cannot directly fund the material needs of antitoxics campaigns, it can fire the imagination and empower the will as members of embattled communities seek to end the inequitable dumping of hazards and toxins in their neighborhoods. The study and use of fact sheets and health reports alone is not enough to enable the struggle over the long term and in the face of overwhelming odds. By motivating all of the participants to better understand their interdependence on one another—to envision the common bond between rich and poor, city folk and suburbanites, anglos and people of color, humankind and otherkind—Green religion provides the attitudinal resources necessary for enduring commitments to combatting environmental racism and injustice.

Deep Ecology and Wilderness Activism

Radical conservationism today is a practical application of the philosophy of Deep Ecology.[24] The goal of neopreservationism is to renew and reconnect endangered bioregions in order to promote ecological richness and diversity. The core insight of Deep Ecology—namely, that all living things are equal in value and possess the inherent right to grow and flourish—provides the underlying warrant for this goal. First formulated by Arne Naess in a 1973 article by that name, Deep Ecology articulates a spiritual vision of nature as a communal exercise in biotic interdependence, where each life-form is a bearer of equal and intrinsic worth.[25] The ethical corollary to this model centers on equal regard for all species populations. Insofar as all life-forms are codependent members of the biosphere, the hierarchical distinctions that prioritize the interests of humankind over otherkind are consistently effaced.

Since Naess's landmark article, current studies in biocentric moral philosophy stress an attitude of equal regard as the *summum bonum* of environmental ethics. Since all organisms, from single-celled bacteria to highly developed mammals, are coequal centers of biological activity, the maintenance of healthy environments in which the realization of a biocommunity's life cycle can be sustained is the primary concern of a nature-based ethic. The moral rule that results from this premise is variously formulated as the "duty of noninterference," the "principle of minimum impact," or the "principle of nonmeddling."[26] This rule, then, entails a hands-off, live-and-let live behavioral norm that would encourage the practice of thoughtful noninterference in various biotic populations. In conflict situations where humans and other life-forms have competing claims to resources and habitats, the ethical goal would be to develop policies that register *no or as little human impact as possible* on the natural world. Practically, this would entail that in situations where nonessential human interests are furthered by the destruction of plants and animals (for example, in the case of the bulldozing of a coastal wetland in order to make room for a housing development), the decision should be to make little or no provision for such environmental impact. On the other hand, however, in situations where the essential integrity and well-being of a species population is at stake, human or nonhuman, more latitude could be given to measures that will benefit the needy population in spite of the negative effects on the populations not benefiting from the measures in question (for example, in cases where the study and use of some organic specimens are necessary for eradicating certain human diseases). Nevertheless, the same rule applies in both situations, namely, the path of minimum impact on other species.[27]

A minimal impact orientation rooted in Deep Ecology philosophy is the mainspring of neoconservationism. The work of Dave Foreman and others with Earth First! in the 1980s and the Wildlands Project in the 1990s represents the leading edge of this movement. Earth First! emerged out of the disillusionment with the protracted environmental policy debates of the 1970s. Wilderness Society staffer Dave Foreman and some of his colleagues broke with a number of the Group of Ten major environmental organizations and founded the direct-action wilderness defense movement Earth First! in the early 1980s.[28] Foremen and other Earth First!ers became well known for highly public, colorful acts of "monkeywrenching" or "ecotage" in their efforts to undermine the industrial exploitation and destruction of unprotected wild habitats. Foreman and associates appropriated the sometimes gnomic ruminations of Deep Ecology and turned this philosophy into an ideological foundation for controversial, often illegal forays into saving wild places. Taking their cues from the Deep Ecology activism embodied in the novel *The Monkey Wrench Gang* by Edward Abbey, Earth First! members style themselves as the final line of defense against a rapacious industrial machine hell-bent on destroying the last undeveloped areas in North America, with special emphasis on the vast frontiers of the American West. Earth First!'s vision of restoring a Green Wild West in the aftermath of a mass ecocide of biblical proportions—a sort of cowboy apocalypticism—is given voice in the figure of George Hayduke in Abbey's novel:

> When the cities are gone, he thought, and all the ruckus has died away, when sunflowers push up through the concrete and asphalt of the forgotten interstate freeways . . . when the glass-aluminum skyscraper tombs of Phoenix Arizona barely show above the sand dunes, why then by God maybe free men and wild women on horses . . . can roam the sagebrush canyonlands in freedom . . . and dance all night to the music of fiddles! banjos! steel guitars! by the light of a reborn moon!—by God, yes![29]

Hayduke is an antindustrial saboteur who prophesies certain eschatological doom; his end-time fantasy provides the master metaphors for Earth First!'s extremist rhetoric. Through vandalizing logging vehicles, spiking trees targeted for logging, and generally playing havoc with wilderness development operations, Earth First! has emerged as the most charismatic, if not always most successful, activist organization for wilderness preservation in the wake of the Reaganesque market-oriented model of "wise use" environmentalism.

In the early 1990s Earth First! split into two factions. Dave Foreman organized the minority faction into a splinter organization that publishes the journal *Wild Earth* and advocates for the Wilderness Project, an ambitious network of activists and scientists working to establish a connected system of wilderness parks and preserves. This rump faction represents a significant change in philosophy and tactics from the larger Earth First! movement: wilderness *recovery* is now the watchword of the minority group instead of wilderness *defense*, and the angry monkeywrenching tactics of civil disobedience have been replaced by the moderate discourse of earth science and public policy studies. Instead of Hayduke-like apocalypticism, the Wilderness Project is seeking long-term solutions to declining biodiversity in wilderness areas; instead of the countercultural youthful hostility to mainstream bureaucratic environmentalism, the Wilderness Project is eager to make common cause with any prowilderness groups, from biocentric grassroots movements to the more conservative Group of Ten environmental organizations, including entities such as the Sierra Club and the World Wildlife Fund.

The central focus of the Wildlands Project is the enactment of a system of nature preserves for the sake of furthering biological growth and diversity. This system would consist of interconnected core reserves that would allow genetically diverse populations to crossfertilize, evolve, and flourish.

> The mission of The Wildlands Project is to help protect and restore the ecological richness and native biodiversity of North America through the establishment of a connected system of reserves. . . . The environment of North America is at risk and an audacious plan is

needed for its survival and recovery. Healing the land means reconnecting its parts so that vital flows can be renewed. . . . Our vision is continental: from Panama and the Caribbean to Alaska and Greenland, from the high peaks to the continental shelves, we seek to . . . restore evolutionary processes and biodiversity.[30]

While this mission statement may appear to hark back to turn-of-the-century conservationism, the goals of contemporary preservationism are different from the ideals of the national parks and related movements that have sought to set aside scenic places for the sake of human recreation and edification. *Today the concern is with the preservation of whole ecosystems in order to sustain the health of the planet in general rather than with the establishment of picturesque sites and outdoor zoos, so to speak, whose purpose is to refresh and uplift the human spirit.* What distinguishes neopreservationism from its conservationist precursors is its plea for the establishment of large nature preserves as nurseries for comprehensive biodiversity without which, its proponents argue, diverse life on the planet as we know it will be seriously eroded—if not extinguished altogether.

What is the relevance of sustainable spirituality to contemporary conservation efforts? Initially it seems that religion and conservationism have little in common. Indeed, one of the sources of disagreement that led to the split among Earth First!ers in the first place was the contention by Dave Foreman and his allies that the movement had been coopted by spiritually oriented, social justice types who were blunting the hard edge of the movement's originally uncompromising anti-industrial message.[31] Foreman's protestations to the contrary notwithstanding, both militant and bureaucratic forms of neoconservationism are deeply spiritual movements at their core. Let me explain. I have argued that grassroots nature activism represents the tactical edge of Deep Ecology philosophy. As such, the expansive vision of a transcontinental wilderness recovery strategy within neopreservationism is animated by a deeply felt spiritual awareness that all life, human and nonhuman, has intrinsic value and should not be subordinated to the growth needs of late capitalist societies. I label this intuitional perspective "spiritual" in this context because its exponents are committed to preserving the integrity of life as such as an ultimate value. Whatever may or may not be said about its scientific merits, Deep Ecology is a spiritual vision of the highest order concerning the organic wholeness and biotic equality of all life-forms on the planet; and insofar as contemporary conservationism is politically applied Deep Ecology, it is a bearer of Green spirituality to a culture that hungers for authentic religion in an age of corporate televangelism and reactionary fundamentalism.

In the same way, then, that Green religion can empower long-term antitoxics commitments in the face of powerful countervailing market forces, it can also engender a comprehensive, emotionally resonant worldview concerning the sacred, inviolable character of every biotic community. Thus the reason for recovering wilderness places is not for the sake of human flourishing—though human flourishing would be a direct consequence of such recovery work—but because all members of the life-web deserve to achieve their full biological potential as much as possible. In short, green spirituality helps to answer the "Why" question for conservationism, namely, Why care about wild places in the first place? The answer is because such places make up the fragile life-support systems that render the earth a teeming biosphere of interconnected living things. Wild places are the nurseries that make biodiversity possible. This understanding of the distinctive role of wilderness in evolutionary processes is both a scientific and spiritual insight: scientific, because it recognizes that wilderness is essential to maintaining diversity at all levels, and spiritual, because this recognition accords to wilderness the supreme value of being essential to the maintenance of life itself.

Mediating the Debate, Green Religion, and Market Values

To this point I have considered the antitoxics movement and conservationism as often opposing factions, albeit factions that share a comprehensive spiritual vision of restored nature. Yet it is the oppositional character of each movement in relation to the perceived concerns of the other group that is so striking and, at the same time, in dire need of mediation. On the one hand, antitoxics leaders like Lois Gibbs sometimes appear to see little relationship between combatting pollutants in the home and workplace and the mainstream environmental movement's interest in protecting plant and animal habitats: "Calling our movement an environmental movement would inhibit our organizing and undercut our claim that we are about protecting people, not birds and bees."[32] On the other hand, Dave Foreman sometimes strikes a misanthropic note in order to underscore the dissimilarities between wilderness protection and fighting against the social causes that force some human communities into toxic environments: "We aren't an environmental group. Environmental groups worry about health hazards to human beings, they worry about clean air and water for the benefit of people and ask us why we're so wrapped up in something as irrelevant and tangential and elitist as wilderness.... [But] wilderness is the essence of everything. It's the real world."[33] To put the differences between the two movements in the most extreme terms, the antitoxics are sometimes derided as anthropocentric and not truly biocentric while the neopreservationists are criticized as antihuman and ecofascist.

The claim has been made that "[a] balance *can* be struck between *preserving* the wild and *reorganizing* our transactions in cities, suburbs, and countryside."[34] But how can such a mediation between antitoxics and neopreservationists be possible if the one appears to prioritize the needs and interests of discrete human populations while the other appears to prioritize the needs and interests of the organic whole? My thesis is that Green spirituality has the resources for forging rapprochement between these two movements by articulating the operative worldview that is logically entailed by both forms of environmental populism. I am not arguing that this worldview is self-consciously understood as such by adherents of both movements, but that it is the mind-set that is implied by the commitment to the integrity and sanctity of life shared by both groups. This shared worldview is holistic in its vision of the biosphere, prophetic in its despair over the earth's declining biological carrying capacity, and interventionist in its struggle against global market forces that have degraded human and nonhuman environments alike. *"Wholeness" is the epithet for a life-centered spirituality adequate to the ecocrisis of our times.* The English word "whole" is a derivative of a constellation of old Teutonic and old English terms that signified well-being, health, and healing. Etymologically, the word "whole" stems from the Germanic *Heil*, which is associated with vitality, integrity, strength, soundness, and completeness. Likewise, the English word "holy"—derived from *heilig* (a cognate of *Heil*)—historically also had the meanings of well-being and integrity in addition to its denotation as consecrated and set-apart. Wholeness, the whole, and the holy, then, are terms that have historically cross-pollinated one another. To uphold, therefore, the integrity of the *whole* is to experience the *holy* or sacred through living a life of personal and communal *healing* and *well-being*.[35]

My suggestion is that sustainable religion enables a mediation between antitoxics and conservationists by explicating the common spiritual-holistic philosophy that is implied by the beliefs and actions characteristic of both movements. It is important, however, to nuance my claim about the joint status of this implied mind-set so that adherents in both groups can recognize their own orientation in what I am labeling a common worldview. At its core this worldview stresses unity and interdependence, but it also carries different valences of meaning for each group: for antitoxics the commitment to ecological unity can still emphasize attention to human needs in systemically unjust situations; for conservationists, the inherent equality between humans and nonhumans means that the question of human welfare is generally subordinated to, or at best addressed indirectly by, the task of preserving the integrity of whole bioregions. Both groups stress biotic interdependence, but for antitoxics this stress need not include the espousal of biotic equality in the Deep Ecology sense. My point is that rapprochement between the two movements need not entail agreement on all issues, including the question of biotic equality. As long as members of both organizations can recognize their tacitly held (if not always explicitly articulated) commitment to the unity and integrity of all living things, then the ground has been laid for mediating the oppositional stances the two groups sometimes take in relation to the interests of the other group. If, therefore, this common ground can be secured—that is, a unitary vision of all organisms and entities as interdependent, if not always coequal members of an organic whole—then the response to the question whether environmental justice or wilderness recovery should be one's primary focus is a response that is tactical, strategic, and contextual—not deep-down philosophical. The problem, then, is not one of disagreement over the fundamental orientation needed to combat further ecocide but over the political focus and practical measures necessary for enacting this core vision of sustainable ecocommunities, human and nonhuman alike.

For those who suffer from the daily onslaught of toxins in the homes and places where people work and play, it is understandable why such communities seek first and foremost to liberate themselves from the killing fields of America's waste industries. To force such communities into the false choice of unsafe livelihoods or chronic unemployment is an unconscionable Catch-22 that results from aggressive industry efforts to dump toxins into neighborhoods that can least afford to house such hazards. Under these conditions it makes tactical sense for antitoxics groups first to labor against the unequal distribution of waste products in degraded human ecosystems close to home before turning to the equally important task of combatting the despoliation of wildland ecosystems in more remote locales. I am suggesting that this decision should be understood in strategic terms. It is not that antitoxics activists do not appreciate the basic connection between human health and the welfare of the biosphere—indeed, as I have argued here, the implied commitment to holism on the part of antitoxics necessitates just such an understanding, at least tacitly—but rather that the direct threat of killer toxins in their immediate neighborhoods should propel antitoxics to organize against these threats first and foremost.

By the same token, the imminent decline and eventual extinction of numerous species and habitats across North America—from large predators and shorebird populations to native forests and tallgrass prairies—understandably shoulders conservationists with a heavy burden for the long-term health and biodiversity of the continent. This burden should not and need not be regarded in opposition to the similar but distinct environmental burden of antitoxics; rather it is one among many counterpoints to the expansive medley of approaches one can take to restoring the harmony among all living things. For embattled citizens of toxic neighborhoods who are fighting the daily struggle for their very survival, it makes sense for such persons to take up the antitoxics cause as their own;

by the same token, for individuals and communities whose survival needs are not as immediately critical, it is equally understandable why such persons privilege the reclamation and rehabilitation of nonhuman nature and only consider the needs of human populations in relation to sustaining the health of the wider biosphere. In spite of these differences, I believe the bedrock commitment to the integrity and inviolability of life as such among antitoxic and biodiversity activists is the common spiritual vision that sustains both movements. While this common vision leads to different strategic interventions on behalf of healing the Earth, the reverence for life at the foundation of each group needs to be recalled amid the welter of the claims and counterclaims advanced by defenders and detractors of both movements.

The debate between antitoxics and conservationists may appear initially irresolvable. But when one considers the lived context of the environmental crisis as understood by the different disputants in the debate—for example, the daily stream of pollutants into minority urban neighborhoods, on the one hand, or the ongoing attenuation of biodiversity in wild habitats, on the other—then the debate becomes one over which tactics and strategies are effective in which particular circumstances and not over which moral claimant is right or wrong. *One's social location—urban/rural, rich/poor, black/white, and so forth—largely determines the appropriate response to the ecocrisis.* "Nature" is not the special preserve of wilderness activists alone; nature is the lived environment common to humankind and otherkind alike wherever both kinds live and work and love and eat. Nature is the lead-filled air breathed in by schoolchildren in toxic urban killing fields; nature is the pristine landscapes and watersheds that still survive in rural parks and wildlands. Whether antitoxics or neopreservationist in orientation, how one responds to the challenges presented by nature in its myriad forms is shaped by the particular places one inhabits. Thus the environmental orientations of both groups—groups whose core philosophy is similar but whose organizational approaches are often different—are equally legitimate and equally dependent upon the social, economic, and ethnic locatedness of the different participants in the common struggle for ecological wholeness and balance.

Finally, it is important to note that sustainable spirituality is not only valuable as a means of forging a common link among radical Green activists who are alternately justice oriented and biodiversity centered, respectively. In turn, it shines a bright spotlight on the exploitative growth philosophy of market individualism that has led to the environmental squalor that characterizes our own time. Even as sustainable spirituality hopes to mediate the dispute between both forms of Green populism by specifying the animating worldview behind each movement, it also seeks to arbitrate this understandable but unnecessary dispute by identifying expansionist market forces as the real culprit in creating both human sacrifice zones and depleted wilderness areas. When everything is a potential commodity for buying and selling—including whole neighborhoods like Chester, Pennsylvania, or America's current and prospective wilderness reserves, as envisioned by the Wilderness Project—human poverty and biological poverty are the inevitable result. *When every organism or entity becomes commodified or thingified, then life and world lose their sacred character and become objects to be bought and sold.* When all life-forms, human and nonhuman, only have meaning as "products" or "resources" to enable the growth of the market state, the prospects for environmental sanity are meager indeed.

Economic competition breeds more competition, market growth breeds more growth, and the needs and values of fragile human and wilderness ecosystems have little hope for survival against these withering assaults. Growth-obsessed market liberalism driven by the "mindless 'laws' of supply and demand, grow or die, eat or be eaten" tears apart the social and ecological fabric that supports life in urban slums and rural bioregions alike.[36] Sustainable spirituality reminds both the advocates of environmental justice and wilderness protection that they share a core vision of healthy and diverse communities living together on a Green planet. This visionary role is the priestly function of sustainable spirituality: to inculcate in all who struggle for a Green future a common worldview and ethic that can sustain the combatants over the long term. But sustainable spirituality performs a prophetic role as well. It decries the rapacious power of the market to undermine our collective ability to grasp the inherent value and worth of Life itself wherever it is found in the biotic communities that make up our planet home. This unitive vision of a Green sacred Earth has the potential to renew and sustain antitoxics campaigners and neopreservationist activists alike in the long struggle against the regnancy of market liberalism—a regnancy that must be overcome if the prospects for life on the planet in the twenty-first century are to improve.

Notes

1. Bob Edwards, "With Liberty and Environmental Justice for All: The Emergence and Challenge of Grassroots Environmentalism in the United States," in Bron Raymond Taylor, ed., *Ecological Resistance Movements: The Global Emergence of Radical and Popular Environmentalism*, ed. (Albany: SUNY Press, 1995), p. 37.
2. "The Wildlands Project Mission Statement," *Wild Earth* 5 (Winter 1995/96): inside front cover, n.a.
3. See this discussion in Charlene Spretnak, *The Spiritual Dimensions of Green Politics* (Santa Fe: Bear, 1986), pp. 25–53; cf. Catherine L. Albanese, *Nature Religion in America: From the Algonkian Indians to the New Age* (Chicago: University of Chicago Press, 1990), pp. 171–78.
4. Edwards, "With Liberty and Environmental Justice for All," p. 37.
5. See Robert Gottlieb, *Forcing the Spring: The Transformation of the American Environmental Movement* (Washington, DC: Island Press, 1993), pp. 184–91.
6. See Carolyn Merchant, *Radical Ecology: The Search for a Livable World* (New York: Routledge, 1992), pp. 162–67.
7. I have drawn this information from "Chester Decides It's Tired of Being a Wasteland," *Philadelphia Inquirer*, July 26, 1994; and Chester Residents Concerned for Quality Living, "Environmental Justice Fact Sheet" and "Pollution and Industry in Chester's 'West End,'" pamphlets. I am grateful to Swarthmore College students Laird Hedlund and Ryan Peterson for making available to me their expertise and research concerning the Chester waste facilities.
8. Maryanne Voller, "Everyone Has Got to Breathe," *Audubon*, March–April 1995.
9. Editorial, "Chester a Proving Ground," *Delaware County Daily Times*, December 8, 1994; and "EPA Cites Lead in City Kids, Bad Fish," *Delaware County Daily Times*, December 2, 1994.
10. Maryanne Voller, "Everyone Has Got to Breathe," *Audubon*, March–April 1995; and Chester Residents Concerned for Quality Living, "Environmental Justice Fact Sheet," pamphlet.
11. "_____," *Delaware County Times*, 1 August 1995.
12. Howard Goodman, "Politically Incorrect," *Philadelphia Inquirer Magazine*, February 11, 1996.
13. The phrase belongs to Merchant, *Radical Ecology*, p. 163.
14. Chester Residents Concerned for Quality Living, "Pollution and Industry in Chester's 'West End,'" pamphlet.
15. Barbara Bohannan-Sheppard, "Remarks," Department of Environmental Resources Public Hearing, 17 February 1994, transcript.
16. Bill Clinton, Executive Order Number 12898, February 1995; cf. Gretchen Leslie and Colleen Casper, "Environmental Equity: An Issue for the 90s?" *Environmental Insight*, 1995.
17. The First National People of Color Environmental Leadership Summit, "Principles of Environmental Justice," in Roger S. Gottlieb, ed., *This Sacred Earth: Religion, Nature, Environment* (New York: Routledge, 1996), p. 634.
18. See, for example, Arthur Green, "God, World, Person: A Jewish Theology of Creation, Part I," *The Melton Journal* 24 (Spring 1991): 4–7.
19. On this point see my *Fragments of the Spirit: Nature, Violence, and the Renewal of Creation* (New York: Continuum, 1996), pp. 133–70.

20. See Tu Wei-ming, "The Continuity of Being: Chinese Visions of Nature," in *Nature in Asian Traditions of Thought: Essays in Environmental Philosophy* (Albany: SUNY Press, 1989), pp. 67–78.
21. On Native American traditions see John A. Grim, "Native North American Worldviews and Ecology," in *Worldviews and Ecology*, eds. Mary Evelyn Tucker and John Grim, pp. 41–54; on neopagan resources see Margot Adler, *Drawing Down the Moon: Witches, Druids, Goddess-Worshippers, and Other Pagans in American Today*, rev. ed. (Boston: Beacon Press, 1986), pp. 372–421.
22. For a collection of source material and analysis on Green religion, see Gottlieb, *This Sacred Earth*; for general analysis also cf. David Kinsley, *Ecology and Religion: Ecological Spirituality in Cross-Cultural Perspective* (Englewood Cliffs, NJ: Prentice Hall, 1995); and Tucker and Grim, *Worldviews and Ecology*.
23. On Edward's spirituality see Jonathan Edwards, *The Nature of True Virtue* (Ann Arbor: University of Michigan Press, 1960), pp. 1–26; cf. William A. Clebsch, *American Religious Thought: A History* (Chicago: University of Chicago Press, 1973), pp. 11–56.
24. I say radical conservationism in order to distinguish this movement from the reformist orientation of mainstream conservationism. The radicals seek to preserve maximum biological diversity in wilderness areas as their goal, while the reformists emphasize responsible development and resource management as their goals. Groups such as Earth First!, Greenpeace, and the European Greens belong in the radical grouping, while entities within the Group of Ten environmental organizations (for example, the National Audubon Society and the National Wildlife Federation) can be grouped under the reformist label. For the differences here see Merchant, *Radical Ecology*, pp. 157–82.
25. See Arne Naess, "The Shallow and the Deep, Long-Range Ecology Movement," *Inquiry* 16 (1973): 95–100.
26. The articulation of this rule is quoted from Paul W. Taylor, *Respect for Nature: A Theory of Environmental Ethics* (Princeton: Princeton University Press, 1986), p. 174; Bill Devall and George Sessions, *Deep Ecology: Living As If Nature Mattered* (Salt Lake City: Peregrine Smith Books, 1985), p. 68; and Tom Regan, "The Nature and Possibility of an Environmental Ethic," *Environmental Ethics* 3 (1981): 31–32.
27. In the vein of the noninterference maxim, Taylor provides a helpful list of five principles—self-defense, proportionality, minimum wrong, distributive justice, and restitutive justice—for resolving conflicting "claims" between human and nonhuman populations. He also provides a number of case-studies illustrating the relevance of these principles to different hypothetical conflict scenarios. See Taylor, *Respect for Nature*, pp. 256–313.
28. On the history of Earth First! see Dave Foreman, *Confessions of an Eco-Warrior* (New York: Harmony Books, 1990), Christopher Manes, *Green Rage: Radical Environmentalism and the Unmaking of Civilization* (Boston: Little, Brown, and Company, 1990); and Bron Raymond Taylor, "Earth First! And Global Narratives of Popular Ecological Resistance," in *Ecological Resistance Movements: The Global Emergence of Radical and Popular Environmentalism* (Albany: SUNY Press, 1995), pp. 11–34.
29. Edward Abbey, *The Monkey Wrench Gang* (New York: Avon Books, 1975), pp. 100–101.
30. "The Wildlands Project Mission Statement," *Wild Earth* 5 (Winter 1995/96): inside front cover, n.a.

31. On this point see Bron Taylor, "The Religion and Politics of Earth First!" *The Ecologist* 21 (November/December 1991): 258–66.
32. Gottlieb, *Forcing the Spring*, p. 318.
33. Ibid., p. 197.
34. Roger S. Gottlieb, "Spiritual Deep Ecology and the Left: An Attempt at Reconciliation," in Gottlieb, *This Sacred Earth*, 529; cf. a similar attempt to resolve the conflicts between Deep Ecology–inspired wilderness advocates and environmental justice proponents in Michael E. Zimmerman, "The Threat of Ecofascism," in this volume.
35. Definitions and etymologies for these terms are drawn from *The Oxford English Dictionary* (New York: Oxford University Press, 1971).
36. The quotation is from Murray Bookchin, "What is Social Ecology?" in Michael E. Zimmerman, ed., *Environmental Philosophy: From Animal Rights to Radical Ecology* (Englewood Cliffs, NJ: Prentice-Hall, 1993), p. 368. Much of my thinking about the relationship between environmental degradation and market liberalism has been inspired by the writings of social ecologists Bookchin, Janet Biehl, and John Clark. For a thoughtful counterpoint to this approach cf. the argument for a modified "social liberalism" in Avner de-Shalit, "Is Liberalism Environment-Friendly?" in this volume.

16

International Justice and Wilderness Preservation

Mark A. Michael

A central tenet of ecocentric environmentalism is that things other than humans can have intrinsic value and that as a consequence the preservation of wilderness areas throughout the world is morally obligatory.[1] But wilderness preservation places a severe strain on economically struggling societies and in turn on the members of those societies least able to bear it, namely the rural poor, and this seems unfair.[2] One obvious solution is to transfer resources from industrial societies to those in the Third World that implement wilderness preservation, thereby obviating the negative economic impact of that policy. Many ecocentrists have, however, been slow to embrace this solution, perhaps fearing that it concedes too much to anthropocentric concerns.[3] In this paper I want to develop two claims. First, if the implementation of wilderness preservation does raise issues concerning fairness, then ecocentrists should not be troubled by a solution that involves a redistribution of assets, since there is nothing in that solution that is incompatible with or inimical to ecocentrism. The second and stronger claim is that in fact wilderness preservation does give rise to a fairness problem; so that given ecocentrists' advocacy of wilderness preservation along with their other commitments, redistribution is the only available solution. Thus the redistribution of assets is not just one of

Copyright 1995 by *Social Theory and Practice*, Vol. 21, No. 2 (Summer 1995).

many alternative solutions to the problem: rather, ecocentrists, insofar as they advocate preservationist policies, are logically committed to a solution that involves some form of redistribution. I will refer to this as the entailment thesis. The moral of the entailment thesis is that ecocentrists cannot remain neutral on issues of international justice and fairness.

I

I want to explore the fairness problem by offering a parable, which hopefully will serve to motivate an understanding of how and why the implementation of preservation policy purportedly leads to such a problem. The setting of the parable is the desert island that by now is overly familiar to moral philosophers. It is an admittedly fanciful scenario, but as the parable is intended as a thought experiment I do not see any reason to find that problematic. The point I want to press here is that what we think about one aspect of the parable will show what we should think about a corresponding aspect of the current environmental situation if it turns out, as I believe it will, that there are important and relevant similarities between the parable and that situation.

A small group of people washes ashore on an island. In their explorations they discover a huge stockpile of goods, apparently left by an earlier and technologically advanced society; other than that the island is largely bereft of resources. Items such as canned goods that will directly satisfy the basic needs of the castaways are found in the stockpile, as are other items such as farming implements, seeds, and construction equipment and materials, which will allow them to farm a wide variety of crops, build fairly sophisticated housing, and do other things that are not directly related to survival but that will lead to a perceived improvement in the quality of life. A quick survey shows that the stockpile is practically inexhaustible relative to the actual needs and desires of those who washed ashore. The stockpile is, however, in a fairly remote location, perhaps at the uninhabitable bottom of a steep canyon, and so any appropriative activity will amount to a form of labor. Some of the castaways, call them the Earlys, begin to take significant amounts of both the canned goods and the tools and materials with which they plan to develop parts of the island. Others, call them the Laters, see no reason to rush into appropriating, as they suspect that they will be stuck on the island for the foreseeable future, and there appears to be plenty for everyone. They take only the canned goods that they need for immediate nourishment. But the castaways discover that the removal of goods creates an unanticipated byproduct. Harm begins to strike the castaways as a result of their appropriations, although the harm does not necessarily strike the appropriator but rather occurs randomly throughout the community. Furthermore, although these harmful side effects are relatively insignificant at first, they increase in seriousness and frequency as more is removed from the stockpile.

The castaways can determine that they have just about reached a critical balance; more appropriation will result in serious harmful effects that will outweigh whatever benefits accrue to the community as a whole from the appropriation. The castaways see the need to either stop or, at the very least, seriously curtail their appropriations. Consequently, all their productive activities from this point on must be performed with those tools and other supplies that have already been removed from the stockpile. But the bundles of goods held by the castaways are vastly unequal. The Earlys, who appropriated right away, have much more than the Laters, who, for a variety of reasons deferred their appropriations. Because of this inequality, the Laters claim that the current distribution is unfair and so resume their appropriating. The Earlys try to convince them to stop. At first they appeal to the Laters' self-interest. They point out that appropriating will have more harmful effects than good effects. The Laters respond that although it would be irrational for the Earlys to engage in more appropriation because of the harm they are risking, this is not the case for the Laters. The Laters have to satisfy some very basic needs as well as some preferences, and so running the risk of harm is a reasonable gamble for them because they have so little to begin with as a result of their late start in appropriating. The Earlys then suggest that as the Laters' actions will put others at risk of unnecessary harm, their appropriative activities are immoral. The Laters agree it is wrong to put others at a risk of harm unnecessarily. But they point out that, given their resource deprivation, the harm they are exposing others to is necessary. The Laters propose a solution: they will curtail or cease appropriation if the Earlys will agree to a more equitable distribution of what the Earlys originally appropriated. The Earlys balk at this, invoking the notion of desert and noting that they invested time and energy in retrieving their goods. But the Laters are willing to donate a comparable amount of time and energy into community projects to compensate the Earlys for the labor they invested, on the condition that some redistribution take place.

Now the foregoing suggests that the Islanders have a fairness problem, and that a redistribution of assets is one reasonable and obvious solution to the problem. On the face of it there is no reason for the ecocentrist to find this answer problematic. The leading idea of ecocentrism is that things other than humans can have value, but generally it does not deny that humans have value as well. The point ecocentrists are anxious to make is that since things other than humans can have value and we can have duties to act toward them in specified ways, sometimes these duties can take precedence over our duties to humans.[4] If this is correct, then ecocentrism will have implications for those issues in which there is a conflict between our duties to humans and our duties to nonhumans, or between acting to advance human interests rather than nonhuman interests. But there seems to be no genuine moral dilemma or tension between preservation and redistribution. These two duties do not somehow preclude one another; we are not forced to choose one or the other but not both.[5] So whether one thinks that nature has some, or even more, intrinsic value than humans is irrelevant; as long as one accords some moral value to humans so that there are duties to treat persons justly and fairly, it is difficult to see why an ecocentrist should be troubled by the claim that preservation should be linked to some kind of redistribution.

Ecocentrists may nevertheless continue to feel some unease over this claim, thinking that it somehow is grounded in an overly anthropocentric view of the world. In the next section I want to look at some of the sources of this uneasiness and show that these concerns should not prevent an ecocentrist from endorsing redistribution. That is, ecocentrists need have no qualms about adopting redistributive policies, for there is nothing in redistributive policies or in what must be assumed in order to justify such policies that would be inimical to or inconsistent with ecocentric principles.

II

We can group ecocentrists' concerns about the compatibility of their theory with redistributive policies into three broad categories. Some of these misgivings are based on the fact that redistribution requires us to treat nature and natural objects as property or resources, and this seems inconsistent with the attribution of intrinsic value to such objects. The second turns on worries that the justification of redistribution minimizes or deprecates the significance of the ecocentrist principle that there is a duty to preserve wilderness areas. And the third rests on consequentialist concerns that the chances that a policy of preservation will be adopted are significantly lowered if it is coupled with a redistributive policy. I will examine each of these in turn.

First, given their view that natural objects have rights or intrinsic value, ecocentrists may be uncomfortable with an assumption that underlies redistributive policies, namely that natural objects can be owned and subjected to property relations. It will be pointed out that while concerns about distributive justice and redistribution are relevant when those things that persons are morally entitled to treat as property are somehow distributed unfairly, things that have intrinsic value or rights are not the sorts of things that should be owned. So the recognition that natural objects or animals have rights or intrinsic value appears to be inconsistent with the claim that there should be some redistribution of assets.[6] But we need to get clear on exactly how this rather general point is supposed to function in an argument, since it could be construed in a number of different ways. For example, the claim may be that since the species and unique ecological areas that are being preserved and protected by the creation of wilderness areas have rights or intrinsic value, the members of these species and inhabitants of these ecosystems should not be treated as property, any more than humans who were enslaved should have been. Claims that some distribution of slaves is fair or unfair have no point, since no one should own slaves. If this is the concern, however, there is no incompatibility between it and the claim that redistribution is a reasonable solution to the fairness problem; to think otherwise involves a basic confusion. The redistributive claim does not endorse the obviously ludicrous policy of somehow dividing more equitably those things that are being preserved. The claim is not that the United States has an obligation to set aside land to insure the survival of wolves, or China should do the same for pandas, and that pandas and wolves should be redistributed to Third World societies. What gets redistributed are financial assets, and so it is difficult to see why the claim that the things being protected have rights or intrinsic value should be thought relevant to the issue of redistribution.

It could be that the ecocentrist is basing the claim of incompatibility on a much more radical thesis, namely, that no living thing or ecosystem may be treated as a means to human ends, so that no such objects may be treated as property. It would follow that no society was ever justified in engaging in appropriative activities, that the holdings of industrialized nations are held illegitimately, and that redistribution would transfer ownership of something that should not have been owned in the first place. Notice that this is not the socialist claim that land and resources should be held in common, but rather that none of it should be the subject of property relations, regardless of whether those are communal or private. But if this is what is intended, it is surely unconvincing. Humans must consume plants or animals to live, so if all natural objects or living things have rights and are not to be treated as means to human ends, then human survival requires a constant violation of the dictates of morality. An ecocentrist might reply that since humans must kill plants or animals for food, there is a conflict between a human's right to life and the animal's or plant's right to life; and in such cases when basic rights conflict some right must give way, so it might as well be the animal's or plant's. This is no more problematic morally than when a person kills another in a legitimate instance of self-defense. If killing in self-defense is justified, then the attribution of rights to animals will not prohibit us from killing them for food, while at the same time it would prevent us from making them the subjects of property relations. But this response misconstrues what is usually meant by a right to life. This right is typically understood as a right not to be killed, and not as the right to do or have whatever I may need to survive.[7] For example, if another person legitimately owns medicine he needs to stay alive, but I also need it, and there is not enough for both of us, I surely could not justify killing him by appealing to my right to life. But this is exactly what happens when I kill a plant or animal to eat it. I am violating its right to life, even though it is not actively threatening my life in return. Thus if the attribution of rights to living things and natural objects prevents us from treating these things as property, it will also prevent us from killing them and treating them as food. So either we eat and survive and give up on morality; or we take the proposed morality seriously and stop treating things as means to our ends and violating rights by consuming things, and we die. Given these alternatives, I take it that ecocentrists will drop the claim that each and every natural object has this sort of strong right that serves as a kind of moral trump. So at least some natural objects can be the subjects of property rights, and consequently the current allocation of resources may be unfair and then a redistribution would be one way to restore fairness.

At this point ecocentrists are likely to contend that I have failed to appreciate the underlying spirit and guiding principle of ecocentrism, which is captured only imperfectly by ascriptions of rights or intrinsic value to nature. Talk of redistribution presupposes that nature is to be viewed as a resource with nothing more than instrumental value. And claims that something ought to be redistributed to those with less of whatever it is that is being redistributed must assume that what is being redistributed is good or valuable, that it somehow contributes to or is a necessary condition of leading a good life. Thus distributive claims only make sense in a context in which it is assumed that an ever increasing amount of material goods is necessary to maintain one's quality of life, and this assumption feeds attitudes such as consumerism and conspicuous consumption. But ecocentrists do not operate in this context. The whole point of ecocentrism is to deny that one can identify quality of life with the possession of material goods. Ecocentrism advocates a nonconsumerist approach to life, whereas anyone who thinks that redistribution is valuable must be assuming that consumerism is a good thing.

Now this involves some large issues that cannot be entered into completely here. But the following observations should go some distance toward dispelling this sort of worry, which I think lies at the bottom of much of the ecocentrist's hostility to the entailment thesis. The first and most fundamental point is that even if it is a mistake to view nature as a resource only, a point with which I am in wholehearted agreement, it is impossible to treat it as something completely other than a resource. Human beings are material, biological creatures with needs that can be met only by interacting with and changing nature. Whether or not we "see" nature as a resource, we must treat it as such. This does not and should not preclude us from seeing nature as something with its own value and beauty. But the fact remains that at least parts of nature must be used as resources, and so how those are divided up will be a matter of moral concern. It must also be pointed out that none of what has been said here presupposes that the more material goods one has, the better one's life must be going, or that the function of humans is simply to produce and consume. It does not entail that one's life is going badly if one does not own the latest technological gadget, and someone who argues for redistribution can easily and consistently repudiate the consumerism and consumption that is responsible for much of the destruction of the environment. She only needs to hold that some resources, such as whatever is required for adequate food, housing, and clothing, are necessary for a good life, and that these may be distributed unfairly. In fact, the case might be made that those ecocentrists who deny this are engaging in a kind of spiritualization of nature and humans that in fact is inimical to another ecocentric thesis, namely that humans are just another part of nature, no different from other animals. We all have material needs that must be met if we are to survive.

Second, rights and intrinsic value might be introduced into the discussion to make a somewhat different point. If nature has intrinsic value, then we can have duties towards nature; there may be, for example, a duty not to harm natural objects, or a duty not to interfere with natural processes. Practices such as clear-cutting tree stands or turning wilderness areas into farmland appear to violate these duties, and so are serious wrongs by the ecocentrist's lights. But then calls for a redistribution of assets to Third-World countries for refraining from these practices would be similar to suggesting that if it is in a person's economic self-interest to own slaves, but she agrees not to, she should be compensated for the loss. Issues about redistribution and compensation evaporate when we realize that slaves possess rights. But redistributive policies seem to be a kind of compensation for losses that Third-World societies would incur in the process of doing nothing more than what is their duty, and surely there is no moral principle that would require compensation for such a loss.

The ecocentrist is drawing a mistaken parallel, however, based on a misunderstanding of the nature of what justifies the payment to Third-World countries. The aim is not to compensate those societies for forgoing gains they might have received had they engaged in some wrongful activity. Certainly compensation under such circumstances is unreasonable and not required by morality. But the payment being called for here is redistributive and not compensatory. That is, the obligation to make payments to Third-World countries rests on the claim that in adopting preservationist policies the initial conditions that were mistakenly thought to be in effect have changed, so that what appeared to be a justified distribution in fact was not. That the payment is redistributive and not compensatory in intent can be seen if we consider another thought experiment. Suppose that some technological advance makes extremely valuable a certain resource that is found only in Third-World countries. Further, this resource may be developed without creating any significant environmental problems or damaging anything that has intrinsic value. This resource becomes a potential source of great wealth for these Third-World societies. Under these conditions, I take it that there would no longer be a need to redistribute material assets, even after those countries decide to undertake preservationist policies. This suggests that the redistribution is not a form of compensation but rather a requirement of distributive justice. For if it were compensatory, the availability of new resources in the Third World would be irrelevant to whether those societies were entitled to compensation. But with the hypothetical introduction of new and extraordinarily valuable resources, the Third World is no longer at a disadvantage due to an unfair distribution, and under these conditions any claim that some redistribution is necessary would be rendered void. As I will argue in part 3, the distribution arrays dictated by distributive justice will change as various material conditions change, and when environmental concerns force the fencing off of a large part of certain countries from the ravages of development those conditions are changed. The call for a redistribution of assets rests on this claim of a flawed distribution.

The moral gravity of our duty to preserve wilderness areas might figure into a somewhat different argument. Anyone who believes that such a failure amounts to the violation of a stringent duty will want to hold that this duty is unconditional, in the sense that the fulfillment of this duty is not and should not be conditional on other people acting in certain ways. Admittedly, acting on this duty will have significant costs for Third-World societies, and those costs could be alleviated by some form of redistribution. But to hold that redistribution is a precondition of wilderness preservation is to fail to treat the duty as a stringent one, and ecocentrists are committed to the view that it is a stringent duty.

But this concern is based on a misunderstanding of the fairness problem and the redistributive solution. The fairness problem arises because acting on the duty to preserve wilderness will change certain material conditions, which in turn alters the moral terrain and gives rise to a new duty. So the criticism misplaces the direction and the terms of the conditional claim. That is, the claim is not that there is some duty to preserve the wilderness only if some action such as redistribution is taken to solve the fairness problem. Rather, there is a duty to preserve wilderness, and if it is acted on it creates the fairness problem, which can be resolved only by meeting a second duty, in this case the duty to redistribute assets. Thus, if Third-World countries fail to engage in wilderness preservation, they are violating a duty, irrespective of how the industrialized world acts. But if Third-World societies do act on this duty, then industrialized nations are under an obligation to find some solution to the fairness problem and to act on that solution. Thus, the claim that wilderness preservation gives rise to a fairness problem, and the proposal that a redistribution would resolve the problem, does not imply that the duty to preserve the wilderness is somehow conditional or trivial.

A third strategy ecocentrists might pursue for showing that there is an incompatibility between preservation and redistribution will rely on a consequentialist type argument. This strategy turns on noting that environmentalists typically are not in a position to implement policy; rather they are advocates for environmental positions. Thus, the question is not whether certain policies or actions are strictly incompatible with one another. The question instead is whether there is some incompatibility in advocating a policy of preservation and redistribution, as opposed to a policy of preservation only. It would seem that under certain conditions there could be. Suppose, for example, that the probability of the implementation of a program calling for both preservation and redistribution is quite low, whereas the probability of the implementation of a program calling for preservation only is substantially higher. Suppose further that the following is the correct ranking of possible worlds in terms of how much goodness each contains: a world in which there is both preservation and redistribution, a world in which preservation only is implemented, a world in which redistribution only is implemented, and finally a world in which neither is implemented. If all the foregoing is correct, then although there is no strict incompatibility between preservation and redistribution, it would seem that the morally correct thing to do would be to advocate preservation without simultaneously calling for redistribution.[8]

But even if all the conditions here hold, the policy recommendation follows only if we are willing to adopt the broadly consequentialist moral principle that one's obligation is to perform the action that has the highest expected utility, where expected utility is taken in a very general way to be a measure of the goodness or value that will result from some action multiplied by the probability that this action will in fact have those consequences. But it hardly needs to be pointed out that this consequentialist principle is highly controversial. One might argue that a person should simply act on his duty, and that the outcomes and their probabilities are beside the point. I think it is equally far from clear that on the basis of this principle environmentalists should advocate preservation only.

Furthermore, we have assumed for the sake of argument that the probability claims are accurate. But surely these assumptions are suspect on a number of counts. First, one wonders how the specific information required to warrant the probability calculations might be gathered. I suspect such detailed information will not be forthcoming, so the plausibility of these assumptions will have to be supported or attacked with acceptable generalizations about how nations and their populations behave with respect to issues such as sovereignty and self-interest. And here the evidence does not appear overly promising for advocates of preservation only. Admittedly they are very likely to be correct about the attitudes and actions of industrialized nations. A policy that links redistribution with preservation will not be as popular as one that calls for preservation only, since the latter represents little or no cost and significant gains for industrialized nations. But having granted that, we need to remember that the policy must also appeal to and be adopted by Third-World societies. And given that for these countries preservation alone represents a significant cost and only a minimal gain, it is unclear why anyone would think that this policy would be more acceptable to them than one that combined redistribution with preservation. If the cooperation of Third-World countries is necessary and if by and large nations do act in their self-interest, why think that the probability that one policy will be adopted is significantly higher than the other? Of course one might opt for the view that unilateral interference with Third-World countries would be justified because in failing to institute preservation they are practicing something morally reprehensible, on the order of genocide. But I think it would be difficult to sustain such a position.

There is a related consequentialist argument that should be noted in passing. The claim is that although in this instance redistribution should be implemented along with preservation, it sets a bad precedent, for it will somehow encourage people to place human concerns above nature, and thereby it will fail to sufficiently reinforce the idea that nature has value. This policy then encourages anthropocentric thinking, and this is dangerous because there may be situations where there is true conflict between duties to humans and to the environment, and people are all too likely as it is to act on human concerns. But this argument suffers from the same defects as the previous one. It is unclear why reflective persons would be unable to see that there is a difference in the cases, that in the present case one should be concerned with fairness as well as preservation because here the two ideals do not conflict, and that one does not preclude the other. Admittedly there may be some future cases in which there is a conflict and human value might not necessarily take priority; but it would seem to be a fairly simple matter to point out to people that there is this relevant difference in the two cases.

III

In the previous section I tried to deflect ecocentric concerns about invoking redistribution as one possible solution to the fairness problem. In this section I want to locate the exact source of that problem, and to show that a redistribution of assets is the only solution available to ecocentrists. This is a much stronger claim than that advanced in the previous section, where I tried to show only that nothing stands in the way of ecocentrists adopting the redistribution answer to the fairness problem. Now I want to argue that they are logically committed to this solution. I will refer to the hypothesis that the implementation of preservation policies gives rise to a duty to redistribute assets as the entailment thesis.

We must first ask why, if wilderness preservation were instituted and a moratorium on the development of ecologically sensitive areas were declared, the current distribution of assets would be unfair. The general point is that whether any given allocation is fair or not will depend, among other considerations, on whether assumptions made about the quantity of what is to be allocated turn out to be correct. Specifically, if what is to be allocated is practically limitless, then the rules that govern its allocation can be very weak, since there will always be more available for persons who do not at the moment have everything they would like to have. A principle such as Locke's might then be applicable in contexts of unlimited supply. But different and stricter rules will apply when what is to be allocated is somehow limited.[9] If an allocation occurs based on the belief that the supply is practically limitless, and the belief turns out to be false, then the allocation will fail to be a fair one. If we return to the parable, the castaways must decide how to allocate a stock of goods on which no one has a prior claim. They ran into trouble because they proceeded to allocate based on the belief that the stockpile was practically unlimited, only to discover later that it was actually limited because of the side effects of removing goods from the stockpile. Thus, the justification for a redistribution is based on the claim that the original allocation, which would have been fair and just if the original conditions and assumptions about quantity had held, is unfair and unjust because those assumptions were mistaken.

Now there appear to be significant parallels between the parable and the current environmental situation. Industrialized nations are similar to the Earlys in that they have been particularly quick and adept at appropriating natural resources, and they have benefited tremendously from this, at least insofar as we can identify benefit with preference satisfaction. However, the discovery that beyond a certain limit there is a tremendous cost to the indiscriminate cutting down of forests, strip mining, and devoting land to productive uses such as farming and grazing strongly suggests that for industrialized nations the costs of continuing such behavior on a large scale outweigh any future benefits. In addition, ecocentrists would argue that nature has some intrinsic value, and we have reached a point where our appropriations amount to harming things that have value. These points converge on one conclusion, namely, that these practices should be stopped immediately or, at a minimum, seriously curtailed and strictly regulated. The problem is that many of the best places to implement wilderness preservation, an end to clear-cutting, and so on, are in underdeveloped Third-World countries, where, precisely because they are underdeveloped, a large part of nature remains in something like its original, pristine condition, which also makes it the most likely arena for the next round of appropriation. Thus, while it seems difficult to deny that policies of preservation are morally justified, the effect of such policies would be to make large areas of nature unavailable for development. So while humans have operated for centuries under the belief that nature and its resources are practically inexhaustible, we now realize this to have been a mistaken assumption.[10] And as we were mistaken about which initial conditions actually held, we were in turn mistaken about which principles should govern the allocation of natural resources. Thus the holdings that resulted from proceeding under the mistaken assumption would fail to be justified. If this justificatory sketch is right, it gives us some reason to think that a moratorium on development in order to preserve wilderness areas, when combined with the current distribution of resources, would be unfair. Furthermore, if the fairness problem has been described correctly, then it is difficult to see how anything other than a redistribution of assets could solve the problem.[11]

These claims are admittedly controversial. Perhaps the current environmental situation and the process by which wilderness preservation gives rise to the fairness problem has somehow been characterized incorrectly. If this were the case, then although there may be a fairness problem, it could have features that allowed the problem to be resolved with less drastic measures than a redistribution of assets, and this would show in turn that the entailment thesis is false. So we need to canvas the more plausible rejoinders to the fairness problem to determine whether this is the case. We might begin by pursuing a possible point of disanalogy with the story. It might be claimed that unlike the Laters in our story, people in the Third World do not really desire the material goods that would come to them as a result of a redistribution. They prefer to live in traditional ways, and so offering them material goods with redistributive intent is pointless; it will not satisfy any of their wants or help them to achieve any of their goals, since their wants and goals lie in very different directions from our western, materialistic ones. If this is correct, then preservationist policies do not require any sacrifice or loss on their part, and thus there is no need to redistribute anything to them. It is as if the Laters in our story were dedicated ascetic monks who neither wanted nor needed a share in the goods the Earlys had already removed from the stockpile.

Now this is an empirical claim about actual preferences of members of Third-World societies, and its truth would have to be decided through empirical methods. But a few observations may be in order that will cast some doubt on the plausibility of this claim. First, the claim appears to involve a highly romanticized version of what life is like and of what people really want in Third-World societies. Whether we like it or not, western values and ideals, including ones about what constitutes a reasonable life and standard of living, have penetrated those societies, so that many members of Third-World societies have wants and goals not dissimilar to those prevalent in industrialized societies. One might try to argue that those wants and values are not genuine or autonomously formed, that they are the results of clever manipulation by multinational corporations. But such an approach treats members of Third-World societies as if they were children incapable of deciding what they truly want and where their true interests lie. Why should a Third-World person's desire for a washing machine or dishwasher be any less rational or autonomously formed than a similar desire in a member of an industrialized society? For both, the machine is a way of escaping drudgery. There are admittedly subcultures within these countries, such as those in the Amazon rain forest, who are largely untouched by western influences, and who would prefer that the rain forests be preserved as they are so that they can maintain their traditional way of life. But here we need to recall that there is no inherent tension between preservation and redistribution. The presence of these groups with wants that are largely unsatisfiable through material goods does not count against the claim that preservation should be linked with redistribution, since preservation of the environment, which is what we are assuming these subgroups want, does not preclude the possibility of redistributing goods to the larger society.

Rather than pursue other insignificant points of disanalogy, I want to turn to a second response to the entailment thesis, which, because it has significant implications for social theory, is more interesting. Its success depends on showing that the change brought about by the implementation of preservationist policies is morally irrelevant to the issue of redistribution. That is, the central contention of the entailment thesis is that preservation alters certain conditions so that what would have been a fair allocation had those initial conditions held in fact is not, and consequently a redistribution is in order. But if this change turns out to be morally irrelevant to the question of redistribution, then the entailment thesis should be rejected. For example, an ecocentrist might concede that wilderness preservation turns what would have been a practically inexhaustible supply, had it not been for the wilderness preservation, into a finite one. This may mean that all future allocations of resources must be based on whatever stringent principles govern contexts of limited supply. But why in addition must there be some redistribution of what was allocated earlier according to looser, Lockean principles, as long as that allocation occurred in the context of a practically inexhaustible supply? If no answer is forthcoming, then the change of context brought on by wilderness preservation appears morally irrelevant to the question of whether *past* allocations that occurred prior to the change were justified.

Here we need to supplement the discussion of the fairness argument by tackling an issue that was dealt with somewhat superficially earlier, namely, why the change in context from one with unlimited resources to one with finite resources is morally significant with respect to the question of what counts as a fair allocation. We begin by noting that the allocation of resources has a moral dimension just because certain allocations will worsen some people's condition unnecessarily, thereby harming them. This worsening of condition occurs because people are no longer at liberty to use those resources that are allocated to others.[12] Thus, a necessary condition for a justified allocation is that it not worsen anyone's condition unnecessarily. In contexts of unlimited supply, the fact that one person is allocated any amount of some resource cannot worsen the condition of anyone else, since these others can get whatever it is they need or want; if it is truly inexhaustible, there will always be more available. Thus there is little reason to have any principle to govern the allocation of unlimited goods. Of course, typically such resources are practically and not literally inexhaustible, and so some weak principle governing their allocation is usually thought to be necessary. One cannot lay claim to an entire continent by saying, "I hereby claim this continent." A Lockean sort of principle is usually considered a sufficient guarantee that no one's condition will be worsened in these contexts.

When the supply of something is limited, however, the likelihood that any particular allocation will worsen the condition of others unnecessarily increases, and so the principle that governs allocation in this context must be stringent enough to ensure that this does not happen. One way to do this is not to allocate at all and instead to keep things available for all to use, such as a commons system. An alternative is to divide the resource equally; here, although each person is no longer at liberty to use whatever has been allocated to others, her condition has not been worsened unnecessarily, inasmuch as she has received as much as is consistent with a non-worsening of the condition of everyone else who received an equal share.[13] Now while it seems that some variation of either of these methods will ensure that the condition of others is not worsened unnecessarily, it is almost certain that a Lockean principle cannot insure this. For under a Lockean principle the complete stock of some necessary or desirable good could come under the control of a small group of persons, thereby excluding all others from the opportunity to use the good. Thus any allocation of limited resources according to a Lockean principle appears to leave open the real possibility that the condition of some will be worsened unnecessarily and so would not be justified.

We can now respond to the ecocentrist's suggestion that the change brought about by wilderness preservation does not require a retroactive redistribution. The salient point here is that for a given time span, part of which is characterized by practical inexhaustibility of resources, call it *t1*, and another part by limited resources, call it *t2*, the entire time span is a context of limited resources. It is so because, while the condition of those alive during *t1* will not be worsened if resources are allocated on a Lockean principle, the condition of persons alive at *t2* will be worsened if the goods that were allocated on Lockean principles at *t1* were then passed along through inheritance to persons existing at *t2*, so that persons' holdings at *t2* are in large part determined by the allocation that occurred during *t1*. The allocation at *t1* worsens the condition of persons who are resource-poor at *t2* because much less is available for their use than there would have been had the allocation at *t1* occurred on the basis of one of the more stringent principles. This shows that the original allocation at *t1* was illegitimate and so a redistribution of what was allocated earlier is in order.

Perhaps this can be seen more clearly if we recur to the parable. Let us suppose that the first generation of castaways will be able to remove as much as it wants from the stockpile without the harmful effects ever becoming too serious. The Earlys bequeath what they have to their children, and the Laters, having taken little or nothing, have nothing to bequeath to theirs. They are nevertheless unconcerned about this, since they believe the stockpile will continue to be inexhaustible so that their children will be able to take from it whatever they might happen to want. But as the second generation removes goods from the stockpile, the seriousness of the harm increases, so that what would have been an otherwise inexhaustible stockpile turns out not to be. Now the appropriation of the first generation of Earlys does not worsen the condition of the first generation of Laters, as the Laters could have appropriated for themselves whatever they wanted. But the second generation of Laters' condition is worsened as a result of the earlier private appropriation in combination with the bequest to the second-generation Earlys. Thus the entire time span that includes both generations is a context of limited supply, and so the principle that correctly applies in that context, in order to make sure that the lot of no one is worsened unnecessarily, must be one of the more stringent ones.

Now historically, or at least since the advent of capitalism, nations have acted on the basis of Lockean principles, on the assumption that the world's resources are practically inexhaustible. But even if that was true at some point in the past, the need to preserve vast tracts of land as wilderness has led to a change so that we are no longer in a context of practically inexhaustible resources. This claim that resources are not practically inexhaustible is of course a central tenet of ecocentrism. Thus, although the allocation that occurred on Lockean principles when the world's resources were practically inexhaustible did not worsen the condition of Third-World people alive at that time, it did worsen the condition of Third-World people now alive, because they are no longer at liberty to use what was allocated to industrialized nations or to attain the resources for themselves, given that development has been curbed. So the resulting allocation fails to be fair, and a redistribution seems in order. Of course, the claim that the condition of members of the Third World, the second generation of Laters, and those alive at $t2$ is actually worsened is subject to an argument I want to look at shortly.

But before doing that we might notice that the above points will be useful in responding to a related concern that might be raised by the ecocentrist. He might wonder whether the quantitative aspect of the change brought about by wilderness preservation really has the impact that is being claimed for it. It might be suggested here that even the most radical proposals for wilderness preservation would make only a relatively small percentage of land and resources unavailable for development, and so the implementation of this policy is unlikely to turn what would have been a practically inexhaustible supply had it not been for the preservation into one that is finite. Whether or not this happens will depend on a variety of factors, such as how much land is being set aside for wilderness and the extent of actual human needs and desires. But suppose we concede for the sake of argument that wilderness preservation alone would not significantly alter a supply that is practically inexhaustible. Still wilderness preservation in combination with other actually occurring factors, such as a rapidly expanding human population with ever increasing needs and rising expectations, will have this effect of turning what would have been a practically inexhaustible supply into a limited one. While wilderness preservation is not the sole factor or a sufficient condition, it is nevertheless a contributing factor in a change of conditions that is morally relevant. The only way to deny that redistribution is in order would be to claim that even with an expanding human population and the need for wilderness preservation, we remain in a context of practically inexhaustible resources. We have already noted that very few ecocentrists will want to claim this. Thus, at a minimum it appears that wilderness preservation in combination with other factors has put humans in a context of limited supply. Suppose, for example, that the castaways in the parable only had to set aside a small portion of the stockpile, a change that, depending on the specific numbers, by itself might not be sufficient to change a practically inexhaustible supply into a finite one. But if this were combined with more people from other shipwrecks washing ashore, then it is not unreasonable to think that making even a relatively small proportion of the resources unavailable for appropriation would have that effect.

There is, however, a final rejoinder that will be familiar to readers of Nozick and should be looked at.[14] If successful, the argument shows that it is unlikely that an allocation based on the Lockean principle will worsen anyone's condition, even in a context of limited resources. Nozick would suggest that while allocations conducted on the basis of the Lockean principle may make that particular lot of some resource unavailable for others, and may, if it is the last lot, eliminate any opportunity for others to appropriate the resource on their own, it does not follow that the resource is unattainable and that the condition of the nonappropriators is worsened thereby. For the person who obtained property rights by appropriating may be an entrepreneur who does the appropriating in an intelligent, efficient, and hard-working sort of way, so that much more of the resource is made available. The idea here is that since what is being allocated are resources that are useless as long as they stay in the ground, there will be more usable resources available to Third-World countries because of the incentive provided by the Lockean principle. In the absence of the Lockean principle those resources would remain in the ground, useless, and so Third-World societies would be even worse off because of the scarcity. If this is correct, then the situation of nonappropriators is not worsened by allocations conducted in accordance with a Lockean principle, for now more of what they really want is available, and they can get it if they can strike some bargain with the owner. For example, they might sell their labor to the owner of resources in exchange for part of what she has appropriated.

Is it the case that the Third World is better off with the allocation of resources that has occurred historically based on the Lockean principle? This question cannot be answered without first answering the question, "Better off than what?" and identifying the relevant and appropriate alternatives for comparison is an extremely difficult problem. As G. A. Cohen points out in a critique of Nozick, if Nozick manages to limit the alternatives for comparison, then the prospects for carrying his point increase significantly, and we might have to conclude that Lockean appropriation did not make nonappropriators worse off, and consequently did them no harm.[15] If we only compare the situation of nonappropriators in the actual Lockean world with what their situation would be like in a state of nature in which nothing is privately owned and people behave in a Hobbesian way, then it may be plausible to hold that nonappropriators actually are better off than they would be if they were in that Hobbesian world. But as Cohen notes, there are alternative principles of appropriation, ownership, and social organization that would seem to offer an even better alternative for nonappropriators than the Lockean one. That is, efficiency and productivity may very well flourish even more under a cooperative system in which allocation is conducted under those stringent principles that require either some kind of equal distribution or common use of resources. If that is the case, then any allocation conducted under Lockean principles will worsen unnecessarily the condition of nonappropriators in the actual world, as there is an alternative system that could be implemented under which they would fare even better.

Now the question of whether some of these alternatives can be rejected because they are based on unrealistic assumptions or outright factual errors goes beyond the scope of this paper. But looking at the situation of the nonappropriators in the parable might provide some clues as to how such a discussion would turn out. It certainly appears that if the Earlys refuse to redistribute a portion of what they have appropriated to the Laters, then the Laters are worse off than they would have been under almost any imaginable alternative. For there is no reason to think the Earlys are somehow more efficient, more intelligent, or better in ways that would make it beneficial to the community as a whole to have most of the available resources monopolized by one group. The Earlys simply made a decision, perhaps because in fact they were unimaginative and couldn't think of anything better to do, to spend their time appropriating. Given the conditions that all the castaways believed to hold at the time, neither this decision nor the actual appropriation demonstrates anything about whether they were somehow more efficient or better appropriators.

This, admittedly, is hardly conclusive. But what must be noted is that however that discussion turns out, this strategy is not one that easily lends itself to adoption by ecocentrists who want to drive a wedge between preservation and redistribution. For this approach will work only if certain possible states of the world are held to be implausibly utopian or otherwise unrealistic. That is, in order to show that the Lockean appropriation does not worsen the condition of nonappropriators, it must be shown that it is implausible to think that common ownership of resources combined with a high degree of social cooperation will be efficient and lead to a dramatic increase in human well-being in a healthy environment. But most ecocentrists are committed to holding not only that such a world is realistic, but also that such a system of social organization is the only one that can provide a basis for an environmentally sound future. Thus, although a Nozickian sort of libertarian might pursue this lead in order to critique the entailment thesis, such a move would be inconsistent with other ecocentric views and hence is not an option that an ecocentrist could pursue comfortably.

IV

What are the ramifications of the entailment thesis? Many ecocentrists think that ecocentrism should stay above the fray on issues of social and political justice, and so will reject the claim that consistency requires ecocentrists to support redistribution or to endorse specific political viewpoints. They will suggest both that ecocentrists should strive for political neutrality and that this neutrality can be maintained without sacrificing consistency. This suggests a "big tent" model of ecocentrism, that holds that ecocentrism is capable of accommodating all sorts of divergent political views. If this were correct, then ecocentrists could adopt whatever views on social and political issues happened to suit best the political theory they happened to subscribe to; conversely, people could adopt ecocentrism without having to be concerned about whether they would have to revise any of their other political beliefs or theories in the wake of this.

But I have argued here that this cannot be done. The claim that our moral views form a complex and interrelated system, so that alterations in one part of the system may have repercussions throughout, requiring the revision of parts of the system that may not at first seem directly related, is hardly controversial. In order to maintain their neutrality, ecocentrists must show that their way of thinking about environmental ethics is unique in this respect, and consequently that ecocentric views somehow fall outside the web of our moral system. This uniqueness thesis is often supported with the claim that ecocentrism offers a new paradigm, and the implication seems to be that this new paradigm is somehow incommensurable with other moral concerns.[16] The lesson to be learned from the entailment thesis is that there is no reason to suppose that environmental ethics is unique in this regard. The entailment thesis serves as a counterexample to the belief that one's views on environmental issues can be isolated from the other beliefs in one's moral system.

Furthermore, if the entailment thesis is a reliable and representative indicator so that we can generalize from it, then the concerns of ecocentrists will generally dovetail with the concerns of those who take what might be described as a broadly liberal/leftist approach to issues of social and political justice. Now one would expect this to be a welcome development for ecocentrism, for it provides the basis of what should be a mutually beneficial alliance. The Green movement is perhaps the best example of the form such an alliance might take. And it is undeniable that many ecocentrists are concerned about issues of justice and do in fact think that it would be a good thing to redistribute assets to Third-World societies. However, many others seem to share the perception that these stands are ultimately unrelated to their ecocentric concerns. Thus the belief that ecocentrists can maintain political neutrality is fairly pervasive. Its pervasiveness explains, for example, why ecocentrism draws such a large part of its core support from what has come to be called the "new class."[17] Wilderness preservation, as well as many other ecocentric proposals, represents a significant benefit to the members of this class. And given the belief that political neutrality can be maintained, this class does not believe it is committed to arguing for and supporting policies such as redistribution that would clearly have a cost and that, given concerns about fairness, would have to be borne in part by them. Thus by the new class's lights, putting ecocentric policies into effect will have no concomitant costs for it, and so ecocentrism can be held painlessly. Similarly, if in fact most ecocentrists do perceive ecocentrism to be politically neutral, we can explain not just the failure of ecocentrists and workers to recognize their common interests, but also the outright hostility that sometimes characterizes the relations between these groups.[18]

The recent dialogues and exchanges between ecocentrists and social ecologists has led to a rapprochement that may be part of a growing recognition that ecocentrism cannot maintain a political neutrality on all issues and be logically consistent.[19] This paper has been an attempt to further dispel that notion. For if the entailment thesis is correct, then ecocentrists will be committed to adopting some fairly specific positions on issues of social and political justice. But again, if I am right, this loss of perceived neutrality is not something that ecocentrists should find disturbing.

Notes

1. In a relatively new field such as environmental ethics, labels are fluid and generally not very helpful at pinning down exactly which movement or thesis is being referred to. My specific concern here is with two influential variants of ecocentrism. The first is Deep Ecology, an account of which can be found in the work of people like Bill Devall and George Sessions, *Deep Ecology* (Salt Lake City: Peregrine Books,1985); Warwick Fox, *Towards a Transpersonal Ecology* (Boston: Shambhala Press, 1990): Arne Naess, *Ecology, Community, and Lifestyle* (Cambridge: Cambridge University Press, 1989); and John Rodman, "The Liberation of Nature?" *Inquiry* 20 (1977): 83–145. The other is indebted to the work of Aldo Leopold and is sometimes referred to as the Land Ethic. Some of the better known exponents of this position are Holmes Rolston III; see his *Environmental Ethics: Duties to and Values in the Natural World* (Philadelphia: Temple University Press, 1988); and J. Baird Callicott, *In Defense of the Land Ethic* (Albany: State University of New York Press, 1989).
2. See, for example, Ramachandra Guha's discussion of this problem in "Radical American Environmentalism and Wilderness Preservation: A Third World Critique," *Environmental Ethics* 11 (1989): 75–82.
3. See, for example, Eric Katz and Lauren Oechsli, "Moving Beyond Anthropocentrism: Environmental Ethics, Development, and the Amazon," *Environmental Ethics* 15 (1993), especially 55–56. Also, David M. Johns, "The Relevance of Deep Ecology to the Third World: Some Preliminary Comments," *Environmental Ethics* 12 (1990): 233–52. Johns's paper is typical in many respects. He sees the force of Guha's complaint, and advocates alliances with certain third-world organizations (238–39), but he never comes out and advocates any concrete redistributive policies. He seems to think that redistribution is only important in the context of

an anthropocentric system of values that holds that "the poor can become better off through economic growth and/or more egalitarian distribution" (240). His nonanthropocentric solution is to adopt an ecocentric view that will lead to a dramatic decrease in consumption; and this will somehow leave more for everyone, which will somehow lead to a more equitable distribution and an improvement in the situation of people in the Third World. But the mechanism by which this is supposed to occur remains vague and unspecified.

4. For an example of how such a moral theory might be worked out see Paul Taylor's *Respect for Nature* (Princeton: Princeton University Press, 1986), esp. chap. 6.

5. Notice the situation under consideration is not one that involves a trade-off so that if Third-World countries agree to preserve some wilderness areas, industrialized nations will provide them with financing and technology that will enable them to exploit alternative, equally sensitive areas. Instead, what is redistributed might be earmarked for projects that would help these nations to convert to cleaner and more environmentally sound technologies. Whether it is legitimate to append conditions such as this to the funds that are redistributed would probably depend on the particular situation, and so I do not go into that issue here. Furthermore, I am assuming that there are sufficient funds for both preserving whatever wilderness remains in industrialized nations and for redistributing assets to Third-World societies, so that we are not in a position where, because of limited financing, industrialized nations must choose between preserving and cleaning up their own environments and redistributing assets toThird-World countries.

6. Katz and Oechsli in "Moving Beyond Anthropocentrism" advance a version of this argument. Their analogy is two would-be murderers discussing how the property of their prospective victim is to be divided between them. Their point is that, given the nature of the proposed act, questions of what counts as a just distribution are irrelevant.

7. The classic discussion of this point can be found on pp. 55–56 of Judith Jarvis Thomson's paper, "A Defense of Abortion," *Philosophy and Public Affairs* 1 (1971).

8. Katz and Oechsli in"Moving Beyond Anthropocentrism," allude to this sort of argument; "If preservationist policies are to be justified without a loss of equity, there are only two possible alternatives: Either we in the industrialized world must pay for the benefits we will gain from preservation or we must reject the anthropocentric and instrumental framework for policy decisions. The first alternative is an empirical political issue, and one about which we are not overly optimistic. The second alternative represents a shift in philosophical world view" (57). Note that the only reason given here for rejecting the first alternative is that the authors are not "overly optimistic," presumably about the likelihood of its being implemented. Thus they opt for the alternative that supposedly is more likely to be adopted.

9. Of course Locke himself recognized this. He thought his principle applied only to those contexts in which there is "enough, and as good, left in common for others." *Second Treatise of Government* (Indianapolis: Hackett, 1980), sec. 27.

10. Locke, for example, thought that there being enough and as good left over accurately described conditions that existed at that time, so that appropriation was being carried out in a context of a practically inexhaustible supply. See his comments about America and Spain in the *Second Treatise of Government*, sec. 36.

11. I understand the idea of a redistribution very broadly here to include any transfer of resources that is unconditional or unidirectional, that is, a transfer that

does not involve an exchange of goods or assets. Thus programs sponsored by conservation groups in industrialized nations such as the Nature Conservancy, whereby tracts of land in Third-World nations are "purchased" and held as wilderness, would appear at first glance to count as a redistribution. So too would a program in which wilderness would be protected in return for a reduction or cancellation of some of the nation's outstanding international debts. However, whether these are the best ways to implement wilderness preservation, or whether they will truly perform the function that is morally required of this sort of redistribution, are issues that must be postponed to a later discussion.

12. And this in turn may lead to the subjugation of nonowners. See, for example, Jeffrey Reiman's account of this process in *Justice and Modern Moral Philosophy* (New Haven: Yale University Press, 1990), pp. 237–43.

13. For one proposal about how this might work see Ronald Dworkin, "What is Equality? Part II: Equality of Resources," *Philosophy and Public Affairs* 10 (1981): 283–92.

14. See Robert Nozick, *Anarchy, State, and Utopia* (New York: Basic Books, 1974), pp. 175–77.

15. G. A. Cohen, "Are Freedom and Equality Compatible?" in J. Elster and K. Moene, eds., *Alternatives to Capitalism* (Cambridge: Cambridge University Press, 1989), pp. 119–21.

16. The following is a pretty clear statement of this view: "Ecological issues are not just one more set of 'social problems' to be appended to the existing anthropocentric 'social justice' political agenda. Of primary importance is a radically new understanding of reality; a 'conversion' to an ecological consciousness." George Sessions, "Ecocentrism and the Greens: Deep Ecology and the Environmental Task," *The Trumpeter* 5 (1988): 66. The same thinking underlies John Rodman's views in "The Liberation of Nature?"

17. For an interesting discussion of the influence of the new class in the radical environmental movement, see Robyn Eckersley's "Green Politics and the New Class: Selfishness or Virtue?" *Political Studies* 37 (1989): 205–23. She suggests that there are a number of reasons for the participation of this class in the environmental movement, only one of which can be traced to self-interest.

18. For example, the hostility of some ecocentrists toward workers in the timber industry, and the need for an alliance between the two, is discussed in John Bellamy Foster's "The Limits of Environmentalism Without Class: Lessons from the Ancient Forest Struggle of the Pacific Northwest," *Capitalism, Nature, Socialism* 4 (1993): 11–41.

19. See, for example, the dialogue between Murray Bookchin and Dave Foreman in Steve Chase, ed., *Defending the Earth* (Boston: South End Press, 1991).

17

Solidarity Across Diversity:
A Pluralistic Rapprochement of Environmentalism and Animal Liberation

Brian Luke

As an animal liberationist with deep environmentalist sympathies, I have been very concerned to find ways to reconcile the apparently intractable conflict between the individualism of animal liberation and the holism of some environmentalisms. The early literature relating these two movements was often exclusionary and somewhat hostile, with, for example, Tom Regan referring to Aldo Leopold's Land Ethic as "environmental fascism"[1] and J. Baird Callicott calling the concerns of animal liberationists "incoherent," "insensitive," "life-loathing," and "anti-natural."[2] On the other hand, those writers who project a more conciliatory tone, and who might be taken to have offered a reconciliation of the two perspectives,[3] have in my view only brought together with environmentalism a rather attenuated animal protectionism, one marked by a lingering ideology of nonhuman moral inferiority,[4] and lacking any explicit mention of the defining goals of animal liberation—the abolition of animal farming, animal vivisection, and sport hunting. In this article, I attempt to bring animal liberation and environmentalism together while doing justice to both.

Copyright 1995 by *Social Theory and Practice*, Vol. 21, No. 2 (Summer 1995).

Animal liberation and environmentalism are often theoretically unified by modeling them as successive stages in the continuing outward extension of the scope of our moral concern, an extension moving ultimately toward a single criterion of value so broad that it includes both individual animals and ecosystems. In this article I draw from ecofeminist theory to argue that the extensionist model is not well-suited for moving us beyond the exploitative tendencies of anthropocentrism. Unity between environmentalism and animal liberation is not to be found in reducing them to the affirmation of a single broadly defined source of value. Rather, I argue that the two movements are affirming distinct values, and we should affirm these values without privileging either over the other in cases of conflict. Finally, I describe how solidarity between the movements can be based on recognizing that each creates social space for the other by undermining a common obstacle, anthropocentric ideology.

From the numerous typologies of environmental ethics, two significant distinctions emerge, that between anthropocentric and nonanthropocentric theories and that between individualistic and nonindividualistic theories. The anthropocentric/nonanthropocentric distinction refers to whether nature is respected for its contribution to human ends or for its own sake; while the individualist/nonindividualist distinction refers to whether the entities respected for their own sake are individuals or not. The various theories regarding our ethical relations to nature can then be organized by these distinctions:

	INDIVIDUALIST	NONINDIVIDUALIST
ANTHRO.	liberal humanism[5]	communitarianism[6]
NONANTHRO.	animal liberation[7] biocentrism[9]	ecocentrism[8] species preservation[10]

This typology is often interpreted by academic ethicists as highlighting points of dispute between various environmentalisms. Rather than seeing here a range of valid but distinct concerns, for beings of various sorts, for individuals and communities, there is a tendency to become territorial, arguing for one criterion of value at the expense of the others. The attempt is to show the invalidity of other criteria, or at least to show that one's preferred criterion of value takes priority over the others in cases of conflict.

Another common response to the typology above is to erase differences altogether by subsuming various environmentalisms under one global criterion of value. My attempt in the following is to retain a recognition of the diverse values affirmed by the various environmentalisms without thereby pitting them against each other. My main focus is on the connections between animal liberation and the other nonanthropocentric environmentalisms—that is, biocentrism, ecocentrism, and species preservationism, the three of which I refer to collectively as "deep environmentalism" to distinguish them from animal liberation and from anthropocentric environmentalisms.

The Ethics of Exploitation

One suggested relationship between animal liberation and deep environmentalism is moral extensionism. This model, developed most explicitly by Roderick Nash in *The Rights of Nature*, places animal liberation, biocentrism, and ecocentrism in a spectrum expressing the continual outward extension of moral consideration, starting with self and moving sequentially through these stages and others: family, tribe, race, humanity, animals, plants, ecosystems. There is some confusion over whether the extensionist relationship is normative or historical, since Nash first states that the model "should be regarded as an ideal type and not as an historical description of the actual thought of any specific individual or group of people,"[11] but then he immediately goes on to relate extensionism to the historical expansion of the concept of natural rights in the West.[12]

There are problems with both the normative and the historical interpretations of extensionism. The historical interpretation is ethnocentric, ignoring nonwestern cultures that embodied environmental consciousness long before the civilized white man began "granting" rights to women, blacks, and some animals and plants. Even within western culture, historical extensionism is problematic, insofar as deep environmentalism is not developing after animal liberation, as an extension of it, so much as the two movements are developing more or less simultaneously and in significant ways independently of each other. Deep environmentalism may even slightly precede animal liberation, thus contradicting the historical model: *A Sand County Almanac* was published in 1949, *Animal Liberation* in 1975; Greenpeace began in 1969, People for the Ethical Treatment of Animals was founded in 1980; the Endangered Species Act passed in 1973,[13] while the first March on Washington for the Animals took place in 1990. Also, if moral consciousness developed stage-wise as suggested by historical extensionism, we would expect to see deep environmentalists as a subclass of animal liberationists, each environmentalist having already developed an animal liberationist consciousness as a propaedeutic to his or her environmentalism. In fact, many deep environmentalists are either ignorant of, indifferent to, or actively against animal liberation.

If extensionism is not an historical thesis but an "ideal type," its point must be normative, that logically one *should* move from one stage to the next, from humanism to animal liberation, animal liberation to deep environmentalism. In other words, humanism implies animal liberation, which itself implies deep environmentalism. Or, more strongly, normative extensionism could hold not only that humanism implies animal liberation and deep environmentalism but also that the best or even the only way to ground deep environmentalism is to start from humanism and draw out its inherent logic in stages.

There are problems with both the weaker and the stronger versions of normative extensionism. The plausibility of the weaker thesis depends on the strength of the attempts to argue that consistent humanists should be animal liberationists or deep environmentalists. Extensionist arguments are used by, among others, Tom Regan and Peter Singer (animal liberation), Kenneth Goodpaster and Mark Bernstein (biocentrism), and Lawrence Johnson (animal rights, biocentrism, species preservation, ecocentrism).

Regan starts with the human right not to be treated as a resource or a receptacle, Singer starts with the human right to equal consideration. Both then argue that to be consistent we must also attribute these rights to animals, because humans and other animals are relevantly similar. The difficulty in showing that there is no relevant difference between humans and animals that would justify denying these rights to animals is that one's criterion of relevant difference depends on one's normative theory (for Kantians, the rational/nonrational difference is always a relevant difference). So to avoid the circularity of simply positing that rationality is not relevant, Regan and Singer are forced to bring in the "argument from marginal cases," pointing out that there is an apparent inconsistency in denying rights to animals because of their supposed lack of rationality while attributing rights to nonrational or "marginal" humans. This argument, however, has nothing to say to those, such as R.G. Frey, who are willing to allow the exploitation of marginal humans, or to those, such as Bonnie Steinbock, who would continue protecting marginal humans but out of nonbinding "sentimental" concern rather than respect for rights.[14]

Extensionist arguments by deep environmentalists fare no better. Kenneth Goodpaster summarizes the arguments for extending the criterion for moral considerability from rationality to sentience, then argues that the sentience criterion still arbitrarily excludes many worthy of consideration—namely, living but nonsentient beings.[15] His main positive argument for rejecting the sentience criterion as arbitrarily narrow is the following:

> Biologically, it appears that sentience is an adaptive characteristic of living organisms that provides them with a better capacity to anticipate, and so avoid, threats to life. This at least suggests, though of

course it does not prove, that the capacities to suffer and to enjoy are ancillary to something more important rather than tickets to considerability in their own right.[16]

Again there is a problem with circularity. Whether the evolutionary explanation of the development of sentience even "suggests" the intrinsic moral significance of all life depends on one's ethical and metaethical perspectives. To one who, like Peter Singer, believes that without sentience there is "nothing to be taken into account,"[17] an evolutionary theory of the development of sentience does nothing more than explain the conditions for the emergence of beings with intrinsic significance.

Lawrence Johnson's ambitious attempt to derive animal rights, biocentrism, species preservationism, and ecocentrism from a common humanistic starting point faces the same problem as the other extensionist arguments. Johnson's crucial premise is the following: "Sometimes our interests do not affect our sentience (or intellectual capacity) or vice versa. Even then they are morally significant in their own right."[18] From this he reasons that:

> To be fair, then, we must conclude that although non-sentient beings clearly have fewer interests, the interests they do have, such as they are, have moral significance as well,to the extent of those interests. There is no significant difference to determine a difference in moral status.[19]

This reasoning does not show that consistent humanists must respect the interests of nonsentient beings. It shows that consistent humanists must either respect the interests of nonsentient beings or disrespect the nonsentient interests of humans (that is, take the interests that do not affect our conscious states to be of no moral significance or of only instrumental moral significance). There is a tendency in western humanism to derogate those of our interests that are labeled "animalistic" or nonrational. This dualistic view of human nature, splitting ourselves into an elevated rational/"human" part and a subordinated nonrational/"animal" part, facilitates the exclusion of animals from moral concern. Given the history of this dualism and its continuation today, I would think that many humanists are prepared and willing to similarly split off and derogate our "plantlike" nonsentient interests as a way to facilitate the exclusion of plants and other nonsentient beings from moral consideration. In doing so they would be denying Johnson's crucial initial premise concerning the moral significance of our nonsentient interests. Since there is no logical invalidity in denying this premise, Johnson fails to show that humanism implies animal liberation or deep environmentalism.

One might use Mark Bernstein's argument here to bolster Johnson's premise concerning the moral significance of the nonsentient interests of humans. Bernstein intuits that preventing a normal adult human from coming into existence would be less wrong than killing an identical human.[20] But *ex hypothesi* the two humans would have lives with identical experiences, so the greater wrongness of killing as opposed to preventing creation cannot come from foreclosing future pleasures. Bernstein concludes that part of the wrongness of killing a human must come from interfering with an organism's natural development. Of course, plants are organisms with natural developmental paths, so the biocentric principle follows from Bernstein's conclusion. But a determined Kantian could avoid these biocentric implications by interpreting Bernstein's thought experiment as showing that the wrongness of killing a person comes not from foreclosing future pleasures but from interfering with a rational agent's plans, plans that an existent human has but that a potentially identical but still nonexistent human does not have.

So there are logical problems with the attempt to argue from humanism to animal liberation to deep environmentalism. Even if these logical problems could be worked out, there are still objections to the strong extensionist thesis (that is, to the idea that successive extensions of humanism are the best way to ground environmentalism). First, the extensionist strategy privileges individualistic environmentalisms from the start.[21] Whether utilitarian, deontological, or contractarian, western humanistic ethics are typically individualistic, granting rights and ceding intrinsic value solely to the individual person or, in the case of utilitarianism, to the affects of individuals (their pleasures and pains). The sequence of anthropocentrism to animal liberation to biocentrism is extensionistic; as the criterion of moral considerability moves from rationality to sentience to conativity, the class of moral subjects expands inclusively, without excluding moral subjects from the previous stages: humans are still included as sentient beings, animals are still included as living beings. On the other hand, the move from rationality or sentience to systemic beauty, stability, and integrity is necessarily a noninclusive move from an individualistic to a nonindividualistic criterion, so it is a shift, not an extension. There is no holistic kernel in humanistic ethics to develop into ecocentrism. Thus it is no accident that species preservationists and ecocentrists, unlike animal rights theorists such as Regan and Singer, generally begin not with humanistic ethical principles but with ecological insights.[22] Those few who wish to support their affirmations of species preservationism and ecocentrism by extending individualistic humanism are placed in the peculiar position of needing to argue that species and ecosystems are, in the end, really individuals, appearances notwithstanding.[23] The peculiarity lies not so much in holding that species and ecosystems are really individuals (this may turn out to be true), but rather in that one's case for respecting species and ecosystems for their own sake should depend on such controversial, abstruse, and (most to the point) *irrelevant* arguments. The features of ecosystems pointed to most typically to communicate their intrinsic value (beauty, stability, and integrity, for instance) are features of ecosystems however the argument turns out over whether ecosystems are individuals. Likewise for the status of species. So the emphasis on such abstrusities should signal us that something has gone awry in our theorizing—perhaps the extensionist model itself?

That the stages of the extensionist model exactly reproduce the levels of value in the hierarchical great chain of being should also give us pause. The great chain of being, in which each existent in the universe is typed, each type has a fixed value relative to the others, and the types of greater value have a right to use the types of lower value, has been identified by ecofeminists as a primary ideological device for legitimating relations of dominance and exploitation in the West.[24] The relative ordering of the rungs of this hierarchical ladder—God, angels, rich white males, other people, animals, plants, matter—matches the order of the successive stages of extensionism. Animal liberationists and deep environmentalists challenge the moral acceptability of exploiting animals and nature, and thus we must at least to some extent deny the success of the great chain of being as a justification of anthropocentric exploitation. But the extensionist model of environmentalist development does not fully reject the great chain of being. Peculiarly, extensionism allows a normative cosmology that functioned to legitimate anthropocentrism to direct our theoretical efforts to overturn anthropocentrism. It does this in two ways, substantively and formally, both of which have unfortunate consequences for environmental ethics.

Substantively, the extensionist model enjoins us to order our expansion of moral concern by following the hierarchical ladder downward, first expanding consideration to all people, then to animals, then to plants, finally to "the land" (that is, to ecosystems and species). But why not consideration for ecosystems, then animals? Or why not different sequences for different people or different societies? Why this particular order? Is it because this order follows our differential capacities to perceive the intrinsic value of various kinds of beings and act respectfully towards them? But it appears to me that the kinds of beings one finds it easiest to respect vary across societies, situations, and individuals (more on this below). Now if a suggestion that we generally find it relatively easy to respect people, less easy for animals, more difficult still for plants, and so forth, were supported by the assumption of a corresponding scale of decreasing intrinsic value, then we would simply be back to presuming the great chain of being. Many animal liberationists and deep environmentalists, notwithstanding their opposition to anthropocentric exploitation, do retain such a scale of decreasing intrinsic value. I criticize this kind of generalism (every plant is of less intrinsic value than any animal, and so forth) below.

Extensionism holds that once the criteria of considerability are sufficiently broadened, the logic of humanism is sufficient to structure a deep environmental ethic. So beyond retaining the substance of the great chain of being, extensionism also retains its formal features. Since the great chain of being functions to legitimate class relations of dominance and exploitation, it is not surprising to find that the logic, or metaethical structure, of extensionist environmentalism also supports dominance and exploitation. It does this in at least these four ways: dualism, rankings, generalism, and privileging sameness over difference.

Dualism

Closely aligned with anthropocentrism and the great chain of being is hierarchical dualism, which is the tendency to polarize entities and properties into pairs of opposites, with one member of each pair designated superior to its opposite.[25] Significant pairs in western thought include (with the first member of each the designated superior): reason/emotion, culture/nature, male/female, rationality/instinct, human/animal, civilized/savage, active/passive, spiritual/corporeal, light skinned/dark skinned. Hierarchical dualism is used to legitimate the rankings of the great chain of being by associating those higher on the chain (that is, rich white men) with the "superior" poles and, conversely, those lower on the chain (all the dominated classes) with the inferior poles. Civilized, rational men deserve consideration and have rights, while others do not. The extensionist environmentalists retain this dualistic, exclusionary thinking. The boundaries shift, certainly, but the logic does not: there are insiders and outsiders, and only insiders are worthy of moral consideration.

For Peter Singer, only those beings capable of feeling pleasure or pain are worthy of consideration, while for Lawrence Johnson, the morally worthy insiders are limited to those entities that can be shown to possess interests. In this way we are theoretically preparing to defend ourselves against charges of exploitation, should advocates arise for beings falling outside the rigorously defined boundaries of considerability.

Rankings

Dualism separates beings into two classes. This crude polarization limits its usefulness as an ideological tool, in that class relations in the West are not simply bipolar, but form more highly articulated chains (for example, men dominating women as a class, but white women in some cases exercising power over black men and women, while all these groups of people exploit animals). The legitimation of these extended hierarchies requires not a polarization but a spectrum of value. This is easily enough done since rationality, the central characteristic used to defend hierarchies, admits of degrees. Thus historically in the West women as a class have not been identified with animals, but placed somewhere in between white men and nonhuman animals in their mental abilities; similarly for nonwhite men.[26] And animals themselves have not been seen as totally mindless,[27] but as slightly "above" plants and the land and therefore deserving a modicum of respect.

The extensionist environmentalists retain this ranking of beings according to their degree of possession of mental qualities deemed intrinsically valuable. The new criteria of considerability—sentience, subjectivity, autopoiesis—also admit of degrees of possession, and this feature has been used by extensionists to specify degrees of intrinsic value among the morally considerable. Moreover, the extensionists, as might be expected from theorists who presume the basic adequacy of humanism, are unable to relinquish humanism's designation of rationality (of the type supposedly possessed by adult white males) as the supremely valuable characteristic. In the following passages by Singer, Regan, and Johnson, we see both privileging of rationality and ranking according to degrees of possession of the criterion of considerability:

> Some non-human animals appear to be rational and self-conscious beings, conceiving themselves as distinct beings with a past and a future. When this is so, or to the best of our knowledge may be so, the case against killing is strong, as strong as the case against killing permanently defective human beings at a similar mental level. . . . When we come to animals who, as far as we can tell, are not rational and self-conscious beings, the case against killing is weaker.[28]

> Recall the [lifeboat case]. There are five survivors: four normal adults and a dog. The boat has room enough only for four. . . . Now, the harm that death is, is a function of the opportunities for satisfaction it forecloses, and no reasonable person would deny that the death of any of the four humans would be a greater prima facie loss, and thus a greater prima facie harm, than would be true in the case of the dog. . . . Our belief that it is the dog who should be killed is justified. . . .[29]

> [S]ome interests have more weight than others. It is a matter of degrees. In arguing that the interests of a mouse are morally considerable, I am not claiming that setting a mousetrap is on a par with the premeditated murder of a human being. Normally, a human being has more interests than does a mouse. Moreover, the interest of a mouse

in continuing to live is not the same as the interest of a human in continuing to live. A mouse only has an interest in continuing a mouse life. That interest counts for what it is, but only for what it is.[30]

Upon reading these passages, with their generalized devaluations of nonhuman life relative to human life, we may need to remind ourselves that these are written by theorists placed on the nonanthropocentric side of the fence today. In many ways Regan, Singer, and Johnson do move beyond traditional anthropocentric thought. But we see here a residual anthropocentric tendency to rank beings by type that, I argue in the next section, is both unnecessary and pernicious.[31]

Generalism

Many contemporary environmental theorists are concerned to specify general rules for "resolving" conflicts between individuals or groups. In fact, these abstract principles do not resolve conflict (that is, indicate a way to accommodate all the involved parties) so much as they pick a small number of properties or relations to determine ahead of time which parties will be sacrificed for which others. This expectation of conflict, and the disposition to manage conflict by an a priori designation of a class of losers (those sacrificed), is a carryover from the humanistic ethics of exploitation. It is striking that we find this sort of generalism even among environmentalists such as Callicott and Taylor, who are not extensionists in other ways.

The expectation of conflict, especially of irreconcilable conflict requiring a sacrificee, is appropriate in societies committed to class relations of dominance and exploitation. Exploitative relations continually generate conflict, and the commitment to the class structure renders a lasting resolution of these conflicts (that is, a social revolution) unthinkable. And of course the deductive formulation of abstract principles of priority makes sense in a class-based society, since ideologically one needs a theory explaining why one group should systematically lose while another systematically wins. But outside a commitment to exploitation, it is difficult to understand either (1) the expectation of irresolvable conflict, (2) the disposition to settle conflict through sacrifice, or (3) the strategy of applying general, predetermined rules to conflict situations.

Those of us committed to a classless society should avoid this sort of generalism. The strategy of applying general, predetermined rules to conflict situations habituates us to thinking in class terms, that is, to focusing on a few general properties, the possession of which separates winners from losers, rather than considering the full complexity and particularity of a conflict situation. Note especially that nonanthropocentric environmental theorists who take a generalist approach are prepared to sacrifice nonhumans (*qua* less rational) to humans (*qua* more rational) in their a priori treatments of conflict (see quotes above). Moreover, the resolve to sacrifice one party for another becomes self-fulfilling, as our expectation of irreconcilable conflict deters us from working sedulously enough to uncover the deep but contingent class structures that often generate conflict. The expectation of irreconcilable conflict blocks the development of the sufficient imagination, patience, cooperation, and determination necessary to preempt conflict. To ask who gets priority when people are in conflict with animals or the land, and especially, to give a general, presumptive answer in favor of people, is to subvert the process of analyzing and opposing the political and economic structures that give rise to the artificial human "needs" to eat meat, to clear forests, to hunt, to pollute, to experiment on animals. On the other hand, once we expect and insist on rapport between humans and nonhumans, and between individuals and their communities (by "rapport" here I mean either harmony or conflicts that are mediable), we position ourselves to perceive and remove contingent sources of pervasive and systematic conflict.

We may have a general expectation of rapport without becoming dogmatically insistent that there will never be irremediable conflicts. Consider Susan Finsen's attempted reconciliation between the holism of ecocentrism and the individualism of animal liberation.[32] Finsen points out that within certain ecological niches it is adaptive for species to develop a reproductive strategy that entails treating species members as individuals—for example, many mammalian species have a small number of offspring and nurture them, the parents and offspring recognize each other, and mates choose each other. Thus there is an ecological basis for granting individuals within these species respect, since the species themselves have found the recognition of individuals to be adaptive. Conversely, many of the beings who have been relatively neglected by animal liberationists, such as flies, have adapted reproductive strategies in which individuality is not respected. In such cases it would be both ecologically disastrous and disrespectful of the species life itself to grant individual rights. While Finsen's points here are very useful for developing our capacity to see how concerns for ecological health and for individuals' well-being can be mutually supporting, it is an unwarranted step from this to Finsen's conclusion that holism and individualism are "entirely complementary."[33] There may be cases of conflict between ecological and animal liberationist concerns that foil all our expectations for rapport. One such possible case is the struggle between The Nature Conservancy and People for the Ethical Treatment of Animals (PETA) over snaring feral pigs in Hawaii. Snaring causes a painful death for these pigs, as PETA points out, but The Nature Conservancy argues that it is the only viable method for protecting the ecosystem from the destruction caused by this nonindigenous species.

Even if some cases cannot be mediated in a way completely harmless to all involved, this does not justify the construction of priority principles that give one type of value general precedence over another. Forgoing the construction of general a priori rules of priority does not mean that when recalcitrant conflict situations arise we will find ourselves with no resources for decision and action. As Elizabeth Dodson Gray points out (by way of analogy), in working with disputing children "parents continually find some grounds for making our decisions, grounds other than ranking our children in some hierarchy of their worth."[34] In our relations with nonhuman nature, as in our intrafamily situations, we can pursue a process of working through conflicts while maintaining consideration and full respect for all the parties, a respect that includes a disposition to avoid general rankings of beings based on type.

Privileging Sameness

[T]he judgment "good" did *not* originate with those to whom "goodness" was shown! Rather it was "the good" themselves, that is to say, the noble, powerful, high-stationed and high-minded, who felt and established themselves and their actions as good, that is, of the first rank, in contradistinction to all the low, low-minded, common and plebeian. It was out of this *pathos of distance* that they first seized the right to create values and to coin names for value.[35]

The power to declare oneself and one's class as good and the efficacy to promulgate this evaluation throughout society is a privilege of class dominance. Rich white men presume their own intrinsic value, use their class power to exact the considerations due to those with intrinsic value, and then have the leisure to wonder which others are sufficiently like them also to deserve these considerations. In this way a morality originating in the self-valorization of the powerful will tend to privilege sameness over difference: all morally considerable beings must have something in common that makes them valuable, since value comes from sufficient similarity to those who are axiomatically valuable.

Extensionist environmentalists follow this pattern. Their ethical theories start with the presumption of the intrinsic value of the human (we "seize the right" to declare ourselves good):

I start with the *assumption* that humans are in the moral universe, that we humans ought to be the objects of moral concern.[36]

In order to explain the basis of the case for the equality of animals, it will be helpful to start with an examination of the case for the equality of women. Let us *assume* that we wish to defend the case for women's rights.³⁷

> [T]he *assumption* will be made that normal adult humans are moral agents . . . Among our *prereflective intuitions* are those that concern when it is wrong to harm a moral agent.³⁸

The extensionist strategy is to declare ourselves good (assume we have certain rights) and then to wonder which beings are sufficiently similar to us also to warrant this privileged standing, defining by contrast the beings so relevantly different that they justifiably miss out on the protections this standing confers. The model is summarized by James Rachels:

> In arguing that animals do have rights . . . we may use the following method. First we select for discussion a right which we are confident that humans do have. Then we ask whether there is a relevant difference between humans and animals which would justify us in denying that right to animals while at the same time granting it to humans. If not, then the right in question is a right possessed by animals as well as by humans.³⁹

The implication that the source of intrinsic value is some property held in common with normal adult humans artificially limits the range of recognized values. The grandeur of a mountain, the evolutionary history and ecological fit of a species, are not attributes of individual humans. Yet such grandeur and ecological fit are seen as the wellsprings of intrinsic value by some environmentalists. Michael Smith rejects extensionism by pointing out that "the reasons for valuing the jungle or the primrose, the desert or Antarctica are manifold and often concerned more with our perception of their disparity from humanity than any affinities."⁴⁰ Such attributions of value may or may not be acceptable in the end, but they at least deserve discussion, whereas the extensionist emphasis on similarity to normal humans precludes any such discussion.

To summarize: extensionism is a superficially attractive model for unifying animal liberation and deep environmentalism. Yet it is historically inaccurate and normatively problematic. Extensionism's normative problems stem from its retention of both substantive and formal features of the western hierarchical worldview. The formal features—dualism, rankings, generalism, and privileging sameness—form a coherent and useful way to defend the systematic exploitation of a range of classes by a ruling elite, but as part of a liberatory theory they are unnecessary and pernicious. The privileging of similarity artificially restricts the range of types of intrinsic value, while the ranking of beings based on general type favors a hierarchical approach to conflict—sacrifice one group's interests to another, according to general rules of priority—rather than a vision of a concerted effort to remove systemic sources of conflict coupled with a decision to mediate whatever conflicts remain with a full recognition of the intrinsic importance of all parties.

In the following section I outline a more deeply liberatory model of the relations between animal liberation and deep environmentalism, one based on moral pluralism and an appreciation of common interests.

Solidarity Across Diversity

Moral pluralism recognizes more than one source of value, more than one reason for treating a being respectfully. What we might call "expansive pluralism" goes beyond moral pluralism by not just recognizing diversity but by welcoming and seeking out diverse types of value. I am suggesting a metaethical framework that understands different environmentalisms as appreciations of different sources of value (for example, rationality, sentience, autopoiesis, ecosystemic integrity, species beauty, and so forth)—values that are not reducible to a single common property, that need not be ranked relative to each other, that are not expected to be in irreconcilable conflict with each other, and that are not understood as an exhaustive list of the reasons to respect a being. We need not expect these diverse values to come to be appreciated according to some predetermined schedule, such as the extensionist paradigm. Individuals are not alike in their sensitivities, and, more significantly, types of value that may be appreciated from one social position may be opaque from another. It will always be difficult for animal farmers to fully appreciate the intrinsic value of their "stock," since their livelihood depends on treating these animals as instruments. Disinterested observers may be best able to perceive some types of intrinsic value (consider the appreciations emerging from ecology and ethology, or from the urban animal liberationists who do not farm), but both disinterest and the luxury to observe are functions of special economic and social positions. Other perceptions of value may best emerge from special types of engagements. Consider, for example, the distinctive appreciation for dogs and horses described in animal-trainer Vicki Hearne's work. We cannot know when and where value will emerge (or disappear) since we do not know ahead of time what social positions and relations will develop (or dissipate).

Individuals with special sensitivities or in favorable positions for perceiving a type of value are often able to communicate their perceptions to others. This greatly increases the possibilities for expansive pluralism, since we can seek out communications with these gifted or privileged individuals in order to increase our own moral capacities. ("Privileged" here does not signify elitism, since moral insight often [typically?] resides with members of socially subordinated or marginalized classes.) This approach reverses the attempt to minimize acknowledged sources of value. The minimization of acknowledged values may be motivated by an expectation of conflict. As discussed earlier, the expectation of irreconcilable conflict is appropriate given a presumption of hierarchical class structure. But those committed to abolishing hierarchies, such as feminists (especially radical and ecofeminists), theorize ethics differently, de-emphasizing conflict, focusing instead on the development of our moral capacities, and more frequently showing an openness to moral pluralism and expansionism.⁴¹

Callicott argues that pluralism is philosophically unsatisfactory; a lack of "theoretical unity" leads to practical difficulties: "When competing moral claims cannot be articulated in the same terms, they cannot be decisively compared and resolved."⁴² It would be more correct to say that under moral pluralism competing claims cannot be decided in an a priori, class-based manner; but, as discussed above, we are still capable of making decisions—better decisions, in fact. I do not share Callicott's belief that when moral claims are made commensurable—"articulated in the same terms"—then we have "discovered something true and deep about morality."⁴³ The reduction of moral considerability to the possession of a single characteristic is a great distortion of the wide diversity of bases we have for respecting species, ecosystems, the land, individual animals, and plants.

Nevertheless, to reject the artificial unification of environmentalisms under a single criterion of value is not to reject the unity of environmentalisms altogether—there are other types of theoretical unity. For example, the acknowledgment of diverse ethical grounds for considerability is metaethically unified through a perspective that respects the abilities of persons to apprehend intrinsic value of many different types. And, as I shall discuss in the remainder of this article, animal liberation and deep environmentalism, though based on different moral sensibilities, are importantly unified in their rejection of anthropocentric ideology.

Animal liberation is based on compassion for individual animals as they are restrained, injured, frustrated, made to suffer, killed, or otherwise harmed through the operations of the major institutions of animal exploitation in the West. Though some theoreticians, such as Regan and Singer, ground animal liberation not in sympathy but in judgments of unfair treatment deduced from abstract argumentation, the written non-theoretical accounts of animal liberationism and my direct experiences with activists lead me to conclude that sympathetic responsiveness to harmed animals is the prevailing motivation for animal liberation.⁴⁴ Other environmentalisms are based on a range of different motivations including self-preservation, concern for future humans, aesthetic appreciation, reverence for life itself, a sense of natural history, and respect for ecological integrity.

Though sympathy for individual animals is quite distinct from respect for ecological integrity, these motivations, and the motivational bases of the other types of nonanthropocentric environmentalisms, are each undermined by the anthropocentric ideology predominating in the West. The notion that animals are "subhuman," for example, renders strong sympathies for nonhumans suspect. Nonhuman animals are not worthy of such attention; those who persist in showing concern for animal well-being are perversely misplacing their compassions, transferring them from their proper sphere, human life, to an inappropriate realm. So in anthropocentric society we learn to hide, disregard, or prevent altogether our feelings for animals. The anthropocentric viewpoint sees ecological systems of soils, waters, plants, and animals as existing to serve man. So a concern for the land for its own sake seems out of place. Ecocentric sensibilities are, like sympathies for suffering animals, suppressed. Concern for ourselves and our children, expression of which is allowed (though highly regulated), develops to the exclusion of our other moral capacities.

So animal liberation and deep environmentalism face a common obstacle. Since anthropocentric ideology forms a coherent whole, a rejection of one part of it brings into question the entire system. As animal liberation develops, anthropocentrism is challenged, which creates possibilities for the development of deep environmentalism. And conversely, the deep environmentalist challenge to anthropocentrism makes space for animal liberation. I will give examples of both these processes.

In my own case I developed an animal liberationist consciousness first. Rejecting the institutionalized exploitation of animals still places one outside the bounds of rational normalcy in this culture. My conviction that the feelings and perceptions leading me to vegetarianism and antivivisectionism were valid allowed me to maintain a sense of myself as reasonable. Knowing that the judgment of irrationality was misplaced in my case gave me the capacity to look at others labeled irrational with an extremely open mind. In particular, anyone who argued for expanding the range of morally considerable entities had my respectful attention as a kindred spirit. As I felt I perceived something significant about individual animals that is routinely set aside in this society, I presumed that those attributing intrinsic value to individual plants, species, and ecosystems may well also be perceiving some neglected significance. Though not particularly well placed or innately suited for developing deep environmental consciousness on my own, I have been able to start moving in this direction by attending to the expressions of those environmentalists challenging anthropocentrism.

Conversely, there is evidence that deep environmentalist challenges to anthropocentrism play a crucial part in the development of animal liberationist sympathies for some activists. This is a two-part process: (1) a sense of ecological crisis undermines confidence in conventional humanistic values, and (2) this weakening of anthropocentric authority allows people to reassert their traditionally derogated sympathies for individual animals. As an example of the first part of this process, consider the following comments on the development of a deep environmental consciousness:

> As environment changes, any species that is unable to adapt, to change, to evolve, is extinguished. . . . The human species is one of millions threatened by imminent extinction through nuclear war and other environmental changes. . . . From this point of view, the threat of extinction appears as the invitation to change, to evolve. . . . The change that is required of us is not some new resistance to radiation, but a change in consciousness. . . . A biocentric perspective . . . may give us the courage to face despair and break through to a more viable consciousness, one that is sustainable and in harmony with life again.[45]

The ideological system supporting the hierarchical political and economic structures in the West has been severely shaken by this system's increasingly evident destruction of the biosphere. When faced with the possibility of human extinction, self-preservation, a disposition highly developed under the humanistic ethos, motivates a thorough reassessment of our practices and beliefs. Institutions and values previously so entrenched as to be unquestionable are brought into doubt as we try to figure out the root causes of this crisis. In such a climate, sensitivities of various sorts long buried by the hegemony of anthropocentrism find political expression, first by a few isolated individuals, then in larger numbers as the establishment of organized movements increases our confidence.

There is an apparent paradox in the desperate, self-concerned promotion of nonanthropocentric environmentalisms—how can we be affirming the intrinsic value of nonhumans (their value independent of our interests) if we are motivated in the first instance by interest in our own continued existence? It might seem that if, on the one hand, the plea is to act merely *as if* nature mattered,[46] lest we destroy ourselves, then we are still thinking anthropocentrically; yet, on the other hand, if we are truly affirming deep environmentalism then we are by definition not acting solely out of human self-interest. Apocalyptic deep environmentalism is common but somewhat puzzling.

The key to understanding the relationship between human self-interest and deep environmentalism, I would suggest, lies in understanding deep environmentalist consciousness (as well as animal liberationist consciousness, for that matter) as a culturally latent human perceptual/motivational capacity. In many cultural and social settings, we are capable of seeing and affirming the intrinsic value of nonhuman nature, yet we may choose not to, be taught not to, be encouraged not to through various sanctions, or be systematically prevented from doing so by the presence of mechanisms that block the development or application of this capacity. A list of some of the social processes that tend to block deep environmental consciousness may communicate my meaning here: sanctioning the consensus that only humans are intrinsically valuable; erasing or ridiculing deep environmental elements of other cultures or of our own minority traditions; promulgating the notion that we must exploit nature to be healthy;[47] preventing the sorts of knowledge of and relations with nonhumans that tend to support deep environmentalism.

This understanding of deep environmentalism as a potential capacity that may or may not be actualized brings up the possibility that we may deliberately choose to pursue those courses that tend to develop deep environmental consciousness. So the path from self-interest to deep environmentalism is not direct and immediate—ending with an enlightened anthropocentrism that is deep in name only—but is indirect by at least one step. Self-interested criticism of anthropocentrism entails that we work to remove the barriers to the development of deep environmentalism (such as the instilled fears and ignorances listed above) and place ourselves in those situations and relations that tend to foster a truly deep environmental consciousness.

Anthropocentric ideology is one of the central barriers to the development of deep environmental consciousness,[48] so a self-interested rejection of anthropocentric society includes a re-evaluation of anthropocentric ideology. The process connecting deep environmentalism to animal liberation can now more fully be articulated as follows: (a) sense of ecological crisis; (b) perception of anthropocentrism as a primary culprit and deep environmentalism as a necessary remedy; (c) self-interested decision to seek means to foster deep environmentalism; (d) identification of anthropocentric ideology as a major barrier to the development of deep environmental consciousness; (e) challenges to anthropocentric ideology; and finally, since anthropocentric ideology is also a major barrier to animal liberationist consciousness, (f) social space is (perhaps inadvertently) created for the development of animal liberation.

The applicability of this framework to some activists is supported by the work of Susan Sperling, though to see this requires a bit of work, since she interprets her data differently than I do. Sperling's understanding of animal liberation is summarized as follows:

> A high degree of concern about pollution and technological manipulation of the environment and living organisms is expressed by most adherents of the modern animal rights movement. Many increasingly

view our species as subject to materialistic technological forces antithetical to nature and as facing a potential ecological apocalypse with other species. Animal rights has emerged in recent years as a response to these anxieties about the human relationship to nature much as occurred in the Victorian era.[49]

Sperling notices millennarian tendencies among today's animal liberationists and among the nineteenth-century antivivisectionists. Activists of both periods frequently speak of the development and application of technology and modern medicine as a callous expansion, unchecked by social morality, that threatens our imminent destruction. In the Victorian period the threat posed is the dismemberment and destruction of people, particularly the politically vulnerable (women, workers, blacks), through human experimentation and the iatrogenic effects of modern medicine, an ascendant healing philosophy conceived as the invasion and manipulation of inert bodies. Today this iatrogenic threat continues but is joined by the threat of the impending collapse of the life-giving biosphere. Nineteenth- and twentieth-century animal liberationists see their work as part of the effort to turn back these destructive processes, an effort that if successful would usher in "the millennium," that is, a period of peace and security organized around compassion and respect for living beings.

The contemporary activists quoted by Sperling suggest a significant connection between concerns over environmental crises and the development of animal liberation:

"The advent of nuclear weapons and environmental pollution have made us realize that science and technology need to be controlled to some extent."[50]

"People are realizing that we can, in fact, destroy the earth and ourselves in the process. . . . I believe that's the arena that was available for the animal rights movement to come forth."[51]

"The specter of a nuclear holocaust, the reality of toxic dumps, of deformed babies, dead rivers, extinct species, have pierced the shroud of numbness and paved the way for questioning authority and wanting to see for oneself. The horrors of the labs and factory farms are no longer hidden."[52]

I am arguing that the sense of imminent ecological crisis destabilizes our confidence in anthropocentric ideology sufficiently to allow the reclamation of compassionate identification with individual animals. Sperling interprets the ecology/animal-liberation relation quite differently—for her the two are linked symbolically through animal experimentation apprehended purely as metaphor. For animal activists, Sperling claims, the vivisection of an animal represents the spoliation of the planet by industrial civilization or the invasion of the human body by medical science: "Animal experimentation is the key metaphor for the abuses by technological society of living organisms and the ecology."[53]

Sperling's idea is that activists oppose animal vivisection not out of concern for the well-being of the individual animal subjects; our real concern is over iatrogenesis or the environmental crisis. Apparently, animal experimentation is such a potent metaphor for these technological manipulations of humans and the land that we come to see animal experimentation *as* iatrogenesis or environmental destruction, thus concluding that we can actually save ourselves and the planet by abolishing animal research: "The redemption of society is believed literally to hinge on the abolition of the use of animals by science."[54]

By explaining antivivisectionism as a symbolically mediated opposition to ecological destruction, Sperling is denying the grounding of animal liberation in responsiveness to the suffering of real individual animals.[55] Similarly, J. Baird Callicott denies the compassionate basis of our outrage at factory farming and vivisection. For Callicott, these institutions are not objectionable because they cause great harm to animals for the material gains of a few people; rather, their immorality lies in the "transmogrification of organic to mechanical processes."[56] Callicott explains:

> The very presence of animals, so emblematic of delicate, complex
> organic tissue, surrounded by machines, connected to machines,
> penetrated by machines in research laboratories or crowded together
> in space-age "production facilities" is surely the more real and visceral source of our outrage at vivisection and factory farming than
> the contemplation of the quantity of pain that these unfortunate
> beings experience.[57]

Callicott here forgoes his capacity to see animals as concrete individuals; for him, they are "emblems." But of course, animals are not merely *emblematic* of delicate, complex organic tissue, they *are* delicate, complex, and organic beings. Callicott, for reasons of his own, is underestimating here the importance of responsiveness to suffering in animal liberationist motivation;[58] even so, I would not simply dismiss his points, since it may be that our capacity for compassionate political responsiveness to suffering is conditioned by the frameworks within which the suffering occurs (including even whether the suffering is framed by the "transmogrification of organic to mechanical processes").

Compare Peter Singer's critique of Sperling's book. Singer supports his statement that Sperling "fails to grasp what the movement is really all about" by arguing that animal liberation organizations are "seeking above all to prevent needless and unjustifiable suffering."⁵⁹ Yes, Sperling unjustifiably minimizes the importance of responsiveness to suffering, but what Singer misses here is that he and Sperling might each have a grasp on one part of what the movement is "really all about": the prevention of suffering, certainly, but also "a response to anxieties about the human relationship to nature." Not either/or, as Callicott, Sperling, and Singer all assume, but both; and not even both accidentally linked but rather the questions about our relationship to nature conditioning and facilitating the political expression of our concerns over animal suffering.

Sperling's theory of a purely symbolic relation between apocalyptic environmentalism and animal liberation is not at all supported by the crucial evidence, the statements of the animal liberationists.⁶⁰ For the activists and writers Sperling quotes in her study, animal vivisection and the environmental crisis are related conceptually, not symbolically. Animal vivisection does not *stand for* the vivisection of the planet; rather, behind both these practices as well as other systems of domination is a common attitude or mind-set. This position is stated explicitly and repeatedly by Sperling's interviewees:

> "We are very concerned about human rights; we are focusing on animal rights. These movements are all related because there's the same basic underlying attitude of supremacy that permits and perpetuates the racist, sexist, speciesist attitudes and concomitant crimes against creatures."⁶¹

> "More people are now aware that the state of mind that would permit the testing of radioactive weapons on monkeys also permits the vivisection of our planet for the same purpose."⁶²

> "The threat of nuclear annihilation and ecological catastrophe have brought home to a lot of people that Western military-industrial policies, practices and attitudes are pathological and aberrant."⁶³

An anthropocentric worldview facilitates the continuation of animal exploitation, so debunking this ideology is one element of animal liberationism. Similarly, deep environmentalism proceeds only to the extent that anthropocentrism is rejected. So the two movements share a common goal, and this relation is reflected in the activists' comments. Sperling's suggestion that activists see animal liberation as sufficient for or the key to human salvation is not supported by any of her subjects' statements. The statement Sperling quotes by Michael Fox, that "human liberation will begin when we understand that our evolution and fulfillment are contingent on the recognition of animal rights,"⁶⁴ might seem to support her thesis, but only because it has been removed from its immediate context, which explains that:

> I am concerned over the suffering of wild and domestic animals; I am equally concerned, however, for the future of my own species, since the values and actions that destroy non-human life and cause untold suffering do no less to humans. . . . Our detachment from nature, our lack of a sense of kinship and continuity with all life underlie the inhumane uses and abuses of animals and the havoc that is being wrought on the natural world.[65]

Here again we see the relation clearly stated: animal vivisection not as a symbol for environmental destruction, but as backed by the same anthropocentric values.

Similarly, Sperling vaguely claims that John Aspinall "links salvation from ecological disaster to recognition of the human-animal bond and animal rights,"[66] leaving out his explicit explanation of the *conceptual* connection:

> The concept of sanctity of human life is the most damaging sophism that philosophy has ever propagated—it has rooted well. Its corollary—a belief in the insanctity of species other than man—is the cause of that damage. The destruction of this idea is a prerequisite for survival.[67]

The key to a rapprochement between deep environmentalism and animal liberation is understanding that (1) potentially conflicting values such as systemic health and individual well-being need not be subordinated one to the other, and (2) the two movements are advocating for diverse values as those values are threatened by structures of human domination backed by a common anthropocentric ideology. So it is in the destruction of this idea: anthropocentrism, the "insanctity of species other than man"—that, notwithstanding the diversity of the values championed, deep environmentalism and animal liberation can find solidarity.

Notes

1. Tom Regan, *The Case for Animal Rights* (Berkeley: University of California Press, 1983), p. 362.
2. J. Baird Callicott, "Animal Liberation: A Triangular Affair," in Eugene Hargrove, ed., *The Animal Rights/Environmental Ethics Debate: The Environmental Perspective* (Albany: SUNY Press, 1992), pp. 37–69.
3. Such as Mary Anne Warren, "The Rights of the Nonhuman World," in Hargrove, *The Animal Rights Debate*, pp. 185–210; J. Baird Callicott, "Animal Liberation and Environmental Ethics: Back Together Again," in Hargrove, pp. 249-61; and Lawrence Johnson, *A Morally Deep World: An Essay on Moral Significance and Environmental Ethics* (Cambridge: Cambridge University Press, 1991).
4. Warren: "Animals have a significant right to life, but one that is somewhat more easily overridden [sic] by certain kinds of utilitarian or environmental considerations than is the human right to life" (p. 202); Callicott: "There is something deeply amiss in the concept of *equal* moral consideration or *equal* moral rights for animals" (p. 255); Johnson: "Some interests have more weight than others. . . . The interest of a mouse in continuing to live is not the same as the interest of a human in continuing to life. A mouse only has an interest in continuing a mouse life" (p. 7).
5. Humanistic environmentalism: nature is significant just to the extent that it impinges on the interests of rational agents.
6. Nature may be considered significant for its effects on human communities as well as on human individuals. (For example, the impending destruction of traditional subsistence fishing in some areas may be of concern even if individual members of these cultures could be integrated into the global economy.)
7. The terms "animal rights" and "animal liberation" are sometimes used interchangeably. They refer, depending on the context, either to an affirmation of the intrinsic moral significance of individual animals, or to the political movement to abolish the major western institutions of animal exploitation, namely animal farming, vivisection, and sport hunting. In this paper I use "animal rights" for the affirmation of moral significance and "animal liberation" for the political movement.
8. Ecocentrism: affirmation of the intrinsic value of ecological systems (Callicott, "Animal Liberation." See also Aldo Leopold, *A Sand County Almanac* [New York: Ballantine, 1949]).
9. Biocentrists, also called reverence-for-life theorists, advocate respecting individual plants, animals, and all living beings for their conativity or autopoietic qualities. See Paul Taylor, *Respect for Nature: A Theory of Environmental Ethics* (Princeton: Princeton University Press, 1986); Kenneth Goodpaster, "On Being Morally Considerable," *Journal of Philosophy* 75 (1978): 308-25; Albert Schweitzer, "The Ethics of Reverence for Life," chap. 6 in *The Philosophy of Civilization* (New York: Macmillan, 1949); and Mark Bernstein, "Towards a More Expansive Moral Community," *Journal of Applied Philosophy* 9 (1992): 45–52.
10. Species preservationism may be based on aesthetic or scientific appreciation of animal or plant species as types; see Eugene Hargrove, ed., *Foundations of Environmental Ethics* (Englewood Cliffs, NJ.: Prentice-Hall, 1989). Or it may be based on the perception of species as genetic lineages having morally significant interests distinct from the aggregated interests of the individual organisms of the species; see Johnson, *A Morally Deep World*.

11. Roderick Nash, *The Rights of Nature: A History of Environmental Ethics* (Madison: University of Wisconsin Press, 1989), p. 4.
12. Ibid., p. 6.
13. Though its passage was undoubtedly due primarily to enlightened anthropocentrism, there were also deep environmentalist views stated clearly in support of the legislation. Nash, *The Rights of Nature*, p. 175.
14. See Brian Luke, "Justice, Caring and Animal Liberation," *Between the Species* 8 (1992): 100-107; R. G. Frey, "Vivisection, Morals and Medicine," *Journal of Medical Ethics* 9 (1983): 94-97; and Bonnie Steinbock, "Speciesism and the Idea of Equality," *Philosophy* 53 (1978): 247–56.
15. Goodpaster, "On Being Morally Considerable."
16. Ibid., p. 316.
17. Peter Singer, *Animal Liberation* (New York: New York Review Press, 1990), p. 8.
18. Johnson, *A Morally Deep World*, p. 117.
19. Ibid.
20. Bernstein, "Towards a More Expansive Moral Community," p. 50.
21. See Kenneth Goodpaster, "From Egoism to Environmentalism," in K. E. Goodpaster and K. M. Sayre, eds., *Ethics and Problems of the 21st Century* (Notre Dame: University of Notre Dame Press, 1979), pp. 21–35; and J. Baird Callicott, "The Conceptual Foundations of the Land Ethic," in Callicott, *In Defense of the Land Ethic* (Albany: SUNY Press, 1989), p. 85.
22. Even individualistic deep environmentalists, such as Taylor, often ground their theories in ecology rather than humanism.
23. See Johnson, *A Morally Deep World*, pp. 152-55.
24. For example, Elizabeth Dodson Gray, *Green Paradise Lost* (Wellesley, MA: Roundtable Press, 1979).
25. See Val Plumwood, *Feminism and the Mastery of Nature* (London: Routledge, 1993), p. 43.
26. See Carol Adams, *Neither Man Nor Beast: Feminism and the Defense of Animals* (New York: Continuum, 1994).
27. At least not until the historically recent development of the notion of animal automatism, or mindlessness, initiated by Descartes and picked up by his vivisectionist contemporaries. See Singer, *Animal Liberation*, p. 201.
28. Peter Singer, *Practical Ethics* (Cambridge: Cambridge University Press, 1979), pp. 103-104.
29. Regan, *The Case for Animal Rights*, p. 324.
30. Johnson, *A Morally Deep World*, p. 7.
31. Michael Smith suggests that such moves back towards anthropocentrism are due to the theorists being unable to face up to the radical egalitarianism entailed by their own extensionist models. Michael Smith, "Letting in the Jungle," *Journal of Applied Philosophy* 8 (1991): 145–54, p. 148.
32. Susan Finsen, "Making Ends Meet: Reconciling Ecoholism and Animal Rights Individualism," *Between the Species* 4 (1988): 11–20.
33. Ibid., p. 14.
34. Gray, *Green Paradise Lost*, p. 148.
35. F. Nietzsche, *On the Genealogy of Morals* (New York: Vintage Books, 1969), pp. 25–26.
36. Johnson, *A Morally Deep World*, p. 6; emphasis added.
37. Singer, *Animal Liberation*, p. 1; emphasis added.

38. Regan, *The Case for Animal Rights*, pp. 152, 258; emphasis added.
39. James Rachels, "Do Animals Have a Right to Liberty?" in T. Regan and P. Singer, eds., *Animal Rights and Human Obligations* (Englewood Cliffs, NJ: Prentice-Hall, 1976), p. 206.
40. Smith, "Letting in the Jungle," p. 151.
41. See Carol Gilligan, *In a Different Voice* (Cambridge: Harvard University Press, 1982); Sarah Hoagland, *Lesbian Ethics: Toward New Value* (Palo Alto, CA: Institute of Lesbian Studies, 1988); and Marti Kheel, "The Liberation of Nature: A Circular Affair," *Environmental Ethics* 7 (1985): 135–49.
42. Callicott, "Animal Liberation and Environmental Ethics," p. 251.
43. Ibid., p. 250.
44. See Luke, "Justice, Caring, and Animal Liberation."
45. John Seed, "Anthropocentrism," in Bill Devall and George Sessions, eds., *Deep Ecology: Living as if Nature Mattered* (Salt Lake City: Gibbs Smith, 1985), pp. 244-45.
46. As in the subtitle of *Deep Ecology*: "living as if nature mattered" (Devall and Sessions).
47. This motivates individuals to block deep environmental consciousness on their own, since, if the premise were correct that we had to exploit nature to survive, a deep environmental consciousness would lead only to death or to a uselessly bad conscience.
48. Cf.: "It is through the denial of human superiority . . . that the biocentric outlook is linked to the attitude of respect for nature." Taylor, *Respect for Nature*, p. 155.
49. Susan Sperling, *Animal Liberators: Research and Morality* (Berkeley: University of California Press, 1988), p. 200.
50. Ibid., p. 95, quoting Peter Singer.
51. Ibid., p. 120.
52. Ibid., p. 121.
53. Ibid., p. 19.
54. Ibid., p. 194.
55. Sperling mentions compassion for animals as a motivation for antivivisectionism *one* time, dismissively (p. 74), while her book contains dozens of references to the vivisection of animals as symbolizing processes directly affecting people. Sperling apparently can conceive of feelings for animals only to the extent that animals approach a humanlike condition: "The image of the chimpanzee in the early Goodall accounts was of an *almost human creature* living in a state of nature, a being toward whom one could feel a comfortable identification, *if not actual affection* " (p. 169; emphasis added).
56. Callicott, "Animal Liberation: A Triangular Affair," p. 58.
57. Ibid.
58. Underestimating compassion, perhaps, even in his own motivations—why are factory-farmed and vivisected beings "unfortunate," if the issue is not about their suffering; similarly, what might be the moral significance of their being "penetrated" and their being composed of "delicate" tissues?
59. Peter Singer, "Unkind to Animals," *New York Review of Books*, February 2, 1989, pp. 36, 37.
60. Her theory does, however, entail the fundamental irrationality of animal liberationists (since animal vivisection is not in fact identical to the devastation of the

biosphere, and the abolition of the use of animals in science is evidently not the crucial determinant in the struggle to save the Earth). This not-so-indirect attribution of irrationality to animal liberationists should be placed within the context of Sperling's evident desire to legitimate her participation in animal research projects specifically criticized by animal liberationists (see Sperling, *Animal Liberators*, pp. 10–14).
61. Ibid., p. 117.
62. Ibid., p. 120.
63. Ibid., p. 121.
64. Ibid., p. 18.
65. Michael W. Fox, *Returning to Eden: Animal Rights and Human Responsibility* (New York: Viking Press, 1980), pp. xiii–xiv.
66. Sperling, *Animal Liberators*, p. 138.
67. John Aspinall, *The Best of Friends* (New York: Harper & Row, 1976), p. 155.

18

The Sustainability Question

Carl Mitcham

According to the contemporary historical account, Socrates' distinction was to bring philosophy down from the heavens by asking questions about piety and impiety, the good, the shameful, and justice and injustice.[1] In the process, philosophy both identified and criticized those ideals that guide the conduct of human affairs. In our time, this identifying and criticizing can appropriately be brought to bear on sustainability, an ideal now widely appealed to in contemporary social and political discussions of the environment.

Sustainability is an ideal, not unlike love or patriotism, that points toward something necessary and even noble, but that can also readily become an ideological rationalization misleading to all who use it. It is important to notice, for instance, that the very word "sustainability," as an abstract noun, did not appear until the 1970s[2]; its very abstractness indicates a peculiarly modern proclivity for high level generalities. As an abstract generality, sustainability is readily subject to disagreements over its precise implications—in this case, disagreements among economic, medical, political, ecological, and related interpretations.[3]

The sustainability question bears, in the first instance, on the issue of whether or not and how our technologically based economy can be sustainable in the face of what are presented as environmental constraints. But in the second instance, there are the less than heavenly philosophical questions concerning the historical and philosophical background of sustainability, the immediate origins of the concept of sustainable development, and relations between sustainable development and its near neighbors. Only against such a background can one begin to appreciate the fundamental critiques of sustainability.

1. The Historico-Philosophical Background of Sustainability

The historical and philosophical background against which the idea of sustainability has emerged can be indicated by contrasting it with two different theories about the basic character of historical change. History is fundamentally either cyclical or progressive. Progress, in turn, can be either toward a perfect future or away from an imperfect past.

The primordial experience of time and history is certainly cyclical. The sun rises and sets, then rises again. The moon returns every twenty-eight days. The seasons come and go, and come again. Any event that first looks like it is not part of some cycle tends to be interpreted as forming part of a larger, perhaps unseen cycle: millennial cycles, cycles of great many-thousand year yugas, maybe even just one great big cycle from creation and fall to redemption as new creation.[4]

The idea of one great cycle is a decisive step toward the theory of progress. When the cycle of history becomes so immense that one only remembers the beginning and hopes for the end, what becomes more important than any return to a previous state is the change taking place in some limited time slice within the cosmic circle. As the past is left ever further and further behind and the future becomes more and more an imaginative construct, attention becomes increasingly focused on the present and its movements. Within the restricted frame of reference thus defined, historical movement toward an imagined future can easily be thought of as progress. The idea of progress originates in and presupposes this two-sided expansion and contraction of perspective.

In Christian theology, for instance, an apocalyptic second coming is conceived of as recovering the state of perfection indicated by the Garden of Eden narrative. But as the second coming of Christ became indefinitely postponed, late classical and medieval Christians began more and more to focus on the practice of the spiritual life in the present—a phenomenon that finds expression, for instance, in the rise of Christian monasticism. When the Christian mythology of progress toward transcendent salvation by means of faith was abandoned, it was easily replaced with a scientific ideology of progress toward happiness in this world and satisfaction by means of power and control over nature. In the political metaphysics of science, the future is projected as realizing that state of perfection implied by natural human potentials for health, well-being, and autonomous freedom. Thus was the idea of a progress toward some final state that is a renewal of a beginning state replaced by progress toward a final state as perfection or realization of potentiality.

Then during the late nineteenth and early twentieth centuries this finality became—not wholly unlike the second coming—increasingly difficult to specify, and the idea of any final state was simply dropped. Progress became indefinite, more away from an imperfect past than toward the perfect future. Originally modern science was presented by its founders as able to provide final and definitive knowledge about the world. In Alexander Pope's heroic couplet, "Nature, and Nature's laws lay hid in night. / God said, *Let Newton be!* and all was light." But with recognition of the inability of any scientific theory to withstand the onslaughts of further investigation or the introduction of further refinements—that is, with recognition that no scientific theory is ever final—the idea of indefinite (if not infinite) improvement was substituted for the realization of an epistemological perfection. As one commentator on Pope has it: "It did not last: the Devil howling 'Ho! / Let Einstein be!' restored the status quo."[5]

This idea that science and technology may get "better and better" in comparison with the past, but not necessarily closer and closer to some definable ideal, is given extended articulation in Karl Popper's philosophy of science. Scientific theories can be falsified but not confirmed.[6] Moreover, although scientific theories are more comprehensive today than yesterday, it remains difficult to imagine what a completely comprehensive theory would be like—and whether it is even possible to strive for such a theory. The ideal of a comprehensive theory or discipline is replaced by one of multiple theories and difference. The leap from Popperian falsificationism to postmodernism is not so great as is often supposed.

Interpretations of evolutionary biology can further exemplify this new idea of indefinite progress away from a past that fails to point toward any definite future. Some animals can smell, others can see. Some can smell and see. The perfect sensorium would be like . . . who knows? But having two senses is nevertheless better or more inclusive than having only one, and three better than two, ad infinitum. A multiplicity of senses, indeed an increasing number of senses enhanced by all available instrumental means, denotes progress.

In like manner, although the perfect human being or society may be impossible to specify, surely we can recognize when one is better than another. This is the implication of Winston Churchill's oft-quoted observation that democracy is definitely not the perfect form of government. In fact, "democracy is the worst form of government except all those other forms that have been tried from time to time."[7] With the fall of communism, the same might be said for capitalist economic and technological development, including its environmental shortcomings. The technocapitalist record of environmental pollution has got to be the worst there is—except for all the others.[8]

It is important to note, however, that this backward-looking or escapist theory of progress entails a stigmatization of the past—a past inherently less than able to defend itself. The motor of progress so conceived can be captured in the fearful question, "Do you want to go back to a time when . . . ?"—with little appreciation of what that "time when" might have really been like. With regard to the sustainability question, it is possible that at some deep level there is simply the fear of an allegedly unsustainable and inhuman past of scarcity that must at all costs be avoided. Sustainability may not be so much a positive ideal as an attempt to escape the unsustainable.

2. Immediate Origins of the Concept of Sustainable Development

Implicit in this typically modern idea of progress as an indefinite and continuous superseding of the past, rather than as an approach toward some definite future, is the notion that progress is without limits. The early modern theory of progress, such as can be found in the Marquis de Condorcet and Immanuel Kant, actually had its own kind of limits: some kind of perfect state. But the theory of indefinite progress lacks any such projected termination point. Perhaps, for example, one could keep adding on senses forever: we certainly create instruments that register more and more aspects of reality through X-ray machines, radio receivers, ultrasound and magnetic resonance imaging devices, and so forth. Such extensions of perception could just go on forever, and even be incorporated into our physiological sensorium, after the manner of the science fiction dreams of some artificial intelligence enthusiasts. Androids and cyborgs become our siblings—if not ourselves.

By contrast, it is the idea of limits to progress that is distinctive late-modern historical consciousness. Or, more cautiously, it is the construction of such limits that is an ardently contested theme in the conflict between modern and postmodern interpretations of history. For late modern theorists, this is an issue closely associated with research sponsored by a group called the Club of Rome and first published in a book entitled, appropriately enough, *The Limits to Growth* (1972).[9]

The Club of Rome is an informal, international association of not more than one hundred scientists, business executives, scholars, and public officials that grew up in the late 1960s and early 1970s around the charismatic Aurelio Peccei and his concerns about the global development system.[10] Its intentionally vague character as a kind of transnational, intellectual mafia may even have contributed to its ability to attract extensive media attention. The fundamental argument of the Club of Rome and that "report to" it called *The Limits to Growth* is that technological development and societal increase simply cannot continue as they have for the last two or three hundred years.

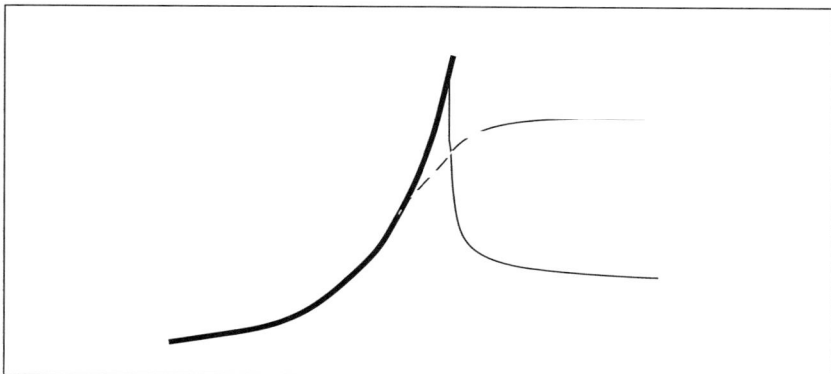

Figure 1.

The theory of progress that compares the present with the past, and that does not consider the future except as an open-ended possibility for further growth and improvement, is a source of increasingly intractable problems. By comparing the present only with the past and noticing how much better things are now than they were then, one fails to notice that in the near future things may well get worse—worse even than they were in the past. Moreover, the catastrophe will occur precisely as a result of the action of the very forces that make the present better than the past, and that continue to be supported because they do so.

To put the point in graphic form (see Figure 1): The simple point of *The Limits to Growth* is that exponential growth cannot continue indefinitely. Exponential increase of one kind or another has been characteristic of modernity since the seventeenth century, whether in population, food, industrial production, energy consumption, CO_2 emissions, etc. (see the dark solid line on the graph). Such growth will either terminate in a catastrophe (thin solid line) or level out into a logistic curve (broken line). Some wits have even described the exponential portion of this graph as the "club" of Rome, as that with which Peccei and his cohorts were wont to hit others over the head.

The argument of *The Limits to Growth* was that we are going to go over the cusp of catastrophe if we do not take conscious action to create a curve of logistic accommodation to resources. Even in the follow-up book, *Mankind at the Turning Point* (1974)—which tried to put forward a more positive vision of the future as one of possible "organic" rather than "undifferentiated" growth—the message continued to be phrased in largely negative terms and tone.[11] Human beings have to *stop* what they are presently doing. The implicit if not explicit message is that we should replace growth with a no-growth or steady state economy.

The shift from emphasizing what *should not* be done to stressing what *should* and can be done is constituted by the shift from a discussion of "limits to growth" to "sustainable development." This shift in the framework of the discussion was decidedly initiated by two other reports:

- the *World Conservation Strategy* (1980) of the International Union for Conservation of Nature and Natural Resources

- *Our Common Future* (1987) of the World Commission on Environment and Development.

Because of their importance for all subsequent discussion, and their influence on such international programmatic documents as *Agenda 21* (the product of the 1992 Earth Summit in Rio de Janeiro), it is worth considering both of these documents in some detail.

The first of these, the *World Conservation Strategy*, is exactly what its name implies, a strategy document aimed at government policy makers, conservationists, and development practitioners.[12] It consists of a spare fifty pages of text divided into an introduction followed by three sections on the general objectives of conservation, a set of priorities for national action, and a set of priorities for international action. Each of the three sections is further broken down into six or seven two-page subsections with from eight to fifteen numbered paragraphs. This is clearly a policy and action manual organized around thought bites if not sound bites.

Development is defined (section 1, paragraph 3) as "the modification of the biosphere and the application of human, financial, living and non-living resources to satisfy human needs and improve the quality of human life." Moreover, "for development to be sustainable it must take account of social and ecological factors, as well as economic ones; of the living and non-living resource base; and of the long term as well as the short-term advantages and disadvantages of alternative actions."

At the same time, conservation is defined (section 1, paragraph 4) as "the management of human use of the biosphere so that it may yield the greatest sustainable benefit to present generations while maintaining its potential to meet the needs and aspirations of future generations." "Conservation, like development, is for people" (section 1, paragraph 5), not for the things conserved.

Such definitions lead readily to the conclusion that "development and conservation operate in the same global context, and the underlying problems that must be overcome if either is to be successful are identical" (section 20, paragraph 1). For the *World Conservation Strategy* recognition of the "limits to growth" is, as it were, the precondition to further and continuing growth. The limits and conservation must be incorporated into further growth or development strategies.[13]

Our Common Future (1987) is a much more discursive text, with an influence that is difficult to overestimate. As the first chapter of *Agenda 21* acknowledges in somewhat breathless prose:

> In 1987, the U.N. World Commission on Environment and Development linked the issue of environmental protection to the seemingly unrelated topic of global economic growth and development. Headed admirably by Norwegian Prime Minister Gro Harlem Brundtland, this commission produced a stunning report entitled *Our Common Future*, which carefully documented the status and future of the global economic and ecological situation. Perhaps the most lasting accomplishment of the Brundtland Commission, however, was to thrust the concept of "sustainable development" into the mainstream of world debate. Although this concept had been the focus of discussion in the world scientific community for some time, its introduction into international dialogue elicited an almost instant response.[14]

Our Common Future was the third in a series of reports from so-called independent commissions headed by prominent European social democratic politicians.[15] Extending the tactics of the *World Conservation Strategy*, this new document emphasizes a shift from limits to sustainability that can be described in the following way. Gro Brundtland, Prime Minister of Norway and chair of the World Commission on Environment and Development, was faced with competing interests. On the one side were environmentalists, who argued for a limits-to-growth or no-more-growth position to meet the threat of pollution, protect natural resources, and respect the rights of future generations. On the other were representative economists, especially of the Third World, who argued the need for development and more growth, to alleviate poverty in the present and to make it possible for these nations to play a larger role in international affairs. The Brundtland report bridging these conflicting interests was to propose neither simply limits nor simply development but *sustainable development*.

To quote the widely influential definition of the Brundtland report, sustainable development "meets the needs of the present without compromising the ability of future generations to meet their own needs." As *Our Common Future* immediately explains:

> The concept of sustainable development does imply limits—not absolute limits but limitations imposed by the present state of technology and social organization on environmental resources and by the ability of the biosphere to absorb the effects of human activities.

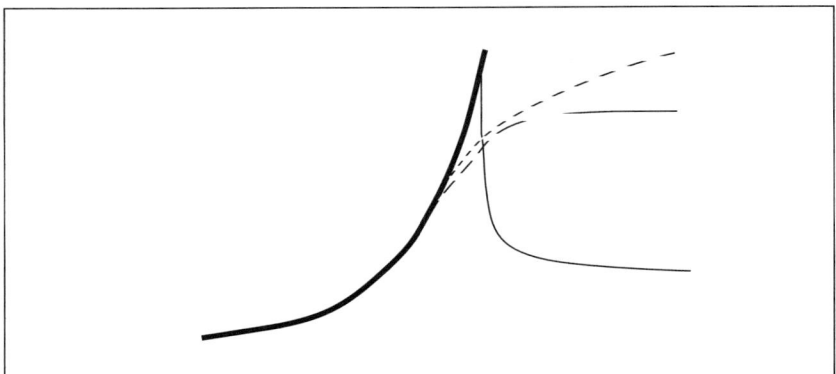

Figure 2.

But technology and social organization can be both managed and improved to make way for a new era of economic growth.[16]

The change that takes place after unlimited growth is thus defined not as a situation of no-growth or no-more-growth but as a situation of sustainable growth. To some extent it seems as if the Brundtland report is proposing a possible third option in the transformation of the exponential curve (see Figure 2): not catastrophe (thin solid line), or leveling off (broken line), but continued more moderate growth (dotted line)—sustainable growth that does not eventuate in catastrophe.

Similar concepts of sustainability appear to animate a host of post–Earth Summit books on sustainable development and sustainability.[17]

3. Sustainable Development and Its Near Neighbors

It is not exactly clear, however, the extent to which sustainability must mean continued growth. As already indicated, the concept entails a creative ambiguity, which is precisely what makes it useful for bridging the gap between no-growth environmentalists and pro-growth developmentalists. In order to explore this ambiguity, it is useful to consider some near-neighbor concepts and their permutations.

In relation to economics, the ideal of sustainability has been strongly influenced by post–World War II theories of economic development. Here, two classic references are President Harry Truman's inaugural address on January 20, 1949,[18] and Walt W. Rostow's influential 1956 article, "The Take-off into Self-sustained Growth."[19] As elaborated in Rostow's *The Stages of Economic Growth: A Non-Communist Manifesto* (1960), economic development is said to go through five distinct stages:

1. traditional society

2. preconditions for take-off

3. take-off

4. drive to maturity

5. high mass consumption

In traditional society there is no dynamism of economic growth, which instead occurs only more or less by chance. In traditional or pre-take-off societies, growth is often the result of outside factors, influxes of capital and external development interests. When these external inputs falter, so does growth.

According to Rostow, for growth to take off and become what he calls "self-sustaining,"

> The forces making for economic progress, which yielded limited bursts and enclaves of modern activity, expand and come to dominate the society. Growth becomes its normal condition. . . . New techniques spread in agriculture as well as industry. . . . [T]he basic structure of the economy and the social and political structure of the society are transformed in such a way that a steady rate of growth can be, thereafter, regularly sustained.[20]

As Hannah Arendt noted at this same time, however, such "unhindered development"—as illustrated by the "economic miracle" of postwar Germany, in which "the expropriation of people, the destruction of objects, and the devastation of cities [turned] out to be a radical stimulant"—undermines the uniquely human world of cultural and political orders in which human beings have traditionally been thought to be at home. In the new order of insistent production and consumption, "not destruction but conservation spells ruin, because the very durability of conserved objects is the greatest impediment to the turnover process, whose constant gain in speed is the only constancy left."[21]

For Arendt the problem of destruction focuses on the cultural-political order, not on nature. Two decades later, however, it was precisely in an effort to respond to a broadened appreciation of the destructive character of self-sustaining technological change that opposing concepts of "sustainability" began to emerge among alternative-technology adherents of the limits-to-growth criticism. The attempt to spell out alternative approaches was especially important in two areas: one dealing with energy, the other with agriculture.

In the energy crisis of the mid-1970s, for example, the argument for the replacement of dependency on nonrenewable resources (coal, oil, and gas) with utilization of renewable resources (water, wind, solar) and the adoption of what Amory Lovins termed a "soft energy path" clearly points toward a kind of sustainability. According to Lovins and his colleagues, energy production and use cannot continue indefinitely at the present accelerating rate; increased pollution and the eventual depletion of resources will not allow it. The self-sustained growth in energy production is not indefinitely sustainable; Rostow's last stage of high mass consumption is a prelude to disaster.

Two noteworthy features of Lovins's argument are that he does not see energy use continuing to grow at an exponential rate, and that he postulates not just a leveling off but an actual decline in energy production and use. These ideas were first presented in *Foreign Affairs* (October 1976), in an article that became the central chapter of *Soft Energy Paths* (1977).[22] Such a position created and continues to create considerable discussion. One of the most controversial implications of Lovins's view concerns the extent to which his soft energy paths and energy conservation proposals call for more general alterations in the lifeways of advanced industrial societies. For Lovins himself, they need not occur; more efficient energy use is presented as actually increasing the sustainability of increases in material standards of living.

It should also be noted that Lovins and the alternative or sustainable energy coalition has also had a direct impact on U.S. governmental policy. Extending the Solar Energy Research, Development, and Demonstration Act of 1974, President Carter in 1977 founded the Solar Energy Research Institute, which subsequently evolved into the National Renewal Energy Laboratories in Golden, Colorado. In relation to energy, "renewable" is another term closely related to "sustainable." Despite erratic funding, this national laboratory has maintained a presence on the political scene through both Democratic and Republican administrations.

With regard to agriculture, the idea if not the term "sustainability" has considerably deeper roots. The idea that farmers should not just use the land, wear it out, and move on, but should instead conserve and sustain it became a major theme in U.S. governmental policy and agricultural education with the establishment of the Department of Agriculture in 1862 and the subsequent founding of the major land-grant universities.

But the first theory about how to achieve this conservation was through "scientific farming" as promoted by agricultural extension agents attached to the new agricultural universities. Building on the work of British and German chemists such as Sir Humphrey Davy (1778–1829) and Justus von Liebig (1803–1873), agricultural science proposed that the regular utilization of inorganic fertilizers, especially nitrogen, would indefinitely increase and prolong land productivity. But nitrogen utilization established a farming dependence on the industrial production of chemical fertilizers, thereby replacing farming with agribusiness.

A different vision of agricultural sustainability developed in conjunction with a critique of scientific farming as tending to promote unnecessary dependence on industrial development. This critical movement originated with Sir Albert Howard, a British agricultural advisor to India in the 1920s, who argued that Indian farmers could maintain their independence and increase yields by means of decentralized fertilizer production through organic composting.[23] But more immediately influential was the U.S. writer and back-to-the-land pioneer, J. I. Rodale (1898–1971), who in 1942 coined the term "organic farming" when he founded what would become the journal *Organic Gardening and Farming*. With arguments that can be seen to foreshadow Club of Rome ideas, Rodale's proposals for organic farming also included criticisms of mono-crop production agriculture. Indeed, this critical or alternative agriculture movement is now widely identified with the work of J. I. Rodale, his son Robert Rodale (1930–1990), the Rodale Institute, and associated publications such as *Prevention* (1954–present) and *New Farm* (1979–present).

But for Robert Rodale, especially, the idea of organic farming needs to be expanded beyond the realm of agriculture.

> In recent years, events have forced the concept underlying organic research to expand. . . . There is an energy crisis, and the nonorganic way of growing food uses too much energy. Our environment has become contaminated by agricultural chemicals, many of which pollute the normal sources of food. The ozone layer of the upper atmosphere . . . is threatened with destruction by the overuse of chemical nitrogen fertilizers. . . . Because of these and similar conditions, there is now a need for a concept of organic research that will support an effort going far beyond the basic goals of natural soil improvement and improved food quality.[24]

As a result, during the latter half of the 1980s Rodale's preferred term became "regeneration," and he worked to define methods of farming that not only sustained the land but actually regenerated it. He argued in the newsletter *Regeneration* (1985–present) that similar attitudes and methods can be carried over into society.[25]

In the 1970s E. F. Schumacher's *Small Is Beautiful* (1973) and the "intermediate technology" movement gave added scope and life to both alternative energy and organic farming arguments.[26] The intermediate technology movement itself devolved into the "appropriate" and then the "alternative technology" movements, which in turn had the effect of spawning the term "alternative agriculture." But during the last half of the 1980s, as illustrated by articles in the academic, peer-reviewed research publication, *American Journal of Alternative Agriculture* (1986–present), alternative agriculture became increasingly associated with what is now explicitly termed "sustainable agriculture."

Since the late 1800s, the primary strategy of North American agriculture had been to increase yields and therefore profits per person-hour of labor by the increased utilization of machines, chemicals, hybrid plants, and animals. Yet this had the effect of making farmers excessively dependent on external, high-technology, and increasingly high-cost inputs as well as on the vagaries of the global economy. As the public grew more concerned about environmental degradation in the 1980s, there emerged influential arguments for adopting alternative, low-technology inputs. It was found that even if high-tech methods reduce yields, net farm income can be maintained or even raised through corresponding reductions in high-cost inputs.

The influence of this contrast between high-tech, high-cost inputs and low-tech, low-cost inputs was reflected in 1988 when the appropriations bill for the Agricultural Productivity Research component (Subtitle C) of the 1985 Food Security Act (which contained the official reference to "alternative farming") explicitly specified "low-input" as an area to be funded for research. Subsequently, agricultural research scientists at Washington State University proposed the term "low-input sustainable agriculture" as the appropriate focus. No doubt aided by the ease with which the term can be turned into the acronym LISA, this became a standardized term in the National Academy of Sciences report on *Alternative Agriculture* and related discussions.[27] Then under Subtitle B of Title XVI of the Food, Agriculture, Conservation, and Trade Act of 1990, LISA was broadened and replaced by the Sustainable Agriculture Research and Education (or SARE) Program, with four regional bases and a diverse membership, including representatives ranging from family farms and agribusiness to nonprofit, governmental, and academic organizations.

Concepts of sustainable energy use and sustainable agricultural production are both more limited concepts than that of sustainable development, although in the hands of Lovins and Rodale a clear if implicit expansion takes place. Another term, which explicitly broadens rather than concentrates the focus, is that of "sustainable society." Here the work of Lester Brown and the Worldwatch Institute has been the most influential. For Lester Brown the issue is not sustainable economic growth or sustainable energy production, but sustainable society. Brown's tactic is to use the notion of sustainability and enlarge its context in such a way as to reform its meaning.[28]

It was in 1978 that Denis Hayes, a cofounder with Senator Gaylord Nelson of Earth Day 1970, wrote the first Worldwatch tract employing the term "sustainable society."[29] Then in 1981 Lester Brown's book *Building a Sustainable Society* made a comprehensive case for sustainability broadly construed as an ongoing social project. Brown wrote, "This book is not an isolated effort to discuss the sustainable society, but rather part of the continuing research program of the Worldwatch Institute." The picture he sought to draw of a sustainable society "has of necessity been painted with a broad brush."[30] But it includes corporations, religion, universities, public interest groups, the communications media—and makes common alliance with such alternative or countercultural ideals as "voluntary simplicity" and "conspicuous frugality." From 1984 on, Worldwatch has issued an annual *State of the World* "report on progress toward a sustainable society" to promote such themes.

The symptomatic character of this transition from sustainable development to sustainable society is highlighted by another book, one not directly associated with the Worldwatch Institute. Although it drew on Worldwatch publications and parallel sources, Lester W. Milbrath's *Envisioning a Sustainable Society: Learning Our Way Out* (1989) makes its own case for why modern society is not sustainable and how the sustainable ideal responds to limits-to-growth ideas.[31] The postmodern society, insofar as it initiates a break with modernity, is, he suggests, better described as the sustainable society.

Complementing Milbrath, Willem Vanderburg further seeks to assess society not just in terms of sustainability with respect to the biosphere but also in relation to human life and culture, giving rise to a notion he calls "human sustainability." For Vanderburg, human sustainability, which he finds lacking in advanced technological society, indicates "the ability of a community to create a way of life which is an expression of [its] values and aspirations and which is able to give meaning, direction and purpose to the lives of its members."[32]

This inflation of the notion of sustainability, when it moves from associations with development to associations with society, deserves to be stressed. As was pointed out earlier, the idea of progress originates in and presupposes a restriction of perspective. Once that restriction is set aside, the idea of progress is no longer nearly as obvious or as easy to defend. Analyses of sustainability may even recover certain cyclical patterns in nature and in history, in a manner not unlike the way discussions of sustainability themselves exhibit some cyclical features. This tendency to break with previous and defining restrictions constitutes the new and characteristic move in sustainability thinking. It also makes "sustainability" even more clearly into a new name for the good broadly construed.

4. Critiques of Sustainability

Nevertheless, one obvious weakness in this characteristic inflation of "sustainability" is that the term easily becomes what Uwe Pörksen has called an "plastic word."[33] Sustainability is now a modular slogan everyone uses, but in this universal approbation people may agree upon nothing. *Science* magazine editorializes about a "sustainable future for planet earth"; *Business Week* does a cover story on "The Push for Sustainable Development"; the latest permutation of *The Limits to Growth* appropriates "sustainable" in its subtitle; and the journal *In Context* bills itself as "a quarterly of humane sustainable culture."[34] Literally hundreds of books and articles on sustainability are published each year. This can be form without substance, word without significance. Efforts to avoid this slide into vacuousness have led to repeated analyses of the meaning, especially of the term "sustainable development."[35]

More substantive efforts to deal with this problem of conceptual inflation take the form of criticisms from both the right and left of the political spectrum. From what might be called the Right comes the argument that "sustainability is not enough." From the Left comes the question, "Sustainability of what?"

For sustainability-is-not-enough critics, sustainability too readily denotes stasis or, at best, dynamic equilibrium. In relation to sustainable agriculture, one critic argues, for example, that "traditional agricultural systems that have met the test of sustainability have not been able to respond adequately to modern rates of growth in demand for agricultural commodities."[36] From this perspective, sustainable development—as slower and more careful development—will be overwhelmed by continuing increases in population and aggregate consumer demand; sustainable development is not itself sustainable.

The most radical representative of this school of thought is Julian Simon. From *The Ultimate Resource* (1981) to *The State of Humanity* (1995), Simon has argued that the limits-to-growth argument is so bad as to "not [even be] worth detailed discussion or criticism."[37] The dynamic of the modern technocapitalist economy, as Arendt rightly observed, depends on the continuous discovery and creation of new conditions for the appropriation and consumption of resources. According to Simon, with nuclear fission and space travel, "we now have . . . the technology to feed, clothe, and supply energy to an ever-growing population . . . to go on increasing forever, improving our standard of living and our control over our environment."[38] Within such a framework, the ideal of sustainability, too, is scarcely worth mentioning. Sustainability exists, as it were, only through attacks on sustainability.

With the "Sustainability of what?" question, a different school of critics inquires more precisely about what is to be sustained. Surely sustainable development is not, as the World Bank defines it, simply "development that lasts"[39]—i.e., that which sustains only itself. Sustainable development must, in sustaining itself, sustain and not destroy something else to which it is related—the natural world, society, a way of life, a culture. The danger, according to those who raise this question, is that sustainable development is simply meant to sustain a way of life in the West that leads to the destruction of other ways of life throughout the world. The suspicion is that sustainability is proposed as a means to sustain, without questioning, what is ultimately an unsustainable way of life.

From this perspective even those who might seem opposed to pro-growth developmentalist assumptions can present a danger. Wolfgang Sachs, for instance, in a provocative article originally entitled "The Gospel of Global Efficiency," suggests that precisely insofar as "Worldwatch alerts us to the need for the efficient use of means [it also elevates] the rules of micro-economics to imperatives for national (and even global strategy)." Sustainable development theorists readily become what Sachs terms "eco-developers" who simply expand the perspective of traditional economics "by surveying the broad range of life-supporting factors in order to assure the sustainability of yields over the long term."

> Worldwatch implies the world-wide victory of [the] specifically modern outlook. . . . The Worldwatch utopia of a sustainable world appears to be peopled by a fairly recent version of homo sapiens: the efficiency-conscious individual.[40]

In pursuit of the ideal of sustainability broadly construed there is the subtle presence of an addiction to management that looks upon the world, in the influential image of Buckminster Fuller, as a spaceship in need of an operating manual.[41] Such an outlook must be recognized as being, at the very least, at odds with much of traditional culture. Perhaps there is a need to critique not just a course of development that may no longer be viable in the face of environmental constraints—and the resultant effort to replace development with sustainable development—but also the very aspirations that have led to such a course and the contemporary debates about the new means that might preserve it.

The ideal of development has its roots not just in Rostow's theories of international economics but also in a typically modern fear of what is presented as its only alternative, scarcity. At the deepest level, criticism of sustainability must address this fear of scarcity and the modern proposal to flee scarcity by means of technological development. But in the face of the empirical evidence of the problem of scarcity in the underdeveloped world even today, how can one possibly object to this proposal?

Consider only one of the most dramatic forms of what is commonly termed scarcity, namely, that which leads to starvation. The developmentalist view is to think that such dire natural or agricultural shortages can only be solved by technological improvements. But as economist Amartya Sen has shown, famines are not nearly so much a consequence of shortages as they are failures in distribution and sharing.[42] Although it is true that drought and flood often precede mass starvation, declines in food production alone cannot explain famine; even in desperate times there is typically enough food in the country to go around or enough money to import it. Disaster strikes, Sen's research shows, not simply when there is no more food, but when what food exists is no longer shared throughout a society. Such failures to share arise today when the poor can no longer afford to buy food following the loss of what Sen variously calls entitlements, capabilities, or functionings—that is, because of a sudden loss of employment or a surge in food prices.

What Sen points to as a loss of abilities to function within the society others have referred to as a loss of solidarity in an economy of subsistence. In the words of two commentators on this phenomenon:

> Ironically, only modern and affluent societies are convinced about the importance of scarcity as a determinant of social behavior, while traditional societies tend to rest on a different conviction. In traditional cultures, wishes and desires are not seen as endless or indefinite but as religiously and culturally constrained. Of course, traditional societies are used to shortages, but precisely because of a commitment to the constraint of desire such shortages are dealt with in ways quite different from those typical of modernity.[43]

The idea of scarcity—from which development and its modification, sustainable development, are proposed as the means of escape—has a history. It is a social construct of the early modern period that has become defining of modern economic life.[44] But at its most profound level, what the modern world calls scarcity might better be called an economy of subsistence—and, indeed, that which is sustainable in the truest sense.

Conclusion

This chapter began with the suggestion that sustainability is an issue about which a Socratic, questioning philosophy might well have something critical to say. It is thus appropriate to conclude by reiterating that point—and to add a word of caution. The danger of critical reflection on various prereflective names for the good is an opening of the door to nihilism or despair. One must beware of calling sustainability a lie.[45] The ideal of sustainability remains a tremendously attractive name coordinate with what in scholastic philosophy were called the transcendentals—those who saw being or reality as manifested in goodness, truth, and beauty. Even when subject to clarifying criticism, that to which such transcendentals ultimately refers must not be rejected as such.

The ideal of sustainability has become a synonym for the first, prereflective name of the good as that which is our own. Just as Socrates taught us that the good that is our own calls for philosophical criticism to save it from popular misunderstandings, so now does the idea of sustainability.[46]

Notes

1. Xenophon, *Memorabilia* I, i, 16. See also Cicero, *Tusculan Disputations* V, iv, 10–11.
2. The *Oxford English Dictionary*, 2d ed. (New York: Oxford University Press, 1992), gives a first citation as Thomas Sowell's *Say's Law: An Historical Analysis* (Princeton, NJ: Princeton University Press, 1972). Michael Redclift, "Sustainable Development: Economics and Environment," in Michael Redclift and Colin Sage, eds., *Strategies for Sustainable Development: Local Agendas for the Southern Hemisphere* (New York: Wiley, 1994), pp. 17–34, includes some remarks on the etymology of "sustainable," but fails to note the conversion from adjective to substantive. Michael Ruse, "Sustainability," in Gunnar Skirbekk, ed., *The Notion of Sustainability and Its Normative Implications* (Oslo, Norway: Scandinavian University Press, 1994), pp. 7–27, too facilely roots the idea in notions of a "balance of nature" and then simplistically identifies it with equilibrium.
3. For a summary review of such disagreements, see Donald Worster, "The Shaky Ground of Sustainability," in Wolfgang Sachs, ed., *Global Ecology: A New Arena of Political Conflict* (London: Zed Books, 1993), especially pp. 134–36. This important essay can also be found in Worster's *The Wealth of Nature: Environmental History and the Ecological Imagination* (New York: Oxford University Press, 1993), pp. 142–55.
4. See, e.g., Mircea Eliade, *Cosmos and History: The Myth of the Eternal Return*, trans. Willard R. Trask (New York: Harper, 1959).
5. J. C. Squire, "In Continuation of Pope on Newton" (1926).
6. See, e.g., Karl Popper, *The Logic of Scientific Discovery* (New York: Basic Books, 1959). Original German: *Logik der Forschung* (Vienna: 1934).
7. Winston Churchill, Parliament Bill speech, November 11, 1947; in Robert Rhodes James, ed., *Winston S. Churchill: His Complete Speeches, 1897–1963*, vol. 7: 1943–1949 (New York: Chelsea, 1974), p. 7566.
8. For a complementary analysis of the theory of progress to that given in section 1, see Carl Mitcham, "Science, Technology, and the Theory of Progress," in Steven Goldman, ed., *Science, Technology, and Social Progress* (Bethlehem, PA: Lehigh University Press, 1989), pp. 240–52.
9. Donella H. Meadows, Dennis L. Meadows, Jorgen Randers, and William W. Behrens III, *The Limits to Growth: A Report for The Club of Rome's Project on the Predicament of Mankind* (New York: Universe Books, 1972).

10. See chapter 4 of Peccei's autobiography, *The Human Quality* (Oxford: Pergamon, 1977).
11. Mihailo Mesarovic and Eduard Pestel, *Mankind at the Turning Point: The Second Report to The Club of Rome* (New York: Dutton, 1974).
12. *World Conservation Strategy: Living Resource Conservation for Sustainable Development* (Gland, Switzerland: International Union for Conservation of Nature and Natural Resources, United Nations Environment Programme, and World Wildlife Fund, 1980).
13. The *World Conservation Strategy* argument is continued in Francis R. Thibodeau and Hermann H. Field, eds., *Sustaining Tomorrow: A Strategy for World Conservation and Development* (Hanover, NH: University Press of New England, 1984).
14. Daniel Sitarz, ed., *Agenda 21: The Earth Summit Strategy to Save Our Planet* (Boulder, CO: EarthPress, 1994), p. 4. One remarkable anomaly in this passage is the idea that environment and development were "seemingly unrelated" until *Our Common Future* made its case, since the Commission was explicitly set up to deal with this issue. Indeed, the relation of environment and development had been a major issue at a previous UN Conference on Human Development organized by Maurice Strong and held in Stockholm in 1972. There, environmentalists had presented development as a threat to the environment, while Third-World developmentalists, led by Indira Gandhi, had responded by criticizing environmentalists as a threat to development. For publications representative of the official "antidevelopment" ideology see, e.g., *The Stockholm Conference: Only One Earth, An Introduction to the Politics of Survival* (London: Earth Island, 1972); Barbara Ward Jackson and René Dubos, *Only One Earth: The Care and Maintenance of a Small Planet* (New York: W.W. Norton, 1972); and Maurice Strong, ed., *Who Speaks for the Earth?* (New York: W.W. Norton, 1973). The Rio Earth Summit of 1992, also organized by Strong, was a twentieth anniversary attempt to bridge the gap that had opened at Stockholm. For an abbreviated proceedings of the Rio Summit see Stanley P. Johnson, *The Earth Summit: The United Nations Conference on Environment and Development (UNCED)* (London: Graham and Trotman, 1993).
15. Simplifying slightly from a rather more complex history, the first of these reports was *Common Crisis* (1980), from the Independent Commission on International Development Issues chaired by Willy Brandt, former Prime Minister of West Germany. The second was *Common Security* (1982), from the Independent Commission on Disarmament and Security Issues chaired by Olof Palme, former prime minister of Sweden. These reports are addressed not only to governmental and developmental specialists but to the general public as well.
16. World Commission on Environment and Development, *Our Common Future* (New York: Oxford University Press, 1987), p. 8.
17. See, e.g., Michael Carley and Ian Christie, *Managing Sustainable Development* (Minneapolis: University of Minnesota Press, 1993); Joroen C.J.M. van den Bergh and Jan van der Straaten, eds., *Toward Sustainable Development: Concepts, Methods, and Policy* (Washington, DC: Island Press, 1994); Michael Redclift and Colin Sage, eds., *Strategies for Sustainable Development: Local Agendas for the Southern Hemisphere* (New York: John Wiley, 1994); Michael Common, *Sustainability and Policy: Limits to Economics* (New York: Cambridge University Press, 1995); Kimio Uno, *Environmental Options: Accounting for Sustainability* (Boston: Kluwer, 1995); and Peter Roberts, *Environmentally Sustainable Business: A Local and Regional Perspective* (London: Paul Chapman, 1995).

18. For reference and analysis of the Truman speech, see Gustavo Esteva, "Development," in Wolfgang Sachs, ed., *The Development Dictionary: A Guide to Knowledge as Power* (London: Zed Books, 1992), pp. 6 ff.
19. Walt W. Rostow, "The Take-off into Self-sustained Growth," *Economic Journal* 50 (4) (October 1956): 150–200. At the time of this article, Rostow was Professor of Economics at MIT; then from 1961–1969 he served as special adviser on development issues to Presidents Kennedy and Johnson.
20. Walt W. Rostow, *The Stages of Economic Growth: A Non-Communist Manifesto* (New York: Cambridge University Press, 1960), pp. 7–9. This volume was reissued in a third edition, 1991. Rostow's theory grew out of an earlier study of *The Process of Economic Growth* (New York: W.W. Norton, 1952; 2d ed., 1962), and has been restated in an edited conference volume on *The Economics of Take-Off into Sustained Growth* (New York: St. Martin's Press, 1963) as well as placed in larger historical context in *Theories of Economic Growth from David Hume to the Present: With a Perspective on the Next Century* (New York: Oxford University Press, 1990).

As an aside, the persistence of Rostow's view of sustainability is promoted by uses such as that provided by an article on "Sustainable Yield" that appeared in *Forbes* (December 14, 1987), the same year as the publication of *Our Common Future*. In the straightforward words of *Forbes*: "Unless you've got an awful lot of capital, living well off it without penalizing your heirs isn't easy" (p. 204). But, of course, for *Forbes* this means increase your capital investment—which is exactly what the transition to economic take-off also requires. There is no recognition that this strategy could, in perhaps another sense, penalize one's heirs.
21. Hannah Arendt, *The Human Condition* (Chicago: University of Chicago Press, 1958), pp. 252–253. The implication of Arendt's analysis for the sustainability question is also pointed out in Hans Achterhuis, "The Lie of Sustainability," in Wim Zweers and Jan J. Boersema, eds., *Ecology, Technology and Culture* (Cambridge, U.K.: White Horse Press, 1994), pp. 199–200.
22. Amory B. Lovins, "Energy Strategy: The Road Not Taken?" *Foreign Affairs* 55 (1) (October 1976): 65–96, and *Soft Energy Paths: Toward a Durable Peace* (San Francisco: Friends of the Earth, 1977). See also Lovins's *World Energy Strategies: Facts, Issues, and Options* (San Francisco: Friends of the Earth, 1975); and Hugh Nash, ed., *The Energy Controversy: Soft Path Questions and Answers* (San Francisco: Friends of the Earth, 1979). As a founder of the Rocky Mountain Institute in Aspen, Colorado (1982–present), Lovins has remained an effective lobbyist for more efficient energy use.
23. Among Sir Albert Howard's many books, see especially *The Development of Indian Agriculture* (London: Oxford University Press, 1927) and *An Agricultural Testament* (London: Oxford University Press, 1940).
24. Robert Rodale, *Our Next Frontier: A Personal Guide for Tomorrow's Lifestyle* (Emmaus, PA: Rodale Press, 1981), p. 14. This book is a rewritten and revised version of *Sane Living in a Mad World: A Guide to the Organic Way of Life* (Emmaus, PA: Rodale Press, 1972). But although the 1972 book makes a case for an expanded idea of organic living, it is not as explicit as the 1981 volume.
25. See also in this same tradition, Wes Jackson, Wendell Berry, and Bruce Colman, eds., *Meeting the Expectations of the Land* (San Francisco: North Point Press, 1984).
26. E. F. Schumacher, *Small Is Beautiful: Economics as if People Mattered* (New York: Harper & Row, 1973). See also Barbara Wood, *E. F. Schumacher: His Life and Thought* (New York: Harper and Row, 1984).

27. *Alternative Agriculture* (Washington, DC: National Academy Press, 1989). For a less official but important complementary volume, see also Clive A. Edwards, Rattan Lal, Patrick Madden, Robert H. Miller, and Gar House, *Sustainable Agricultural Systems* (Ankeny, IA: Soil and Water Conservation Society, 1990).
28. The Worldwatch Institute was founded in 1974 by Brown, who had worked for ten years (1959–1969) as a Department of Agriculture international development specialist and then for five years (1969–1974) for the Overseas Development Council. From the beginning Worldwatch has been a no-frills, savvy organization bucking some of the trends of traditional think-tank culture. Worldwatch markets its research as much as it goes grant hunting. See, for reference, Bob Cohen, "Think Tank, Think Profits," *Newsweek*, April 28, 1986, p. 70.
29. Denis Hayes, *Repairs, Reuse, Recycling—First Steps Toward a Sustainable Society*, Worldwatch Paper 23 (September 1978).
30. Lester Brown, *Building a Sustainable Society* (New York: W. W. Norton, 1981), pp. xi and xii.
31. Lester W. Milbrath, *Envisioning a Sustainable Society: Learning Our Way Out* (Albany: SUNY Press, 1989).
32. Willem H. Vanderburg, "Can a Technical Civilization Sustain Human Life?," *Bulletin of Science, Technology, and Society* 15 (2–3) (1995): 92.
33. Uwe Pörksen, *Plastik Wörter: Die Sprache einer Internationalendiktatur* (Stuttgart: Klett-Cotta, 1989). English version: *Plastic Words: The Tyranny of a Modular Language*, trans. Jutta Mason and David Cayley (University Park, PA: Penn State University Press, 1994).
34. Philip H. Ableson, "Sustainable Future for Planet Earth," *Science* 253, whole no. 5016 (12 July 1991): 117. "Growth vs. Environment: In Rio next Month, a Push for Sustainable Development," *Business Week*, 3265 (May 11, 1992): 66–75. Donella H. Meadows, Dennis L. Meadows, and Jorgen Randers, *Beyond the Limits: Confronting Global Collapse, Envisioning a Sustainable Future* (Post Mills, VT: Chelsea Green, 1992), in which the "limits-to-growth" argument nevertheless remains essentially in tact. *In Context* (from Bainbridge Island, WA 98110) began publication in 1981.
35. Among the more useful examples of this literature are Michael Redclift, *Sustainable Development: Exploring the Contradictions* (New York: Methuen, 1987); Clem Tisdell, "Sustainable Development: Differing Perspectives of Ecologists and Economists, and Relevance," *World Development* 16 (3) (1988): 373–84; and Sharachchandra M. Lélé, "Sustainable Development: A Critical Review," *World Development* 19 (6) (June 1991): 607–21.
36. Vernon W. Ruttan, "Sustainability Is Not Enough," *American Journal of Alternative Agriculture* 3 (2–3) (Spring–Summer 1988): 128.
37. Julian L. Simon, *The Ultimate Resource* (Princeton, NJ: Princeton University Press, 1981), p. 286. See also Julian L. Simon and Herman Kahn, eds., *The Resourceful Earth: A Response to Global 2000* (Oxford: Blackwell, 1984), pp. 34–37. Julian L. Simon, ed., *The State of Humanity* (Oxford: Blackwell, 1995), is an update of *The Resourceful Earth* that does not even bother even to mention, much less discuss, the limits to growth thesis—or the arguments for sustainability of any kind.
38. Norman Myers and Julian L. Simon, *Scarcity or Abundance: A Debate on the Environment* (New York: W. W. Norton, 1994), p. 65.
39. *World Development Report 1992* (New York: Oxford University Press for the World Bank, 1992), p. 34. Quoted from Wolfgang Sachs, "Global Ecology and the

Shadow of Development," in Sachs, ed., *Global Ecology: A New Arena of Political Conflict* (London: Zed Books, 1993), pp. 3–21.

40. The quotations are from an unedited version of a paper subsequently published as "A Critique of Ecology," *New Perspectives Quarterly* (Spring 1989): 16–19.

41. R. Buckminster Fuller, *Operating Manual for Spaceship Earth* (New York: Simon and Schuster, 1969). Explicit proposals for the high-tech management of the environment are exemplified in the NASA "Mission to Planet Earth"—see, e.g., Thomas F. Malone and Robert Corell, "Mission to Planet Earth Revisited: An Update on Studies of Global Change," *Environment* 31 (3) (April 1989): 6–11 and 31–35—and the special "Managing Planet Earth" issue of *Scientific American* 261 (3) (September 1989), which includes William D. Ruckelshaus's "Toward a Sustainable World," pp. 166–170, 172, and 174.

42. Amartya Sen, *Poverty and Famines: An Essay on Entitlement and Deprivation* (New York: Oxford University Press, 1981).

43. Pieter Tijmes and Reginald Luijf, "The Sustainability of Our Common Future: An Inquiry into the Foundations of an Ideology," *Technology in Society* 17 (3) (1995): 328.

44. On this point, the most comprehensive study is Hans Achterhuis, *Het Rijk va de Schaarste* [The reign of scarcity] (Baarn, the Netherlands: Ambo, 1988). For an English precis, see Achterhuis, "Scarcity and Sustainability," in Wolfgang Sachs, ed., *Global Ecology: A New Arena of Political Conflict* (London: Zed Books, 1993), pp. 104–116. Cf. also Nicholas Xenos, *Scarcity and Modernity* (New York: Routledge, 1989).

45. Hans Achterhuis, "The Lie of Sustainability," in Wim Zweers and Jan J. Boersema, eds., *Ecology, Technology and Culture* (Cambridge, UK: White Horse Press, 1994), pp. 198–203.

46. This paper grew out of a presentation at the Consejo Latino Americano de las Ciencias Sociales (CLACSO) in Santiago, Chile, in late 1991. A less complete version of this article appeared as "The Concept of Sustainable Development: Its Origins and Ambivalence," *Technology in Society* 17 (3) (1995): 311–26.

Contributors

AVNER DE-SHALIT is a professor of political science at Hebrew University, Jerusalem. Among his many publications are *Why Posterity Matters: Environmental Policies and Future Generations*.

GUS DIZEREGA's research focuses on modern society's relationship with the natural world and on the nature of liberal democracy. His work has appeared in *Critical Review, The Review of Politics, The Western Political Quarterly, Telos, Environmental Ethics, The Trumpeter, Social Theory and Practice, Shaman's Drum, Gnosis,* and elsewhere. He received his Ph.D. in political theory at the University of California at Berkeley, and is presently a research associate at Berkeley's Institute for Governmental Studies. He resides in Sebastopol, California, where he teaches part-time, writes, sells his art work, and is a student of shamanic healing and related animistic studies.

ROGER S. GOTTLIEB is professor of philosophy at Worcester Polytechnic Institute. He has written *Marxism 1844–1990: Origins, Betrayal, Rebirth* and *History and Subjectivity: The Transformation of Marxist Theory*; and edited *Thinking the Unthinkable: Meanings of the Holocaust, This Sacred Earth: Religion, Nature, Environment, A New Creation: America's Contemporary Spiritual Voices, Radical Philosophy,* and *An Anthology of Western Marxism*. He directs series for Routledge, SUNY Press, and Humanities Press, and is on the editorial boards of *Social Theory and Practice* and *Capitalism, Natural, Socialism: A Journal of Socialist Ecology*. He is currently writing a book on the subjective experience of social, moral, and spiritual life in an age of ecocide.

ERIC KATZ is an associate professor of philosophy and director of the Science, Technology, and Society program at the New Jersey Institute of Technology. In addition to several dozen scholarly essays, he is the author of two annotated bibliographies in the field of environmental ethics. For Routledge he has coedited (with Andrew Light) the collection *Environmental Pragmatism* and has recently completed *Deep Green: A Critical*

Introduction to the Philosophy of Deep Ecology. He is also the author of *Nature as Subject: Essays in Environmental Ethics and Policy.* Katz was the founding vice-president of the International Society for Environmental Ethics.

ROBERT KIRKMAN received his Ph.D. in philosophy from the State University of New York at Stony Brook, where he pursued an interdisciplinary project in philosophy and ecology. A significant portion of his research into the history of ecology was conducted in Paris, France, with the support of a Chateaubriand Scholarship from the French Embassy to the United States. He currently lives and teaches in the New York City area.

ANDREW LIGHT is an assistant professor of philosophy at the University of Montana. In addition to a number of articles on environmental philosophy, philosophy of technology, and political theory, he is coeditor of the recent Routledge collection *Environmental Pragmatism* (1996). He is also editor of the forthcoming volumes: *The Environmental Materialism Reader* and *Anarchism, Nature, and Society: Critical Perspectives on Social Ecology.* He is coeditor of the annual *Philosophy and Geography* and is currently at work on a book on urban environments.

BRIAN LUKE is an assistant professor of philosophy at the University of Dayton, Ohio. His primary writing and teaching interests are in animal rights theory, environmental ethics, feminist theory, and radical political philosophy. Luke is active in grassroots animal rights activism and in coparenting his two sons, Alex and Adam.

DAVID MACAULEY presently teaches philosophy and literature classes at St. John's University and Marymount Manhattan College. He is the editor of *Minding Nature: The Philosophers of Ecology* and has published articles in philosophy, ecology and political theory. He is completing his Ph.D. at SUNY–Stony Brook.

MARK A. MICHAEL is assistant professor of philosophy at Austin Peay State University. He has published papers in journals such as *American Philosophical Quarterly, Environmental Ethics,* and *Journal of Social Philosophy,* and he is currently working on a book on liberalism and environmentalism. His research interests are environmental ethics and social and political philosophy.

CARL MITCHAM is a member of the philosophy department and of the Science, Technology, and Society Program at Pennsylvania State University. His most recent book is *Thinking through Technology: The Path between Engineering and Philosophy.*

JOHN O'NEILL is a lecturer in philosophy at Lancaster University. His publications include *Ecology Policy and Politics: Human Well-Being and the Natural World* and *Worlds Without Content: Against Formalism*, as well as numerous papers in journals and collections. His research interests include environmental philosophy, the foundations of economics, and social and political thought. He is currently working on a book critically examining defenses of market economies.

HOLMES ROLSTON III is University Distinguished Professor of Philosophy at Colorado State University, where he teaches environmental ethics. He is the author of *Environmental Ethics, Philosophy Gone Wild, Science and Religion: A Critical Survey,* and *Conserving Natural Value.* He was the founding president of the International Society for Environmental Ethics, and has been, for almost two decades since its inception, the Associate Editor of *Environmental Ethics.*

DAVID SCHLOSBERG is Assistant Professor of Political Science at Northern Arizona University, where he teaches political theory and environmental politics. He is completing a book project on *Environmentalism and Difference.*

WILLIAM THROOP is the Jefferson-Pilot Associate Professor of Philosophy at St. Andrews College in Laurinburg, NC. He writes on topics in epistemology, metaphysics, value theory, and environmental ethics, and applies insights from pragmatism to the solution of problems in these areas.

STEVEN VOGEL is professor of philosophy at Denison University in Granville, Ohio. Educated at Yale and Boston Universities, he is the author of *Against Nature: The Concept of Nature in Critical Theory.* An earlier version of his essay was presented as a working paper at the Humanities Research Center "Menneske og Natur" at the University of Odense, Denmark. He is grateful to the center for its support.

MARK I. WALLACE is associate professor and co-chair in the department of religion and a member of the environmental studies concentration at Swarthmore College, Pennsylvania. He is the author of *Fragments of the Spirit: Nature, Violence, and the Renewal of Creation* and *The Sacred Naïveté: Barth, Ricoeur, and the New Yale Theology;* editor of *Figuring the Sacred: Religion, Narrative, and Imagination;* and coeditor of *Curing Violence: Religion and the Thought of René Girard.*

PETER S. WENZ, professor of philosophy and legal studies at the University of Illinois at Springfield and adjunct professor of medical humanities at Southern Illinois University School of Medicine, is one of those aging, leftist, feminist, vegetarian environmentalists that one often sees jogging near universities. He teaches a variety of courses in applied ethics, including Moral Issues in the Law, Biomedical Ethics and the Law, and Environmental Values. His single-authored books are *Environmental Justice, Abortion Rights as Religious Freedom,* and *Nature's Keeper.* He and Laura Westra coedited *Faces of Enviromental Racism.* His current interests include exploring social and environmental dimensions of technological change.

MICHAEL E. ZIMMERMAN is professor of philosophy at Tulane University. Author of three books, an anthology, and numerous scholarly articles, his special interests are in Heidegger, environmental philosophy, and transpersonal psychology. As clinical professor of psychology at Tulane Medical School, he regularly teaches a course on philosophy and psychotherapy to psychiatry residents.